WENNER–GREN CENTER
INTERNATIONAL SYMPOSIUM SERIES

VOLUME 42

QUANTITATIVE NEUROANATOMY IN TRANSMITTER RESEARCH

QUANTITATIVE NEUROANATOMY IN TRANSMITTER RESEARCH

Proceedings of an International Symposium held at
The Wenner–Gren Center, Stockholm, May 3–4, 1984

Edited by

Luigi F. Agnati
Department of Human Physiology, University of Modena, Modena, Italy

Kjell Fuxe
Department of Histology, Karolinska Institutet, Stockholm, Sweden

PLENUM PRESS • NEW YORK AND LONDON

First published 1985

Published in Great Britain by
THE MACMILLAN PRESS LTD
Houndmills, Basingstoke, Hampshire RG21 2XS
and London

Published in the United States of America by
PLENUM PUBLISHING CORPORATION
233 Spring Street, New York, NY 10013, USA

ISBN–13: 978–1–4612–9263–0   e–ISBN–13: 978–1–4613–2139–2
DOI: 10.1007 / 978–1–4613–2139–2

CONTENTS

SESSION III: QUANTITATIVE IMMUNOCYTO-
CHEMISTRY AND AMINE FLUORESCENCE
HISTOCHEMISTRY
Chairman: C. Owman

THE PARTICIPANTS

Luigi F. Agnati
Department of Human Physiology
University of Modena
Via Campi 287
I-41100 MODENA
Italy

Kurt Andersson
Department of Histology
Karolinska Institutet
S-104 01 STOCKHOLM
Sweden

Fabio Benfenati
Department of Human Physiology
University of Modena
Via Campi 287
I-41100 MODENA
Italy

Robert Benno
The New York Hospital
Cornell Medical Center
1300 York Avenue
NEW YORK, N.Y. 10021
USA

Theodor Blackstad
Department of Anatomy
University of Oslo
Karl Johans gate 47
OLSO 1
Norway

Tim Cowen
Department of Anatomy & Embryology
University College London
Gower Street
LONDON WC1E 6BT
U.K.

Luis-M. Cruz-Orive
Department of Anatomy
University of Bern
P O Box 139
CH-3000 BERN
Switzerland

Annica Dahlström
Department of Histology
University of Göteborg
S-400 33 GÖTEBORG
Sweden

Pier Luigi Fabbri
The Center for Instruments
University *of* Modena
Via Campi 213
I-41100 MODENA
Italy

Kjell Fuxe
Department of Histology
Karolinska Institutet
S-104 01 STOCKHOLM
Sweden

Dwight German
Department of Physiology
University of Texas Health
  Science Center
DALLAS
Texas 75235
USA

Hans Jørgen Gundersen
Department of Medicine
University of Aarhus
DK-8000 AARHUS
Denmark

Vyvyan Howard
Department of Anatomy
University of Liverpool
P O Box 147
LIVERPOOL L69 3BK
U.K.

Tomas Hökfelt
Department of Histology
Karolinska Institutet
S-104 01 STOCKHOLM
Sweden

Olle Johansson
Department of Histology
Karolinska Institutet
S-104 01 STOCKHOLM
Sweden

Gösta Jonsson
Department of Histology
Karolinska Institutet
S-104 01 STOCKHOLM
Sweden

Madhu Kalia
Department of Neurosurgery
Thomas Jefferson University
1025 Walnut Street
PHILADELPHIA
Pennsylvania 19107
USA

John Morrison
Scripps Clinic and Research
  Foundation
LA JOLLA
California 92037
USA

Lars Olsson
Department of Histology
Karolinska Institutet
S-104 01 STOCKHOLM
Sweden

Christer Owman
Department of Histology
University of Lund
Biskopsgatan 5
S-223 62 LUND
Sweden

Jose Maria Palacios
Preclinical Research
Sandoz Ltd
CH-4002 BASEL
Switzerland

Thomas Rainbow
Department of Pharmacology
Medical School
University of Pennsylvania
PHILADELPHIA
Pennsylvania 19104
USA

Grayson Richards
Pharmaceutical Research Department
F. Hoffman-LaRoche & Co. Ltd.
CH-4002 BASEL
Switzerland

Bernhard Scatton
Department of Biology
Synthélabo - L.E.R.S.
31 Avenue Paul Vaillant Couturier
F-92220 BAGNEUX
France

Jacques Schipper
Department of Pharmacology
Free University Medical Center
Van der Bocchorststraat 7
AMSTERDAM
Holland

Lewis Tucker
The New York Hospital
Cornell Medical Center
1300 York Avenue
NEW YORK, N.Y. 10021
USA

Donald Woodward
Department of Cellbiology
University of Texas Health
  Science Center
DALLAS
Texas 75235
USA

George Wooten
Department of Neurology
University of Virginia Medical
  Center
P O Box 394
CHARLOTTESVILLE
Virginia 22908
USA

OPENING ADDRESS

Ladies and Gentlemen!

On behalf of the organizing committee which consists of Professor Luigi F. Agnati and myself I wish you most cordially wellcome to Stockholm and to this international Wenner-Gren symposium on "Quantitative neuroanatomy in transmitter research". We are very happy that so many of you have been able to come to this symposium in spite of your many other commitments and that many of the leading scientists in the world in this area are with us today at this conference. Ever since our discoveries of the monoamine neuron systems 15 to 20 years ago we have become increasingly aware of the importance of developing methods to quantitatively describe the morphofunctional characteristics of transmitter identified neurons in the brain. In 1974 and 1978 we tried to perform microdensitometry and morphometry on catecholamine (CA) terminals demonstrated by means of Falck & Hillarp technique by analyzing the density distribution of CA profiles in smears by means of grids (Agnati and Fuxe, 1974) and by using optical filters (Agnati et al. 1978). In 1977 we performed a morphometrical evaluation of the distribution of tyrosine hydroxylase immunoreactive nerve terminal networks and their possible codistribution with peptidergic networks (Agnati et al. 1977). A substantial achievement was the transformation of the formaldehyd technique from a very sensitive tool in chemical neuroanatomy into a tool to perform functional studies (see Andén et al. 1969). This was achieved thanks to the work of our group in Stockholm (Einarsson et al. 1975; Löfström et al. 1976), which led to the development of semiquantitative measurements of catecholamine stores in brain. Due to the complaints of many referees we developed by means of quantitative histofluorimetry and appropriate CA standards present in the sections a method to determine CA levels in discrete CA nerve terminal populations within the brain expressed in absolute terms (nmol/g of tissue; Agnati et al. 1979). The rate constants and the turnover rates in these individual CA nerve terminal populations could then be determined from measurements of CA levels at different time intervals following tyrosine hydroxylase inhibition.

With the development of the image analyzers the possibilities to perform otherwise time-consuming morphometrical and micro-densitometrical analyses of transmitter-identified neurons markedly increased. We therefore felt it necessary to bring scientists specialized in morphometry and stereology together with chemical neuroanatomists to discuss the various new methods available for computer-based morphometry and microdensitometry.

'Nimium ne crede colori'

Virgilius, Eclog. 2, 17
70 – 19 B.C.

Fig. 1.

In this way we hope to have a very friendly and fruitful discussion on how to use quantitative neuroanatomical methods in the morphofunctional characterization of transmitter-identified neurons. We hope to show their drawbacks and advantages so that we can use these new approaches in the best possible way to solve our various neurobiological problems. The explosion is already on its way and we should at this meeting be able to learn how to use image analysis. We hope to overcome too enthusiastic and sometimes also too esthetic attitudes towards the "products" of the computer elaborations, which are so beautiful that they could be shown in exhibitions of modern art! It may at this point be useful to quote the latin poet Virgilio (fig. 1): "The colours cannot be fully trusted". I am sure that we will see many beautiful colours at this symposium, which will make a number of morphologists happy. I am also convinced that we will also hear a lot about the theory behind the colours, which will make the biomathematicians happy. We will also discuss the biochemical counterpart of these new approaches to quantitate chemical neuroanatomy. On behalf of the organizing committee I now have the pleasure of opening this multifacit symposium which may satisfy research workers of many different fields.

We are very grateful to Mrs. Gun Hultgren at the symposium secretariat for her excellent assistance.

We are very much indepted to the following sponsors, who made this symposium possible:

The Swedish Medical Research Council
Astra Pharmaceuticals, Sweden
CIBA-GEIGY AB, Sweden
Ferring AB, Sweden
AB Ferrosan, Sweden
AB H. Lundbeck & Co, Sweden
E. Merck, FRG
Merck Sharp & Dohme Research Laboratories, England
Roche-Produkter AB, Sweden

Sandoz AB, Sweden
Sandoz Ltd, Switzerland
Schering AG, FRG
Schering Corporation, USA
The Wellcome Research Laboratories, England

K. Fuxe and L.F. Agnati
Department of Histology,
Karolinska Institutet,
Stckholm, Sweden and
Department of Human Physio-
logy, University of Modena,
Modena, Italy

References

Agnati, L.F. & Fuxe, K.: Quantitative comparisons of amine
    fluorescence in cortical noradrenaline terminals using smear
    preparations. J. Histochem. Cytochem. 22, 1122-1127, 1974.
Agnati, L.F., Fuxe, K., Hökfelt, T., Goldstein, M. & Jeffcoate,
    S.L.: A method to measure the distribution pattern of speci-
    fic nerve terminals in sampled regions. Studies on tyrosine
    hydroxylase LHRH, TRH and GIH immunofluorescence. J. Histo-
    chem. Cytochem. 25, 1222-1236, 1977.
Agnati, L.F., Benfenati, F., Cortelli, P. & D'Alessandro, R.: A
    new method to quantify catecholamine stores visualized by
    means of the Falck-Hillarp technique. Neurosci. Lett. 10,
    11-17, 1978.
Agnati, L.F., Fuxe, K., Andersson, K., Wiesel, F. & Fuxe, K.: A
    method to determine dopamine levels and turnover rate in
    discrete dopamine nerve terminal systems by quantitative use
    of dopamine fluorescence obtained by Falck-Hillarp method-
    ology. J. Neurosci. Methods 1, 365-373, 1979.
Andén, N.-E., Corrodi, H. & Fuxe, K.: Turnover studies using
    synthesis inhibition. In: Metabolism of amines in the brain,
    (ed. G. Hooper), MacMillan, London, pp. 38-47, 1969.
Einarsson, P., Hallman, H. & Jonsson, G.: Quantitative
    microfluorimetry of formaldehyde induced fluorescence of
    dopamine in the caudate nucleus. Med. Biol. 53, 15-24, 1975.
Löfström, A., Jonsson, G., Wiesel, F.-A. & Fuxe, K.:
    Microfluorimetric quantitation of catecholamine fluorescence
    in rat median eminence. II. Turnover changes in hormonal
    states. J. Histochem. Cytochem. 24, 430-442, 1976.

MORPHOMETRICAL AND MICRODENSITOMETRICAL STUDIES ON NON–TRANSMITTER– IDENTIFIED NEURONS

Chairman: G. WOODWARD

QUANTITATIVE ANALYSIS OF THREE-DIMENSIONAL STRUCTURES IN NEUROANATOMY

H.J.G. GUNDERSEN

Stereological and Electromicroscopical Diabetes Research Laboratory, University Institute of Pathology and Second University Clinic of Internal Medicine, Aarhus University, Denmark

INTRODUCTION

It is almost certain that a complete understanding of the nervous system requires a thorough knowledge of all its qualitative, 'dimensionless' aspects (which are its components, how are they connected, what is the transmitter involved etc.) and a good deal of quantitative, 3-dimensional information about number, length, sizes, connectedness etc. of all these cells, nuclei, and major or minor bundles of connections, which are the constituents or structural components of the nervous system. It is also rather obvious that the discovery-phase is the primary one - but discoveries of new cells, transmitters and nuclei are only first steps towards understanding functions - quantitation of all these components at the cellular or at the ultrastructural level is almost inevitable for understanding fully the complex function of their aggregate.

STEREOLOGY

The set of methods by which 3-dimensional information about structures is obtained by way of making 2-dimensional sections trough specimens or organs and then analyzing these sections is known as STEREOLOGY.
In order to get such sections and to transform the 2-dimensional, observed quantities to unbiased 3-dimensional structural quantities a number of problems have to be solved. A number of these problems are of physical/chemical nature (chemical fixation, physical embedding, mechanical sectioning, staining by chemical or immunological methods) others are of statistical/geometrical nature, they are the scope of this text. For a very recent and comprehensive review of stereological methods in a biological context, see Loud (1984). Weibel (1979) has written a complete textbook meant for biologists.

3

Sections and the information in them.

It is both wellknown and intuitively obvious that the way a section or a set of sections is made determines which information is obtainable from it: on a perpendicular section of e.g. a nerve fibre one can determin the cross-sectional area, the shape and the diameter(s) of the nerve fibre, on an oblique section one can only measure the diameter (if the fibre is cylindrical) and only on a section perfectly parallel to the fibre axis can one measure its length (if the fibre is straight) - but not the diameter or the shape. For all practical purposes one can distinguish between just two types of sections: random, independent sections and systematic sections.

A set of random, independent sections is obtained when an organ of known volume V(ref) is cut into a number of slices and small blocks are punched out from each slice and these small blocks are embedded with a random orientation and one section is taken from each block. On such a set one can obtain information about total volume V and total surface area S of a given structure which is identifiable on the sections. If the structure is composed of a large number of elements that are very long compared to their diameter one can also estimate their total length L, see Gundersen (1979). When dealing with cells, nuclei, mitochondria etc., all countable, isolated elements denoted particles, their total number N cannot, however, in general be estimated from such sections, it is only when they all have the same, simple and known shape that an approximate estimate of their total number is obtainable.

Systematic sections are quite often made in neuroanatomical research, for two good reasons: they are easy to make and it is often possible both to get a 'feeling' for the overall architecture of the organ and to identify anatomically defined regions on them. From a stereological point of view, systematic sections have the additional advantage that on such a set one can indeed estimate the total number of particles. As illustrated in Fig. 1 it is often not necessary actually to use true serial sections (when all consecutive sections in a series is used). Depending on the size of the particles two sections to be used may be spaced a known distance by considering e.g. only every third of a series. For reasons of sampling efficiency it is also often an advantage to space the pairs or the sets of more than two sections a large distance so that the sets of pairs sample the whole nucleus, say, which is under study.

Using stereological methods on systematic sections the number N, the total volume V and the total height Σh of all particles can be estimated; The mean particle volume $\bar{v}$ and the average particle height $\bar{h}$ (perpendicular to the sections) is of course estimated from the corresponding total quantities divided by N. The method for estimating particle number is outlined below and fully described in Sterio (1984). It should be emphasized that in general one cannot estimate the total surface area S of particles from simple systematic sections, see also the paper by Cruiz-Orive in this volume.

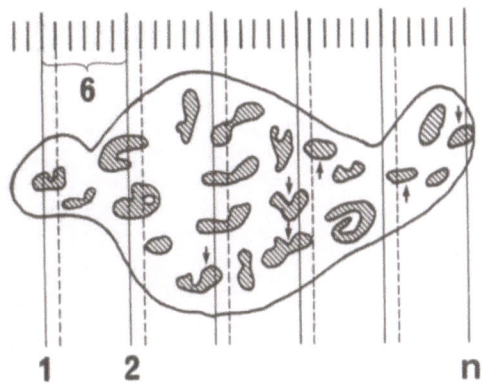

Fig. 1
    Two-dimensional schetch of a small organ or CNS-nucleus con-
taining N (=21) particles and cut with sections that contain the
total cross-sections of the organ, there are therefore no particles
on the edges of the sections. As indicated, 32 sections hit the
organ, but only every k'th (k=6) section (full drawn) together with
its neighbour (dashed) is used for analysis, a total of n pairs of
sections is thereby selected. The number of the very first section
to be used (=3) is selected at random among the first k-1 sections.
An unbiased estimate $\hat{N}$ of the number of particles is obtained by
counting in a section all the Q^- particles <u>not</u> hit by the neighbouring section in the pair, summing Q^- over all sections used and
multiplying ΣQ^- by k: $\hat{N} = k \cdot \Sigma Q^-$. When both sections in a pair is in
turn used as counting section the efficiency is doubled. In the
figure, $\Sigma Q^- = 4$ and 2 when the full drawn and the dashed sections
are used, respectively, wherefore $\hat{N} = 6 \cdot (4 + 2)/2 = 18$. In
practice two modifications may be needed. If the particles are
large compared to the section thickness neighbouring sections in a
pair may be spaced one or several sections. If a section hits many
particles and/or is very large a counting frame may be necessary,
see Sterio (1984). In all events, the number n should probably
always be from 3 to 10 per organ, depending on the homogeneity of
the spatial distribution of the particles.

<u>Stereological sampling probes.</u>

    The most significant single fact about all stereological
procedures is that the whole operation is carried out <u>in three</u>
dimensions: it is only when we combine the information on the sections (obtained with the stereological probes to be described now)
with our knowledge about how the sections were made that the results
are valid for the 3-dimensional structure.

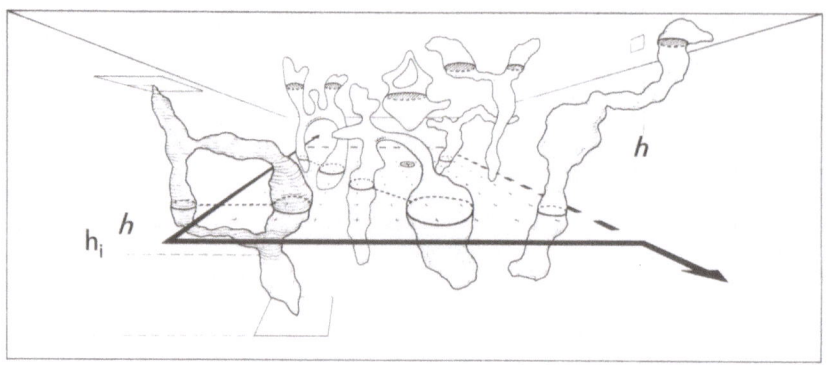

Fig. 2

A disector of distance $\underline{h}$ generating planar transects by cutting
through particles in a containing or reference space. One planar
transect of a particle is the union of all its profile(s) and the
drawn line(s) connecting them in the section plane. On the lower
section an unbiased counting and selecting frame is shown, see
Gundersen (1978). In planar sampling one counts all transects which
are completely or partially inside the frame provided they are not
in any way intersected by a fully drawn edge or an (infinite) exten-
tion line, a total of $Q = 4$ transects is counted in the example.
Two of the particles are not cut by the parallel plane, therefore
$Q^- = 2$. The frame is part of an integral test system where the encircled point is used for estimating the area of the reference
space within the frame, and the 'fine' points are used for estimating the volume of particles, a total of 9 of these hit particle
profiles in the figure. The height h_i and the upper-$\underline{h}$-part of a
particle in the direction $\underline{i}$ perpendicular to the planes of the
disector are shown at the front of the figure. That particular particle is not counted in the frame because the fully drawn line in-
tersects the (dashed) line connecting the two profiles in the
transect.

    There used to be just three stereological probes: 0-dimension-
al points, 1-dimensional lines, and 2-dimensional planes. By using
these probes it is in principle possible to obtain structural information in terms of 3-dimensional volume, 2-dimensional surface,
and 1-dimensional length or height, respectively; as already mentioned only some of these quantities are accessible on certain
types of sections. The precise methodology by which volumes are
estimated by counting points that hit structures, surfaces are
estimated by counting intersection-points between test lines and
surface traces on sections, and lengths or heights are estimated

by counting profiles profiles within test areas are all quite
simple and can be looked up in the general references already
mentioned. What is not quite so trivial, however, is the relationship between the dimension of any of the probes and that of the
structural quantity it estimates: they always add up to the dimension of the sampling space, 3 in this case. It is due to this
simple but strong fact that there exist no direct estimator of
number - a O-dimensional quantity - when using only points or
lines or independent sections. In order to estimate number without
making what is usually quite unrealistic assumptions about particle
shape a 3-dimensional probe must be used - and systematic sections
with a known distance between them may in effect be turned into
such probes. As mentioned by Cruz-Orive in this issue one can invert
the relationship $\bar{h} = \Sigma h/N$ into an estimator of N in sets of systematic sections the total height of which each must be greater than
twice the maximal particle height. The sampling of individual particles and the estimation of their mean height $\bar{h}$ requires special
attention as also described by Cruz-Orive (1980).

DIRECT COUNTING OF PARTICLES USING THE DISECTOR

As described by Sterio (1984) one can count particles directly
in the three dimensions using ordinarily just an unbiased 2-dimen-
sional counting frame on one section and a second 'look-up' section
which is parallel to the first one at a known distance h, see Fig.
2. The complete probe is called a disector and the estimate of
particle number is unbiased under the condition that one can identify in the disector all isolated profiles of a particle. In addition, the sections must be translucent, as most biological sections are, or h must be less than the minimal height of a particle.
A number of further practical details are discussed in the paper by
Sterio, only a few need to be mentioned here.
In order to count e.g.cells, one need not necessarily count the
whole cell in the disector, only some identifiable unit inside a
cell - under the obvious condition that each cell always has one
and only one such unit. The unit to be counted may be the nucleus
in mononucleated cells or a certain part of the cell that is stained with a specific stain. If such stains are used it is also worth
noticing that if stains A, B and C are used in sequence of every
third section then the disector may be formed by taking every third
section. By keeping track of cells through the sections with different stains additional information regarding coexistence of different components may evidently be obtained.

SAMPLING EFFICIENCY

The key to efficient sampling in biology is captured in a
phrase due to Weibel:'Do more less well', Gundersen and Østerby
(1981). The essential point is to put the work and the emphasis in
sampling at the highest sampling level where only rather few items

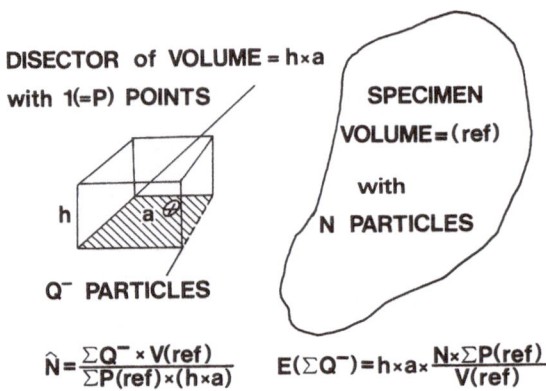

DISECTOR of VOLUME = h×a

with 1(=P) POINTS

SPECIMEN

VOLUME=(ref)

with

N PARTICLES

h   a

Q⁻ PARTICLES

$$\hat{N}=\frac{\Sigma Q^- \times V(ref)}{\Sigma P(ref)\times(h\times a)} \qquad E(\Sigma Q^-)=h\times a\times\frac{N\times\Sigma P(ref)}{V(ref)}$$

Fig.3

A disector of known volume = h · a is used for estimating the number N of particles in a specimen of known volume V(ref). The disector with distance h between sections and area a has a single (encircled) test point. ΣP(ref) is the total number of times this point hit the reference space, a number roughly equal to the number of disectors cut in the organ. As indicated by the formulae, one can manipulate the number of disectors used and the constants h and a of the disector such that the number ΣQ⁻ of particles counted in the specimen approximates a certain, predetermined value, 100, say. In that way sampling efficiency may be kept close to an optimum.

are studied. In biology that means that as many animals, say, as possible are studied by means of a number of sections or blocks in each animal. On each section a number of fields is studied, but within each of the now many fields only rather crude but fast observations are made. There is no need to measure or observe in each field with great precision since fields of necessity are quite different, instead one must observe many fields (and always use unbiased methods for the observation, if at all possible; biases are not eliminated by studying many fields).

Taking the example of estimating the number N of certain identifiable cells in a nucleus sampling may be approximately optimized as follows. The volume of the nucleus must be estimated by using e.g. Cavalieri's principle on systematic sections, see the paper by Cruz-Orive in this issue, and one must have an idea about the order of magnitude of N - if not, a pilot study in a few animals can be performed. The number of disectors cut should most likely be around 5. Using values of these parameters one is now in a position to generate the disector with constants h and a, see Fig. 3, such that the number of cells counted per animal is 100, for most studies a reasonable value. Assuming that N = 4500, V(ref) = 10 mm³

$(10 \cdot 10^9 \mu m^3)$ and $\Sigma P(ref) = 6$ and $\Sigma Q^- = 150$ we have
$\hat{N} = \{\Sigma Q^- \cdot V(ref)\}/\{\Sigma P(ref) \cdot h \cdot a\}$ from which we get $h \cdot a =$
$100 \cdot 10 \cdot 10^9/6/4500 \sim 0.04 \cdot 10^9 \mu m^3$. Taking a reasonable
value of $h = 20 \mu m$ the total area a to be used for counting in
each pair of sections is therefore roughly $2 \cdot 10^6 \mu m^2 \sim 1500 \cdot$
1500. Depending on the magnification this area may be suitable, if
not one can use an area of e.g. $400 \cdot 400 \mu m^2$ distributed over 16
fields of observation on each pair, to give just an example of a
sensible sampling scheme.

   None of the values stated above should be taken too seriously,
only their order of magnitude matters. It is, however, not likely
that the number of section pairs should have optimum far from 5 or
ΣQ^- far from 100.

SELECTION OF PARTICLES

   The disector has an important role for another use than just
counting particles: the unbiased selection of particles for further
study (serial sectioning followed by estimation of individual particle volume, shape, connectivity number etc. etc.) No matter the
reason for selecting particles from sections, it should never be
overlooked that particles cannot be selected just based on the fact
that they were hit by a section – such a sample is heavisy biased,
see e.g. the paper by Cruiz-Orive in this issue and Sterio (1984).
To put it short, only unbiased counting probes are unbiased selecting probes, and since the disector is the only known unbiased
counting probe in three dimensions it is the only known probe for
the unbiased selection of 3-dimensional structures.

REFERENCES

Cruz-Orive, L.M. (1980). On the estimation of particle number.
J.Microsc. 120, 15-27.

Gundersen, H.J.G. (1977). Notes on the estimation of the numerical
density of arbitrary particles: the edge effect. J. Microsc. 111,
219-223.

Gundersen, H.J.G. (1979). Estimation of tubule or cylinder L_V, S_V
and V_V on thick sections. J. Microsc. 117, 333-345.

Gundersen, H.J.G. and Østerby, R. (1981). Optimizing sampling
efficiency of stereological studies in biology or ' Do more less
well ! '. J. Microsc. 121, 65-73.

Loud, A.V. and Anversa, P. (1984). Biology of disease. Morphometric
analysis of biological processes. Lab. Invest. 50, 250-261.

Sterio, D.C. (1984). The unbiased estimation of number and sizes of
arbitrary particles using the disector. J. Microsc. 134, 127-136.

Weibel, E.R. (1979). Stereological Methods. Academic Press, London.

ESTIMATING PARTICLE NUMBER AND SIZE*

LUIS-M. CRUZ-ORIVE

Anatomisches Institut der Universität Bern, Bühlstrasse 26, Postfach 139, CH-3000 Bern 9, Switzerland

0. INTRODUCTION AND SUMMARY

This paper is mainly concerned with the stereology of particles. It has been written bearing the interests of quantitative synaptology in mind, but the results can be applied elsewhere.

The first section deals with the assumption-free estimation of geometric particle properties from serial sections, notably particle number. A ready-to-use version of an earlier method (Cruz-Orive, 1980a) is given which is now valid under arbitrary overprojection (due to section thickness) and truncation (namely unobservability of grazing particle fragments). The recent method of Sterio (1984) is also worth considering, especially for not too complex and not too small particles. In addition, relevant considerations on the proper use of quantitative information from serial sections are brought forward (section 1.4). It is shown that serial sectioning does not always guarantee the unbiased estimation of particle properties.

In general, an optimal combination of serial section and random section measurements should be sought. This is the aim of sections 1 and 2 taken together. The models in section 2 have to be rather restrictive, however, because satisfactory corrections of random section measurements affected by overprojection and truncation are known only for spheres and disks.

Finally, section 3 contains two models for 'platelets with dark spots' with the purpose of estimating the mean number of dense projections sitting on a presynaptic grid of arbitrary shape and connectivity.

_____

* Research supported by the Swiss National Science Foundation grants 3.762.80 and 3.524.83

11

Most of the methods described here were tried on extensive data kindly supplied by Didima M.G. De Groot (TNO Rijswijk, NL). Related results have been reported (De Groot & Bierman, 1983, De Groot, 1984). For reasons of time and space the present paper does not contain any numerical comparisons or cross-checks, however. This was planned for a future joint paper.

The skilful assistance of Mr K. Babl, Ms R. Fankhauser and Ms M. Schweizer in preparing the typescript is gratefully acknowledged.

1. ESTIMATION OF PARTICLE PROPERTIES FROM SETS OF SERIAL SECTIONS

1.1. Model

We consider a population of N disjoint particles of arbitrary size, shape and orientation within a solid Ω of volume V. We are interested in the number N and in geometric properties of an individual particle, such as volume, surface area, linear projection or 'caliper height' in a given direction, etc. We restrict ourselves to the most common case in which the particles, and hence the sections used, are very small in relation to Ω, so that edge effects can be ignored. Nevertheless, the essence of the methods is not affected by this assumption.

In order to estimate N, for instance, we use the identity

(1.1) $N = V \cdot N_V$

where $V = specimen\ volume$ and $N_V = N/V$, the numerical density of particles, have to be estimated separately. For the single specimen Ω at hand,

(1.2) $CE(estN) \simeq \{CE^2(estV) + CE^2(estN_V)\}^{1/2}$

where $'est'$ denotes $'estimate'$ and $'CE'$, $'coefficient\ of\ error'$ namely standard error divided by mean. For a collection of r specimens, the sample $mean(estN)$ and its estimated coefficient or error are computed using conventional statistics.

1.2. Estimating Specimen Volume

If V cannot be measured directly by fluid displacement (Weibel, 1979, p. 239), it can be estimated by systematic sectioning as follows.

(1.3) $estV = T \cdot g^{-2} \cdot (X_1 + X_2 + \ldots + X_m)$, or, if a coherent test system (Weibel, 1979; Cruz-Orive, 1982) is used to analyse each section

(1.4) $estV = T \cdot d^2 \cdot g^{-2} \cdot (P_1 + P_2 + \ldots + P_m)$

where

m :Number of systematic sections analysed.
T :Distance between the upper faces of two consecutive séctions analysed (not to be confused with section thickness.)
g :Magnification.
X_i:Area of the i-th section of the specimen.
d^2:Area per test point of the coherent test system.
P_i:Number of test points in the i-th section.

   If the recommendations (1) - (2) below are respected and if section thickness is negligible, the estimators (1.3), (1.4) are unbiased for arbitrary Ω and their precision is very considerable. If Ω is a triaxial ellipsoid arbitrarily orientated, then for (1.3), $CE(estV) = (1/\sqrt{5})m^{-2}$, (Cruz-Orive, 1984), Thus, in many cases $m = 5$ or 6 will give sufficient accuracy.

Notes

(1) The position of the upper face of the first section should be uniform random between 0 and T, (see Cruz-Orive & Weibel, 1981, Fig. 3a).

(2) If point counting is used, each section should be uniform random within a larger test system (Cruz-Orive, 1982, Fig. 2g)

(3) If the thickness of each section analysed is t, there remain $m-1$ slices of thickness $T-t$ from the material which can be used for subsequent analyses.

1.3. Estimation of N_V

   The material not used to estimate V (see (3) above) is sliced further and laid arbitrarily on a flat surface. The vertices of a coarse square test system may now be used to sample M small blocks of material. This procedure usually copes well with inhomogeneities within Ω. Each of the M blocks is cut into serial sections of a known thickness. A given section toward the middle of the series is taken as *reference section*. An approximately unbiased estimator of N_V is computed as follows:

(1.5) $estN_V = \sum\limits_{i=1}^{M} N_i n_i^{-1} T_i^{-1} \sum\limits_{j=1}^{n_i} m_{ij}^{-1} \Big/ \sum\limits_{i=1}^{M} A_i$

where

N_V = Number of particles per unit volume of the reference solid Ω.
M = Number of independent blocks cut into series
A_i = Reference area, namely the sampled area of the reference sec-
       tion from the i-th block ($i = 1,2,\ldots,M$).
N_i = Total number of particle sections (not just particle
       "fragments") counted in A_i, according to an unbiased counting
       rule (see Cruz-Orive, 1980a, Fig. 1 and Gundersen, 1977)
n_i = Number of particles in A_i tracked through the series.
T_i = Distance between the upper faces of two consecutive serial
       sections of the i-th series. See Fig. 1.
m_{ij} = Number of serial sections from the i-th block in which the
        j-th tracked particle appears. Note that j runs from 1 to n_i.

   Under certain assumptions (see Cruz-Orive, 1980b) a good estimate
of the standard error of $estN_V$ for the single specimen Ω is:

$$(1.6) \qquad SE(estN_V) = \left[(M-1)^{-1} \left\{ \sum_{i=1}^{M} y_i^2 x_i^{-1} \bigg/ \sum_{i=1}^{M} x_i - (estN_V)^2 \right\} \right]^{1/2}$$

where we have put $y_i = N_i n_i^{-1} T_i^{-1} \sum_{j=1}^{n_i} m_{ij}^{-1}$, $x_i = A_i$

Notes

(1) The orientation of each series may be chosen at will.

(2) The distance between the first and the last sections of a series
    must be at least twice the longest linear projection of a par-
    ticle in a direction perpendicular to the sections.

(3) If the particles have a simple shape then only every k-th sec-
    tion needs being observed (e.g. every 3rd in Fig. 1b). If the
    section thickness is t, then $kt = T$. If the particles have a
    complex shape then it may be necessary to observe every serial
    section whereby $k=1$ and $T = t$, as in Fig. 1a.

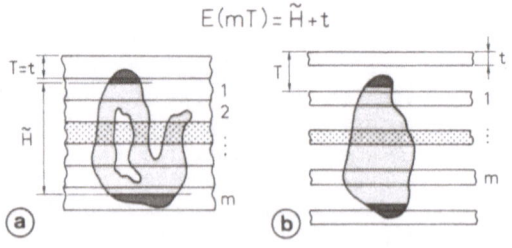

$$E(mT) = \tilde{H} + t$$

Fig. 1. Lateral views of particles hit by serial sections. The re-
     ference section is dotted. See text, 1.3.

(4) <u>Sample size</u>. Here, and elsewhere in this paper, the number of
blocks M per specimen should be established in terms of the variation between and within specimens, relative costs and goals
of the experiment. The reader is referred to Gundersen and
Østerby (1981), Cruz-Orive & Weibel (1981) and references there-
in. To give an orientative figure, $M = 4$ or 5 blocks per specimen (animal) often proves sufficient in electron microscopic
morphometry, provided that the number of animals is properly
chosen according to the biological variation of the parameter
at hand.

1.4. General Considerations on the Estimation of Geometric Particle Properties from Serial Sections

Let ψ be a geometric particle property (e.g. volume v, surface
area s, linear projection H on a given direction, etc.), and $E\psi$ its
population mean over Ω. In order to estimate $E\psi$ suppose that we take
a small block at random from Ω and we cut it into serial sections
of thickness t. As usual, a section toward the middle of the series
is taken as *reference section*. The particles hit by the reference
section do not constitute a simple random sample from the population,
however. Failure to appreciate this will yield biased results. In
fact the sample is $(\tilde{H} + t)$ – *weighted* because a particle is hit and
observed with a probability proportional to $\tilde{H}+t$. See Fig. 2. Here $\tilde{H}$
is the *effective linear projection* of a particle, namely its linear
projection H minus the linear projection of the truncated part.
Therefore

(1.7) $E(\psi|\dagger) = (E\tilde{H}+t)^{-1} E\{(\tilde{H}+t)\psi\}$

where $(\psi|\dagger)$ denotes the property ψ for a *hit* particle. Note that
what can be measured with the aid of serial sections is $(\psi|\dagger)$ but
not ψ. Unfortunately the identity (1.7) shows that $E\psi$ cannot be
estimated in general using the present serial section approach, con-

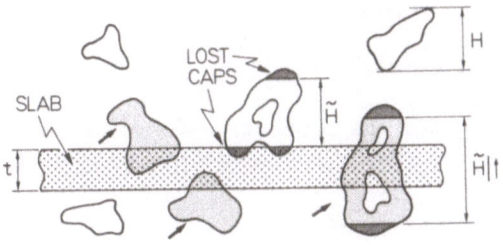

$(\diagup)$: PARTICLES IN $(\tilde{H}+t)$ – WEIGHTED SAMPLE

Fig. 2. Lateral view of a population of arbitrary particles hit by
a slab. See text, 1.4.

trarily to what it is often believed, because ψ enters as a product $\tilde{H}\psi$ in the right hand side of the identity.

An important and general result in this context, based also on the fact that the probability of hitting a particle and observing the profile is $(\tilde{H}+t)A/V$, is the following

(1.8) $N_V = EN_A (E\tilde{H} + t)^{-1}$

where EN_A is the expected number of hit and observed particles in a section of area A divided by A. Fortunately, putting $\psi = (\tilde{H}+t)^{-1}$ in (1.7) we see that $(E\tilde{H}+t)^{-1}$ can indeed be estimated as follows:

(1.9) $est(E\tilde{H}+t)^{-1} = n^{-1} \sum_{j=1}^{n} \{(\tilde{H}_j|\dagger) + t\}^{-1}$

where n is the number of particles hit by the reference section whose effective linear projection $(H_1|\dagger)$, $(H_2|\dagger)$, ..., $(H_n|\dagger)$ can be estimated. The latter step is carried out by means of the following result (which can be proved by integral geometry methods, see Santaló, 1976, chapter 8 for excellent reference):

(1.10) $est\{(\tilde{H}_j|\dagger) + t\} = m_j T,$

see Fig. 1a,b. Thus, it suffices to count m_j and to know T. Combining the results (1.8) – (1.10) we obtain an estimator of N_V for a single block. If M independent blocks with reference areas A_1, A_2, ..., A_M are available, then the i-th N_V estimate may be weighted by $A_i/(A_1 + \ldots + A_M)$, (preferably to $n_i A_i/(n_1 A_1 + \ldots + n_M A_M)$ as proposed in Cruz-Orive, 1980a), whereby the final result (1.5) is obtained.

Suppose now that a coarse test system is superimposed on a reference section of negligible thickness and consider the sample of particles hit by the test points only. The particles hit by the reference section *and* by a test point constitute a *volume-weighted* sample (Cruz-Orive, 1980a, p.27). Thus,

(1.11) $E(\psi|\dagger,p) = (Ev)^{-1}E(v\psi)$

where $(\psi|\dagger,p)$ denotes the property ψ for a particle hit by the reference section $(\dagger)$ *and* by a test point (p). Setting $\psi = 1/v$,

(1.12) $\{E(1/v|\dagger,p)\}^{-1} = Ev$

which means that Ev can be estimated by the *harmonic mean* of the volumes of n particles from the volume-weighted sample. The formula (1.12) is relatively expensive to apply, however. In practice it seems preferable to estimate Ev via the identity

(1.13) $Ev = V_V/N_V$

where $V_V = $ *(Total particle volume in* Ω*)/V* can be estimated simply by point counting on random reference sections only (namely one section per block) whereas N_V is estimable via (1.5) from serial sections.

Finally, *if the particles are convex* and we consider those particles hit by a reference section *and* by the test lines of a test system superimposed on the reference section we have a *surface-weighted* particle sample, whereby

(1.14) $E(\psi|\dagger,\ell) = (Es)^{-1}E(s\psi)$

where $(\psi|\dagger,\ell)$ denotes the property ψ for a particle hit by the reference section $(\dagger)$ *and* by a test line (ℓ). Setting $\psi = 1/s$

(1.15) $\{E(1/s|\dagger,\ell)\}^{-1} = Es.$

Unfortunately, even under ideal conditions ($t = 0$ and no capping) the measurement of particle surface area from serial sections cannot be recommended. Any algorithm used must critically depend upon particle shape. Thus, in order to estimate Es there seems to be little choice but to use

(1.16) $Es = S_V/N_V$

where $S_V = $ *(Total particle surface area in* Ω*)/V* is estimable from random sections by intersection counting with the lines of a coherent test system, whereas N_V is estimable via (1.5). Moreover (1.16) is valid for arbitrary particles whereas (1.15) is not.

1.5. Estimation of Particle Properties in Special Cases

We consider the estimation of the population mean EH of particle linear projection in a given direction u. Clearly the serial sections used must be perpendicular to u. For arbitrary particles the quantity EH is not a very powerful size descriptor. However, let us suppose that:

(i)   The particles are convex platelets.
(ii)   They are isotropically orientated.

Then H becomes equivalent to the mean linear projection of a particle over isotropic directions in space, EH is its population mean and the following relations hold:

(1.17) $EL = 4EH,$ $ED = \pi^{-1}EL = (4/\pi)EH$

where:

L : Boundary length of a platelet
D : Mean platelet diameter, that is the mean linear projection of a
      platelet within its support plane.

Returning to arbitrary particles we have, from (1.8):

(1.18) $est(E\overset{\rightharpoonup}{H}) = N_A/estN_V - t.$

However, the problem remains of relating $E\overset{\rightharpoonup}{H}$ to EH. A useful approach
is to assume $\tilde{H} = KH$, where K is some constant, $0 \leq K \leq 1$. Unfortunately, estimating K requires a precise modelling of particle shape
and of capping mechanism.

For the disk model illustrated in Fig. 3 we have $K = cosZ$. For
arbitrarily orientated disks Z can be estimated from serial sections
as follows. For the i-th block,

(1.19) $est(sinZ_i) = \sum\limits_{j=1}^{ni} min\ B_{ij} / \sum\limits_{j=1}^{ni} maxB_{ij}$

where n_i is as in (1.5) whereas $min\ B_{ij}$, $max\ B_{ij}$ denote respectively
the shortest and the longest intercept length observed in serial
sections for the j-th tracked disk. Formula (1.19) is reasonable if

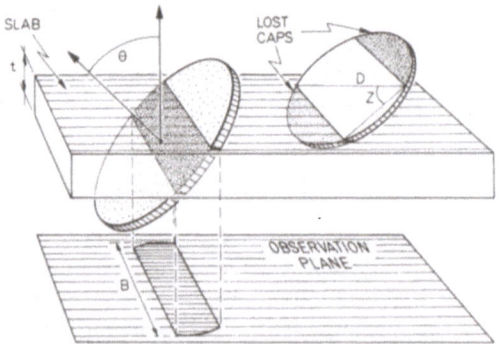

Fig. 3. *Disk model* with the following assumptions. (1) The particles
       are disjoint opaque disks of variable diameter D but of neg-
       ligible thickness. (2) The embedding matrix is translucent
       and there is overprojection effect. (3) *Capping Z-model:* a
       disk segment subtending an angle of less than $2Z$ from the
       disk centre is not observable. Z is independent from disk
       diameter (D) and orientation (θ) but not necessarily from
       section thickness t.

the overprojection effect is not too acute, otherwise Z_i will tend to be slightly overestimated. Now if M blocks are analysed,

(1.20) $est(sinZ) = \sum\limits_{i=1}^{M} n_i est(sinZ_i) / \sum\limits_{j=1}^{M} n_j$

(1.21) $est(EH) = \{est(cosZ)\}^{-1} (N_A/estN_V - t)$.

The preceding formulae assume that the section thickness t is approximately constant for all blocks. Oterhwise, Z_i is also expected to vary and the analysis becomes more complicated. In fact in this case instead of the simple equation (1.21) we have:

(1.22) $estN_V = \sum\limits_{i=1}^{M} N_i (K_i.EH + t_i)^{-1} / \sum\limits_{i=1}^{M} A_i$

which can be solved numerically for EH after inserting $K_i = est(cosZ_i)$.

Notes

(1) If the disks are isotropic then by (1.17) we see that $est(EH)$ suffices to estimate ED but not $Es = (\pi/4)ED^2$ because $ED^2 \neq (ED)^2$. For Es the identity (1.16) should be used.

(2) It can be shown that formulae (1.19) - (1.21) are valid also for *elliptical platelets* if $\tilde{H} = KH$ and if K is constant in all directions. More precisely, in this case $K = \{1-(minB/maxB)^2\}^{1/2}$ for any direction of sectioning.

2. ESTIMATION OF PARTICLE PROPERTIES FROM RANDOM SLABS.

From each of M blocks from Ω, taken for instance as indicated in section 1.3, a single section is taken for analysis. Let A be the total section area. Under proper sampling and ideal conditions, namely $t=0$ and no truncation, the following estimators can be used (summations are over i or j and run from 1 to M):

(2.1) $estV_V = estA_A \quad\quad = P_P = \Sigma(a_1/p_1)g_i^{-2}P_{1i}/\Sigma(a_2/p_2)g_j^{-2}P_{2j}$

(2.2) $estS_V = (4/\pi)estB_A = 2I_L = 2\Sigma(a_1/\ell_1)g_i^{-1}I_i/\Sigma(a_2/p_2)g_j^{-2}P_{2j}$

(2.3) $estN_V = N_A/est(EH) = \{\Sigma N_i/\Sigma(a_2/p_2)g_j^{-2}P_{2j}\}/est(EH)$

where

A_A = (Observed particle section area in A)/A, which is estimated by
      P_P.
B_A = (Observed particle section boundary length in A)/A, which is
      estimated by $(\pi/2)I_L$.
N_A = (Observed particle profile number in A)/A.
a_k, ℓ_k, p_k, $(k = 1,2)$: Area, test line length and number of test
      points, respectively, of the fundamental 'tiling' region of
      the coherent test system used (see Cruz-Orive, 1982, Fig. 4).
g_i : Magnification of the i-th section.
P_i,I_i : Number of test points and of test line intersections, re-
      spectively, with the features of interest.

      If $t > 0$ and there is truncation (hardly avoidable if the parti-
cles are very small) then formulae (2.1) – (2.3) will yield biased
results. Overprojection causes these formulae to yield overesti-
mates, whereas truncation acts in the opposite direction. Unfortunately, the net amount of bias can only be estimated under rather
severe model conditions. Here we consider briefly the cases of
spheres (treated in more detail in Cruz-Orive, 1983) and *disks* (Fig.
3, also treated in the latter paper, sect. 9.6), under the Z-capping
mechanism. For additional models and methods see Weibel,(1980,
chapter 4).

Spheres

$$(2.4) \qquad estV_V = 2(3cosZ-cos^3Z)^{-1}\{P_P-(t/4)estS_V\}$$

$$(2.5) \qquad estS_V = \{1-(2/\pi)(Z-sin2cosZ)\}^{-1}\{2I_L-4t(EH)estN_V\}$$

$$(2.6) \qquad estN_V = N_A\{t+(EH)cosZ\}^{-1}$$

where P_P is given in (2.1), $estS_V$ by (2.5), I_L in (2.2), $estN_V$ by
(2.6) and N_A in (2.3). Yet, of course, there remains to estimate Z
and the mean sphere diameter EH. In order to do this from random
sections the unfolding procedure described in Cruz-Orive (1983,
(4.10)) must be carried out. A good alternative is to use the serial
section methods described in section 1 to estimate N_V, EH and Z,
then to apply (2.4) and (2.5).

Disks

      The model is illustrated in Fig. 3. The following estimators can
be used if the disks are isotropically orientated.

$$(2.7) \qquad estS_V = \{1-(2/\pi)(Z-sin2cosZ)\}^{-1}\{2I_L-(4/\pi)t(ED)estN_V\}$$

$$(2.8) \qquad estN_V = N_A\{t+(EH)cosZ\}^{-1} = N_A\{t+(\pi/4)(ED)cosZ\}^{-1}$$

where ED = *mean disk diameter*. In (2.7), $2I_L$ can be replaced with

$$(4/\pi)B_A = (4/\pi)\Sigma(\textit{Observed disk intercept lengths in A})/A.$$

The equation (2.7) uses the following identity

(2.9) $EB^k = \{(4/\pi)t+(ED)cosZ\}^{-1}\{(4/\pi)t(ED^k)+\Phi_{k+1}(Z)(ED^{k+1})\}$,

valid for $k>-2$. In (2.9), B denotes individual disk trace length in a section (Fig. 3) and $\Phi_n(Z) = \int sin^n x dx$, the integral being from Z to $\pi/2$. The identity (2.9) is analogous to (7.9)of Cruz-Orive (1983) for spheres, and it can be exploited to estimate ED if Z is known (for details see the latter paper, section 5.) If Z is not known, then either the unfolding procedure for disks has to be carried out, or serial sections have to be used (section 1.5). The unfolding procedure uses a sample histogram of B as input and yields an estimated histogram for D. For details see Cruz-Orive, (1983, section 9.6).

3. MODELS FOR THE NUMBER OF DENSE PROJECTIONS ON A PRESYNAPTIC GRID

3.1. The Tile Model

Consider an unbounded lattice of *fundamental regions*, coincident up to a translation and tiling the plane without overlapping. In Fig. 4 the fundamental region is an hexagon. Each fundamental region contains a *convex fundamental domain*, (a circle in Fig. 4). The fundamental domain will play the role of a presynaptic dense projection. Now consider a plane figure of arbitrary connectivity, bounded by a

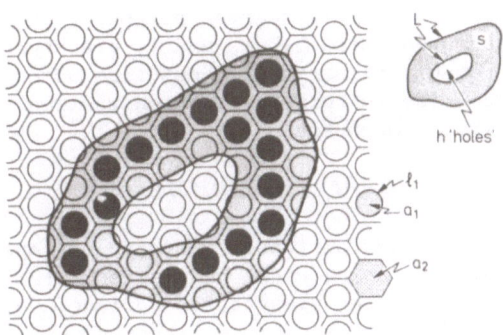

MEAN NUMBER OF 'BLACK DOMAINS' = $\dfrac{s+(1-h)a_1}{a_2} - \dfrac{\ell_1 L}{2\pi a_2}$

Fig. 4. A model for the number of dense projections sitting on a presynaptic grid of arbitrary connectivity. See text, section 3.1.

finite number of piecewise smooth closed curves, placed with isotropic uniform randomness on the lattice. This 'mobile' figure will
play the role of a presynaptic grid. The problem is to determine
the expected number of fundamental domains completely inside the
mobile figure.

Assumption. For any position of the mobile figure, the intersection
between its boundary curves and the boundary of a fundamental domain cannot consist of more than two points.

Classical methods of integral geometry (e.g. Santaló, 1976,
chapter 8) yield the following result:

$$(3.1) \qquad En_0 = \{s + (1-h)a_1\}/a_2 - \ell_1 L/(2\pi a_2)$$

where

n_0: Number of fundamental domains completely inside the mobile
      figure.
s : Area of the mobile figure, i.e., area enclosed by the outer
      boundary minus area of holes.
L : Total length of boundary curves of the mobile figure.
h : Number of holes of the mobile figure ($h = 1$ in Fig. 4).
a_1: Area of the fundamental convex domain within a fundamental re-
      gion.
ℓ_1: Boundary length of the fundamental convex domain.
a_2: Area of the fundamental region.

Estimation. In practice the quantities a_1 and ℓ_1 depend on dense
projection size, whereas a_2 can be calculated from the distribution
pattern of the dense projections (e.g. equilateral triangular lattice in Fig. 4) and from their nearest neighbour distance. If for a
given specimen the quantities a_1, a_2 and ℓ_1 are approximately constant, they can probably be estimated best from orthogonal views of
a few presynaptic grids in relatively thick sections. The population version of (3.1) now reads

$$(3.2) \qquad En_0 = \{Es + (1-Eh)a_1\}/a_2 - \ell_1(EL)/(2\pi a_2).$$

If a_1, a_2, ℓ_1 are not constant, then they may be replaced with their
respective mean values in the right hand side of (3.2), but this introduces some bias. The mean presynaptic surface area Es can be
estimated via (1.16), even though the necessary equation (2.5) is
not exact for non-circular presynaptic grids. On the other hand, EL
includes outer and inner mean boundary length. If the presynaptic
grids are isotropic, approximately convex and non-perforated then
$EL = 4EH$ (see (1.17)) and in turn EH can be estimated via (1.21).
For isotropic grids perforated by approximately convex holes, the

contribution of the latter to EL can be estimated by the same approach. Finally Eh is the ratio of the numerical densities (N_V) of holes to that of presynaptic grids, both estimable at least by serial sections via (1.5). Note that if $Eh \approx 1$ then a_1 does not need to be estimated.

3.2. The Conic Model.

Here we regard the dense projections sitting on a given type of presynaptic grids as a population of conic particles in space with a numerical density of n_V per unit volume of specimen. Then we have

(3.3) $En_0 = n_V/N_V$

where N_V is the numerical density of presynaptic grids of the chosen type in the specimen. Here N_V can be estimated from serial sections via (1.5) whereas n_V can be estimated either from serial sections as well, or via the following analogue of equation (1.8):

(3.4) $n_V = n_A (E\tilde{H}_0 + t)^{-1}$

(3.5) $E\tilde{H}_0 = K_0 EH_0$, $(K_0$: truncation constant),

(3.6) $H_0 = (r/2)\{\pi + a/r - arctan(a/r)\}$,

the latter expression being the mean linear projection of a cone of radius r and height a. Whereas assessing r,a may not be too difficult, estimating K_0 remains a problem. Yet, since H_0 is here often less than t, the value of K_0 is not too critical. $K_0 \approx 0.4$ was a reasonable try for the data we came across.

REFERENCES

Cruz-Orive, L.M. (1980a). On the estimation of particle number. J. Microsc. 120, 15-27.

Cruz-Orive, L.M. (1980b). Best linear unbiased estimators for stereology.   Biometrics, 36, 595-605.

Cruz-Orive, L.M. (1982). The use of quadrats and test systems in stereology, including magnification corrections. J. Microsc. 125, 89-102.

Cruz-Orive, L.M. (1983). Distribution-free estimation of sphere size distributions from slabs showing overprojection and truncation, with a review of previous methods. J. Microsc. 131, 265-290.

Cruz-Orive, L.M. (1984). Estimating volumes from systematic sections. Universität Bern, Anatomisches Institut Internal Report Nr. 115/LC18. Submitted for publication.

Cruz-Orive, L.M. and Weibel, E.R. (1981). Sampling designs for stereology. J. Microsc. 122, 235-257.

De Groot, D.M.G. (with an appendix by L.M. Cruz-Orive)(1984). Improvements of the serial section method in relation to the estimation of the numerical density of complex-shaped synapses. In: Stereology in Pathology (eds. A. Reith and T.M. Mayhew). Hemisphere/ McGraw-Hill, New York, (to appear).

DeGroot, D.M.G. and Bierman, E.P.B. (1983). The complex-shape 'perforated' synapse, a problem in quantitative stereology of the brain. J. Microsc. 131, 355-360.

Gundersen, H.J.G; (1977). Notes on the estimation of the numerical density of arbitrary profiles: the edge effect. J. Microsc. 111, 219-223.

Gundersen, H.J.G. and Østerby, R. (1981). Optimizing sampling efficiency of stereological studies in biology: or 'Do more less well!' J. Microsc. 121, 65-73.

Santaló, L.A. (1976). Integral Geometry and Geometric Probability. Addison-Wesley, Reading, Massachusetts.

Sterio, D.C. (1984). Estimating number, mean sizes and variations in size of particles in 3-D specimens using disectors. J. Microsc., (to appear).

Weibel, E.R. (1979/80). Stereological Methods, Vols. 1 and 2. Academic Press, London.

TASKS IN COMPUTER–ASSISTED NEUROANATOMY: DATA ACQUISITION, IMAGING AND DATABASE

DONALD J. WOODWARD, WADE K. SMITH,
DANIEL S. SCHLUSSELBERG, S. AUSIM AZIZI
and JOHN K. CHAPIN

Department of Cell Biology, University of Texas Health Science Center at Dallas, Dallas, Texas 75235, USA

INTRODUCTION

There is a growing use of computer technology for neuroanatomical studies. First uses began nearly as soon as small laboratory computers became available in the late 1960's and early '70's. The benefits of being able to quantitate what could only be photographed and verbally described were obvious to many workers. It was also a hope that machine storage of two and three dimensional information would lead to new methods of imaging of what could only be represented numerically. However, it was not clear at the outset, perhaps due to the novelty of the concept, how long these procedures would take to develop, and exactly what power and capacity the computing machines would be needed in order to bring the concepts into regular use in the neuroanatomical laboratory.

BACKGROUND

One of us (DJW) (with Dr. Alan Selverston, at University of California) had the pleasure of organizing a planning session on Computer Assisted Neuroanatomy for the Division of Research Resources in 1976 in San Diego, California, USA. Seventy scientists gathered to debate the needs thought to exist for applications of computers in anatomy. It became clear that many parallel, but independent efforts were underway to study biological tissue at the gross, light or electron microscopic levels. Data acquisition, either manually or through automated techniques, was a major concern. Alignment of serial section data required inventive new strategies. Imaging and plotting of two and three dimensional data arrays was recognized as a universal problem. Numerical analysis was needed for all types of data. It was also recognized that data archiving and sharing between laboratories could become commonplace. The cost of the labor involved in preparing programs even then seemed greater than the cost of the equipment, and that

25

distribution of software, if possible, would greatly benefit the neuroscience community.

One product of the planning meeting was a design for a comprehensive neuroanatomical computer-based analysis system which would satisfy a wide range of needs. In our laboratory at Dallas we have directed our efforts toward creating a multipurpose system. From the outset we hoped to design the software programs and hardware with the maximum versatility for different applications in anatomy. The CARP system (Computer Aided Reconstruction Program) has been developed from this point of view. In many ways the basic host computer and programming language has not changed in the past decade. The computer memory and hardware components have become far less expensive but programming costs are greater, so that total costs probably remain similar. The major technical advance has been the introduction of the image analysis and graphics computers specifically designed for acquisition and synthesis of video images of a high resolution.

The development of CARP began in earnest with the acquisition of the first commercially produced high resolution raster graphics system from Ikonas Inc. (later acquired by Adage Inc.) The unique combination of features included video digitization, a flexible color look-up table and graphics controller and a special purpose bit-slice image processor designed to allow a range of image analysis and sophisticated graphics operations to be done at a speed which made practical a wide range of applications in anatomy.

Three major tasks have emerged as functions of the CARP system. Morphological information is: 1) acquired by a variety of strategies, 2) imaged by an assortment of algorithms, and 3) manipulated by a complex data base structure.

HARDWARE COMPONENTS

As shown in Figure 1. the core of the system (Smith, et. al., 1981 and Schlusselberg, et. al., 1982a) is a host computer time shared between the input systems. We thought at the start that one-half megabyte of main memory and a 200 megabyte disk drive would be more than adequate. However, in time the programs grew increasingly complex to accomodate the variations requested by many individual users. Our current version of CARP will require a minimum of 2 megabytes of main memory with a virtual memory operating system and a minimum of 300-500 megabytes of disk space. To a great extent computers with the Unix operating system with Fortran and C languages may soon emerge as a commodity in which very complex software systems can be run by machines from many manufacturers. The graphics computer is a newer concept for which few standards have emerged. The Adage 3000 raster graphics system provides for video digitization, 4 megabytes of image memory

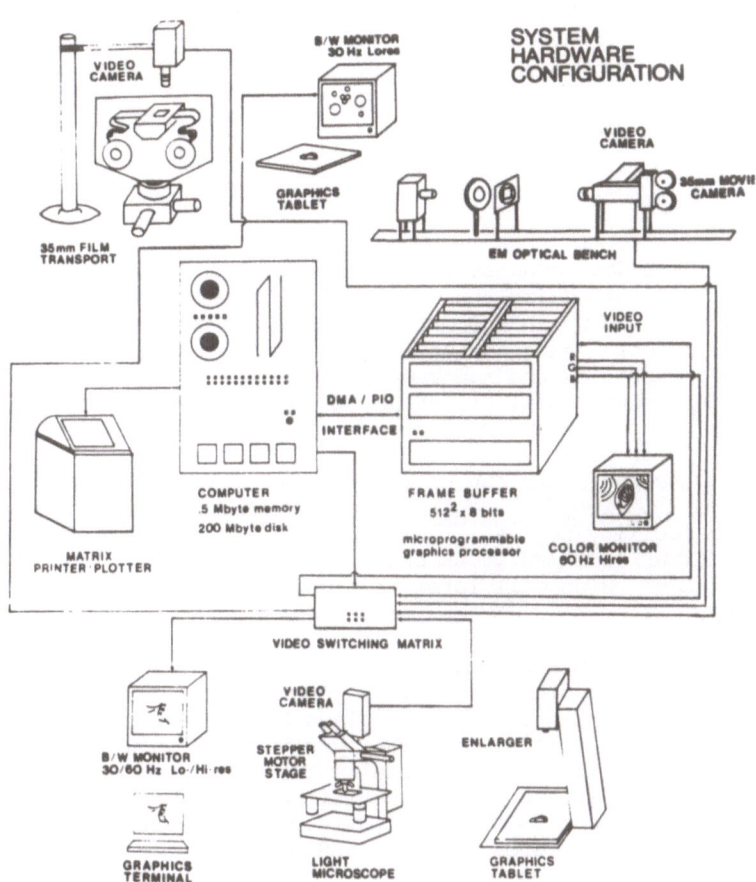

Figure 1. This computer system is built around a DATA GENERAL
Eclipse S/130 16-bit minicomputer with 0.5 megabytes of memory and
200 megabyte disk drive, and an ADAGE 3000 Raster Graphics System
with high-speed microprogrammable graphics processor.

(displayable as 512 x 512 or 1024 x 1024), color coding of video
output, and the microprogrammable graphics processor. Data input
modules include tablet with hand-held cursor for drawing lines and
points of projected images. A tablet, stepper motor stage, drawing
tube, and bit map graphics terminal are used to manually input data
while viewing microscope slides. Finally, a high-precision film
transport with movable platform, a high resolution video camera,
and microfilm viewing system allows for analysis of serial electron
microscopic sections.

SOFTWARE COMPONENTS

Data Input

     The input data acquisition routines are divided into modules
with commands suited for the operation of given hardware. TRACE is
a set of routines which support drawing diagrams consisting of
lines and points. Axes and scales are specified. Labels are
applied to each line and subgroups of points. The TRACE routine
allows morphometric analysis of any image represented as a
histologic or photographic slide projected onto a tablet.
Photographic or other images on paper can be traced. Files
containing segment data can be plotted and information may be
printed on line lengths, centroids and areas of closed contours.
The TRACE facility has been used extensively to determine areas of
cross-sections of human gross brain material or to outline brain
cross-sections which will later be studied under higher power
microscopy.

     MICRO contains routines for plotting data with a light
microscope, tablet drawing tube, and bit map graphics terminal.
The host computer issues command pulses to drive a stepper motor
stage to specified positions. The operator inputs an origin and
axis for each section of biological material. The stage can be
moved randomly, directed toward positions in biological
coordinates, or made to move in a grid-like pattern for systematic
examination of large areas. Lines and points drawn on the tablet
are converted in the host computer into floating point numbers in
the biological coordinate system. Conversion routines plot lines
and points on the graphics terminal. The bit mapped screen is
viewed through the drawing tube. Calculations result in lines and
points being superimposed on the image viewed through the
eyepiece. This arrangement allows easy comparison between visual
images and overlays of what has been stored in the computer
memory. Data input from sequential sections can then be aligned to
create a structure in three dimensions. The TRACE and MICRO
routines seemed simple in concept but have grown to a system of 900
subroutines to accomodate all of the convenience features suggested
by the users.

MICRO is currently being expanded to incorporate a cell recognition subroutine (Schlusselberg, et. al., 1982b) which runs in the high-speed microprogrammable processor. Cell fields with distinctly different grey scale intensities are video digitized. The routine, CSCAN, searches the field for seed points within a pair of intensity thresholds. Dark spots are filled with a scanning routine which determines areas, centroids, and pixel contents. An outline around the blob can be drawn. Areas too large or small can be ejected or marked for operator decision. The video field illustrated in Figure 2. can be scanned in about one second and a list of cell locations and cell descriptions is made available to the host computer on each field. A wide range of quantitative information can potentially be obtained from regions manually outlined or areas determined automatically with CSCAN.

The input mode RAD provides for analysis of film autoradiographs placed on a light box and video digitized (Schlusselberg, et. al., 1983). Video addition is used to average noise. A background image is subtracted to correct for nonlinearities in the light source and electronics of the Newvicon camera, Subroutines of the TRACE mode can be employed to outline subregions of a brain cross section for densitometer analysis. The unique feature of CARP-RAD is an extension of a two dimensional to a three dimensional autoradioraphic analysis system. In our preliminary work subregions of serial sections through the cerebral cortex have been scanned to determine average densities along that axis. Side views of the cortex can be generated showing spatial regions of hightened metabolic activity using the Carbon-14 2-deoxyglucose method for measuring glucose uptake. We anticipate that such techniques will find applications in examining spatial patterns in autoradiographic data from metabolic or receptor binding data.

FILM is another input routine currently being constructed. It will control a 35mm precision film transport and a high resolution video camera for viewing the microfilm copies of aligned serial electron micrographs. FILM consists of many of the functions of TRACE, MICRO and RAD with special routines to control the film advance and film transport micropositioning system. In this way the serial electron microscope reconstruction methods pioneered by Levinthal and associates (Macagno, et. al., 1979) can be included with minimal special purpose programming.

Finally, a future planned acquisition mode will be GEL, a system for study of the spots visualized by two dimensional gel electrophoresis method for separation of proteins. Many of the tasks needed to digitize gels and quantitate spots already exist in TRACE, MICRO and RAD.

D.J. Woodward, et al.

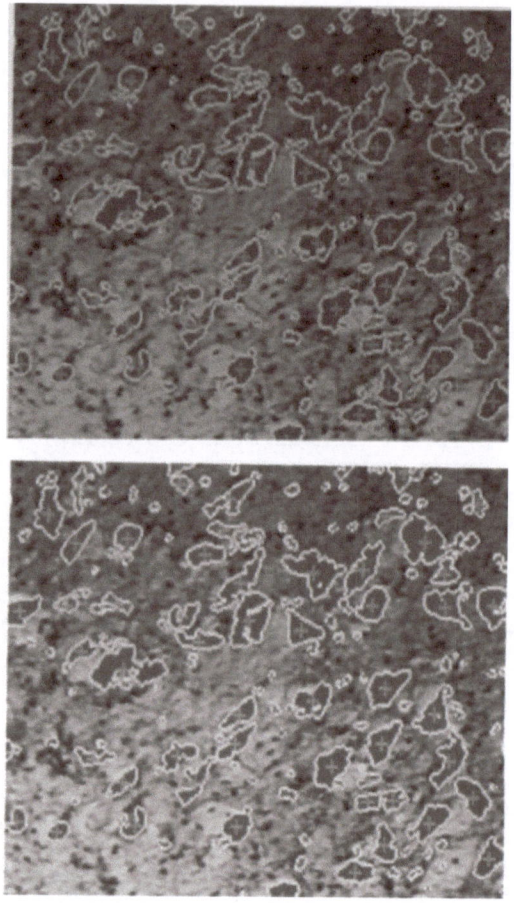

Figure 2. (Upper) Video digitized view of neurons in the substantia
nigra of human.
(Lower) Microcoded cell recognition has located, color coded and
outlined regions within a specified range of areas.

Data Display

The most common two-dimensional representation of an anatomical object is a line traced around its external boundary as viewed in a serial section. These line drawings are stored as ordered vertex lists, where the first point in the list is graphically represented as a "move" (pen up), and subsequent points are "draws" (pen down). Similarly, graphical symbols are used to represent single point data such as cell positions, autoradiographic grains, synapses, etc. Symbol definitions are normalized coordinates stored in the program or data file and are translated to an anatomical coordinate position for final display. A filled region, such as a pigment-containing cell body, can be represented by center of mass, area, and intensity, when using digitized video images for data acquisition.

The graphical display of digitized information requires the definition of an anatomical window through which a part of the anatomical coordinate area is viewed. This requires clipping out portions of data that are outside of the window. The anatomical window is then mapped to a physical display device by using display coordinates which define a viewport. The window can be combined with images of objects seen in data input devices or manipulated by the user to focus in on certain regions and create summary diagrams of all data entered. The current parameters used to define this window are stored in the data file, so that it is not necessary for a user to redefine them each time a data file is examined.

There are several strategies for creating three-dimensional models of previously digitized anatomical data. By specifying certain viewing parameters and applying a perspective transformation, display coordinates are generated from vertex lists to create three-dimensional line drawings. Depth perception in these displays can be enhanced by using line intensity or thickness to represent distance from the eye or by hidden-line algorithms, which assume that each vertex list represents an opaque polygon.

Our efforts have been directed toward exploring new techniques for computer reconstruction of three-dimensional surfaces of objects which have been serially sectioned. Triangulation algorithms are used to generate an ordered polygon list from two vertex lists representing the points defining two "contours" of an object (Fuchs, et. al., 1977). An "object" in this case is usually defined as a surface bordering an anatomic region. A contour represents the external boundary of an object which has been sectioned by a plane. Contours are derived from segments in sequential sections.

There are several requirements of triangulation algorithms which must be satisfied by the database. For any given section, several segments can be digitized, representing different objects

that have been sliced in that section. Each segment that belongs
to one object will generate one contour. Techniques have been
developed to specify which contours belong to the same object in
different sections. If an object branches, it will generate more
than one contour in each section. Special triangulation algorithms
are needed to generate a proper set of polygons for branching
surfaces when given more than two contours as input. It is also
necessary to keep track of the branching hierarchy in biological
objects, since this may contain important structural and functional
information. The simplest way to accomplish this is to store a set
of sequential contours as one branch, which represents one portion
of a three-dimensional surface. Each branch serves as input to a
triangulation algorithm, since it only contains one contour per
section. By manipulating portions of surfaces in this way it is
possible to represent surfaces of considerable complexity by raster
scan computer graphics.

     For data display, there exists a package of graphical output
routines that operates on the data structure, as described below,
to output graphical primitives. In 2D mode, vector lines, points
and symbols are plotted on a variety of devices. In 3D mode, lines
can be assigned color and intensity and line width to achieve depth
shading. As plotting progresses in the frame buffer the Z depth
positions can be stored and compared pixel by pixel. This
technique allows calculation of hidden surfaces and transparencies
(Figs. 3,4). Illumination can be calculated from information
about normals to the polygons which define the surface. The
Gouraud (1971) method of producing shading and illumination
optimizes use of pseudocolor and high speed of microcode programs.
The Phong (1975) method uses a full twenty-four bits for full color
representation. A more satisfactory view of a solid plastic-like
surface is produced but at the expense of cpu time. To produce a
high quality image, a user needs to position calculated light
sources, determine reflectance values and set many parameters,
including eye position and distance, much as with a conventional
photographic or illustration process (Figs. 5,6).

Data Management

     Our current programming environment for efficient data
manipulation is in the 'C' programming language (Kernighan and
Ritchie, 1978), under the AOS Operating System, for a Data General
Eclipse S/130 16-bit minicomputer with a 200 megabyte disk and
magnetic tape backup. We have developed a virtual file system for
flexible interaction with data entered into the computer. This
allows programs and data to vastly exceed host memory limits and
simplifies I/O management. All data is physically stored in a
binary file which is treated as an extension of the program's
address space. The virtual memory software package can address any
element of the file through 32-bit pointers to byte locations.

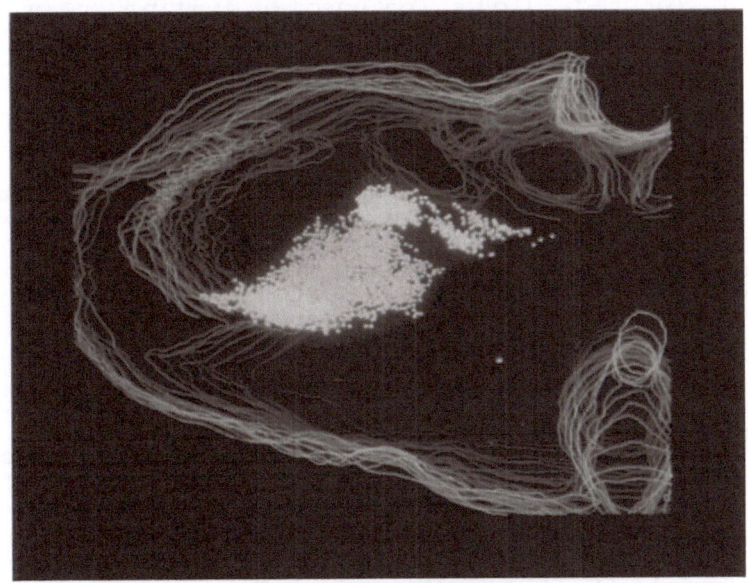

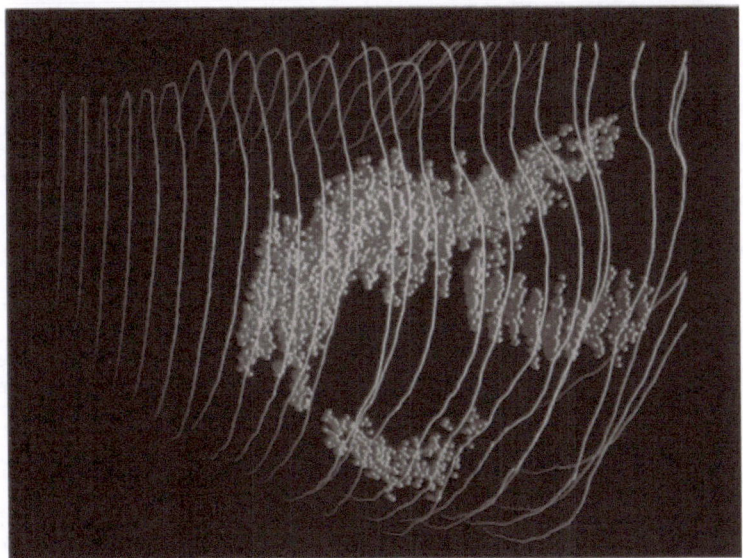

Figure 3. (Upper) Three dimensional distribution of thalamic cells projecting to SI peri-oral regions in rat. View looking into thalamus from behind. Outlines of thalamus (above) and hypothalamus (below) in green (50 um per section.) HRP labelled cells (blue) shown at lower left in VentroPosteroMedial (VPM) nucleus. Additional spots of labelling in Posterior (PO) and Intralaminar (IL) nuclei in upper right. (continued on next page.)

Figure 3. continued. (Lower) Comparison of distributions of
thalamic neurons projecting to SI (yellow) and MI (blue) cortical
forepaw areas. Side view of thalamus; Rostral = left of picture;
Dorsal = top of picture. Outer thalamic boundaries of each section
(with 100 um spacing) depicted with green lines. Two separate HRP
injected brains used. Cells projecting to SI forepaw area (yellow)
occupy VentroPostroLateral (VPL) nucleus caudally, but extend
rostrally into VentroLateral (VL) nucleus. Cells projecting to MI
forelimb area (blue) seen in VL rostrally, where it overlaps with
SI projecting cells. Caudo-dorsally they extend back from VL into
PO (posterior) nucleus, and Caudo-ventrally they extend back into
VentroMedial (VM) nucleus.

Overall, the programs have been designed to optimize the use of new
32-bit microcomputers which are now becoming available.

Data records are organized into nodes which contain
information about the type of data stored and address pointers to
lists of graphical data and other nodes. The structure of the
graphical data list depends upon its type; for example, a segment
consisting of a line points to a list of data triplets. Each
triplet uses 32-bit floating point numbers to represent X, Y and Z
coordinates in 3-space. Segments can include lines (e.g., tracings
around the border of a sectioned object), cell positions,
autoradiographic grain positions, or fiducial marks (landmarks with
relatively constant positions in serial sections which are used for
alignment.)

Our data file representation (Smith, et. al., 1983a and Smith,
et. al., 1983b) of a complicated neuroanatomical object is
organized in a tree-like structure which is similar to the Knuth
transform of an n-ary tree (Pfaltz, 1977). A tree is defined as a
system of the nodes, or data packets. The tree-like structure
comes from the address pointers, within the packets, which are used
to determine where to find in the file the other nodes in the
tree. At the top of the tree is the ROOT node which contains
general information and pointers to family modes. A FAMILY node
might specify, for example, how to find the location in three
dimensions of all cell bodies which are part of a labeled nuclear
region. Alternately, a different category of FAMILY node might
point to all contours or sets of polygon strips which define a
surface around a nuclear region. A variety of different node types
are employed to define all the types of information encountered in
neuroanatomical studies.

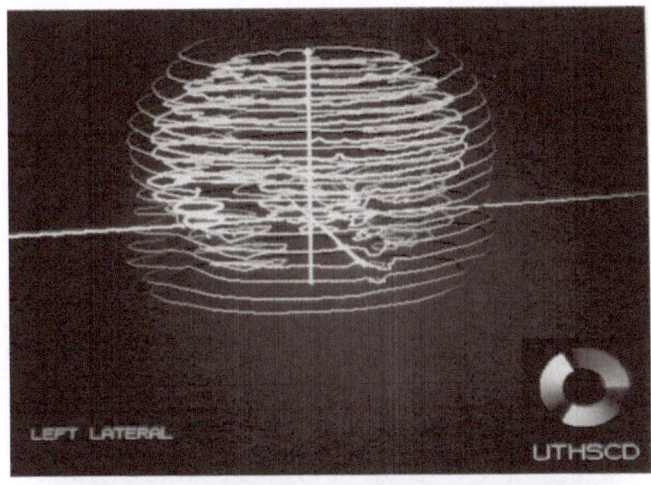

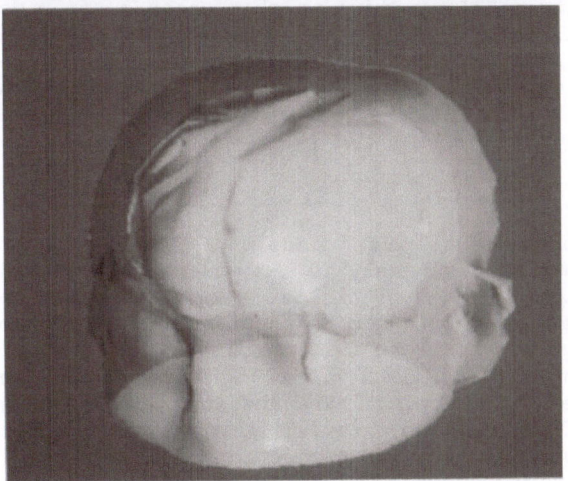

Figure 4. (Upper) A view of contours of cross-sections of brain and skin obtained from a series of slices from an NMR imaging system. (Lower) Surfaces of brain and skin generated from contour data above. The skin is represented as a transparent film.

A major graphical problem exists when an object branches. In this case, it is necessary to construct a hierarchy in the database to represent the relationship between branches. Special FAMILIES of nodes are used for temporary storage and manipulation. If one branch splits into two branches, the three branches will share a contour in one of the sections. To satisfy the triangulation requirements this shared contour must be present in all three branch definitions, but should be split in two of the branches. In our data manipulation procedure, the user specifies where to split the contour because of the unpredictability of contour complexity in serially sectioned biological material. Branches exist in the representation of the nuclei of the inferior olivary complex (Figs. 5,6).

The data management routines allow manipulation of data with use of the tree and node structure. All data is input initially from a sequence of sections and stored in portions of files located by SECTION nodes. The BUILD facility allows sections to be aligned and edited. New nodes are created to define OBJECTS which define ways of gaining access to all contours in sequence which define a continuous surface or a cluster of points. Sequential contours can be subjected to a TILE routine which creates surface definitions and determines normals of adjacent polygons to permit graphical shading routines. A QUANT facility operates on the tree-like data structure to determine volumes, surface areas, cell counts, pixel counts, graphs, bar charts and histograms.

DISCUSSION

Our major goal has been to design a flexible general-purpose system for quantitative analysis and generation of high resolution three-dimensional images of biological material that has been serially sectioned. Data can be input from a variety of devices, and because we use absolute (biological) coordinates stored as floating point numbers, data values can have a wide range (sub-micron to macroscopic). The absolute coordinate system also allows data from the same material to be analyzed with different modalities such as magnified projections onto a tablet or views from a light microscope.

Because of the complex and unlimited nature of biological objects, the database design had to be tailored to provide efficient storage and access. The database routines also had to be adaptable to a wide variety of experimental paradigms such as surface reconstruction, cell and grain counting, morphometric analysis, and regional density. Database design has constituted a major programming effort, at times consuming 90% of our effort, and has proven valuable to the flexibility of the system.

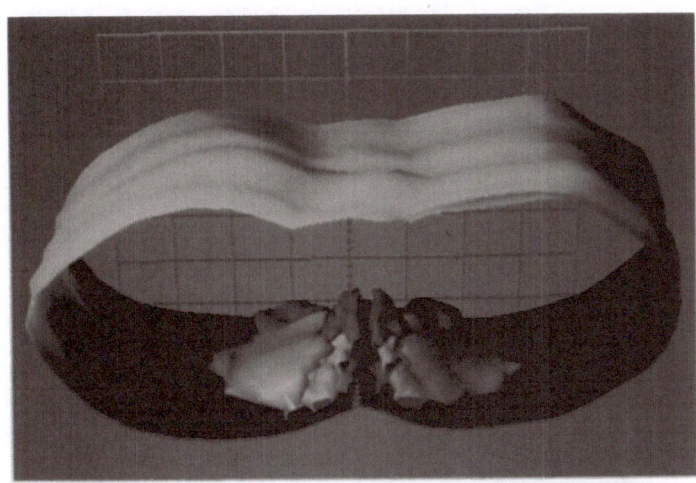

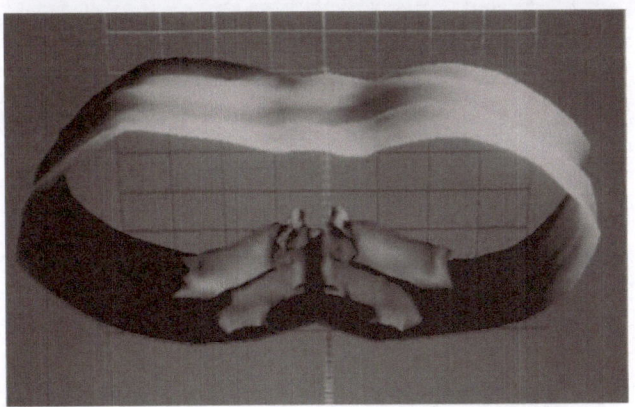

Figure 5. Anterior (Upper) and posterior (Lower) views of the
inferior olivary complex within the outline of the brain stem of
the rat. The eye of the observer is anterior (Upper) and posterior
(Lower) to the nucleus and about 30 degrees above the horizontal
plane.
This nuclear group is composed of three major and several minor
subnuclei. The structure is dominated in size by the medial
accessory olive (Red). The dorsal accessory olive (green) forms
the dorsolateral tier and the principal olive is sandwiched between
the two subnuclei.

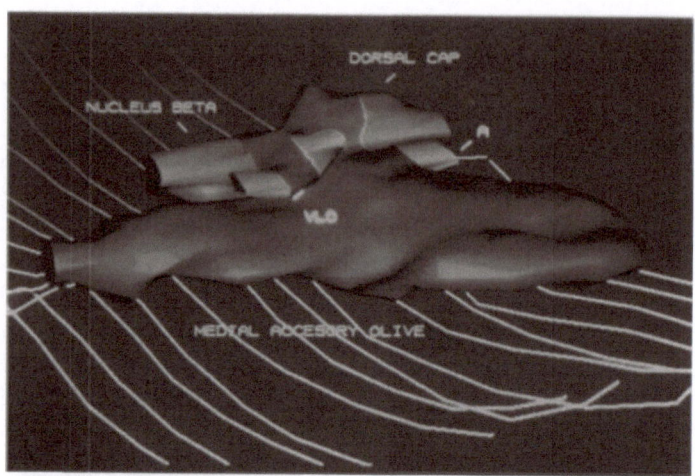

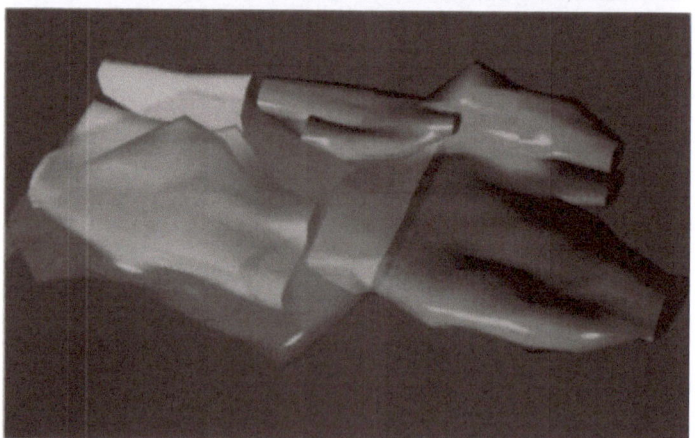

Figure 6. (Upper) A lateral view of the inferior olivary complex of the rat. The observer's eyes are located in the horizontal plane and about 20 degrees posterior. This reconstructed image reiterates the fact that the medial accessory olive is the largest subnucleus of the complex. Other subnuclei are labeled and shown in different colors.
(Lower) Another lateral view of the nuclear complex, where the eye of the observer is located 20 degrees anterior from the midline in the horizontal plane. Using a complex transparency algorithm the dorsal accessory olive (transparent blue) has been generated such that it allows observation of the other subnuclei of the olivary complex

Another valuable asset has been the implementation of various imaging strategies which can utilize separate components within the hierarchical database to allow visualization of relationships within complex biological objects. These routines are highly interactive, allowing the user to display the specific portions of three-dimensional objects of interest. The same routines are used to condense large amounts of quantitative information from multiple three-dimensional objects into color-coded 3D graphical images, which correspond to distributions of densities within 3D objects. This allows direct comparison of asymmetries within a single object, paired objects, or objects from different subjects.

Our view is that one cannot underestimate both the importance and difficulty of producing computer generated high resolution images of rich complexity. Such procedures as we have outlined could soon play a major role in our conceptual approach to anatomical questions.

ACKNOWLEDGEMENTS

We acknowledge the support from grant numbers AA-390, NIAAA 1F32-5198, DA-2338 and the Biological Humanics Foundation.

D.J. Woodward, et al.

REFERENCES

Fuchs, H., Kedem, Z.M. and Uselton, S.P. (1977). Optimal Surface
Reconstruction from Planar Contours. Commun ACM., 20, 693-702.

Gouraud, H. (1971). Continuous Shading of Curved Surfaces. IEEE
Trans Comput., 20, 623-629.

Kernighan, B.W. and Ritchie, D.M. (1978). The C Programming
Language. Prentice-Hall, New Jersey.

Macagno, E.R., Levinthal, C. and Sobel, I. (1979).
Three-dimensional Computer Reconstruction of Neuronal Assemblies.
Ann Rev Biophys Bioengng., 8, 323-351.

Pfaltz, J.L. (1977). Computer Data Structures. McGraw Hill, New
York.

Phong, B.T. (1975). Illumination for Computer Generated Pictures.
Commun ACM., 18, 311-317.

Schlusselberg, D.S., Smith, W.K., Lewis, M.H., Culter, B.G. and
Woodward, D.J. (1982). A General System for Computer-Based
Acquisition, Analysis and Display of Medical Image Data. ACM 1982
Conf Proc., ACM Annual Meeting.

Schlusselberg, D.S., Smith, W.K., Culter, B.G. and Woodward, D.J.
(1982). A Computer System for Semi-Automatic Cell Recognition in
Neuroanatomic Studies. Soc. Neurosci. 1982 Abstract 182.9.

Schlusselberg, D.S., Smith, W.K., McEachron, D.L. and Woodward,
D.J. (1983). A Computer System for Three-Dimensional Analysis of
Regional Brain Activity Using 2-deoxyglucose Autoradiographs. Soc.
Neurosci. 1983 Abstract 106.4.

Smith, W.K., Schlusselberg, D.S., Woodward, D.J. (1981). A
Computer System for Neuroanatomical Data Acquisition, Analysis and
Display. Soc. Neurosci. 1981 Abstract 135.18.

Smith, W.K., Schlusselberg, D.S., Culter, B.G., Woodward, D.J. and
Lacy, E.R. (1983). Hierarchical Database Design for Biological
Modeling. NCGA 1983 Conf Proc., 106-116.

Smith, W.K., Schlusselberg, D.S., Culter, B.G. and Woodward, D.J.
(1983). A Database Structure for Three-Dimensional Reconstruction
of Neuroanatomical Objects. Soc. Neurosci. 1983 Abstract 106.5.

POPULATION CHARACTERISTICS OF NERVE CELL BODIES ILLUSTRATED BY THE POSTNATAL DEVELOPMENT OF CEREBELLAR GRANULE CELLS IN THE RAT

VYVYAN HOWARD*, RHIAN LYNCH*
and PETER WOODHAMS†

*Department of Anatomy, University of Liverpool, P.O. Box 147, Liverpool L69 3BX, UK
†MRC Developmental Neurobiology Unit, Institute of Neurology, London WC1 2NS

INTRODUCTION.

Until very recently research with transmitter identified neurons (TINs) using immunofluorescent techniques has largely been qualitative in nature. The change in emphasis to quantitation in TIN studies, for example Fuxe et al (1983), may be hampered by lack of objective data on various aspects of the technique. The "totality" of staining of target cells under any particular experimental circumstance is yet not fully understood. For meaningful stereological estimates to be made it is self evident that all the cells of a particular population should be equally identifiable from the point of view of contrast in the microscope.

The problems of identifiability of TINs can be looked at in several ways. If the concentration of precipitate in the nerve cell is inadequate then that cell will not be visualised. Furthermore, if the arrangement of precipitate is not homogeneous then those cells that are sectioned through poorly stained areas of the cytoplasm will give rise to false negatives. The nucleus is not stained normally in these preparations and there are indications that the cytoplasm is not equally stained. For example in the paper by Williams and Dockray (1983) the distribution of enkephalin related peptides in the granular layer of the rat cerebellum by the P.A.P. technique shows that in Golgi cells there is a granular staining close to the nucleus. Such an inhomogeneity is likely to give rise to serious stereological problems, unless it is constant and predictable. This is not likely to be the case and therefore would probably lead to unpredictable bias of unknown magnitude. Another factor likely to cause problems in morphometric studies is the reproducibility of experimental conditions between one animal and another. If there is a large inter-animal variation in staining by these techniques then major problems in interpretation of results will ensue. Dahlstrom (1967) was the first to use colchicine to block axonal transport and cause an increase in amine storage granules in the soma of nerve cell somata. This dramatically increases the identifiability of TINs but once again, for predictable quantitative analysis, the various questions about dose dependent effects and inter-animal variation that need to be

41

known have not yet been fully addressed.   Matthews et al (1982) have
shown that certain effects of colchicine <u>are</u> dose related and there seems
to be room for further research into that effect upon TINs.

Overall population characteristics of groups of neurons are of some
importance.   The fact that size distributions of neuronal somata, dendrite
internode lengths and synaptic areas all appear to be highly positively
skewed has been reported by Howard and Scales (1982) and Howard
(1983).   The theoretical evaluation and functional significance of highly
asymmetric size distributions in biological systems has been dealt with by
Howard (1981a, 1981b and 1983).   When highly restricted patterns of
biological variation are found, then they usually indicate strong selection
pressures acting upon the system under consideration.   And in the papers
mentioned above, Howard has come to the conclusion that the positive
skewing found among population characteristics of many of the anatomical
features involved in information processing is an indication that there is a
selection pressure for them to progressively diminish their size throughout
phylogeny.

The study of population characteristics can also indicate the presence
of mixtures of populations of nerve cells.   This is well illustrated in studies
on motoneuron populations by Howard et al (1980) and Howard and Scales
(1981).   The predictable bimodality of profile size distributions from
measurements made in motor nuclei has led to a method for the calculation
of relative and absolute numbers of alpha and gamma motoneurons, using
stereological techniques.   Another strength of studying population
characteristics is that systematic variations in the overall mean size of
groups of neurons of the same type may be found in different anatomical
locations.   This is illustrated by the discovery that Purkinje cell diameter
appears to vary with position within the cerebellar folia.   Purkinje cells at
the top of folia have been found to be consistently smaller than those at
the bottom.   This has been reported by Howard et al (1984).

Methods that are essentially qualitative in nature can be tailored for
use in quantitative studies.   At present the reasons why certain neurons
take up HRP, while others of the same population do not, is not
understood.   In a study of the motor nucleus of the rat trigeminal nerve
Lynch (1982) and Lynch and Howard (1983) have demonstrated that the use
of HRP, injected into individual muscles innervated by the fifth nerve, can
be used to delineate areas within the motor nucleus.   After mapping has
been performed then, by counterstaining with a "total" dye, it is possible to
count every profile within the mapped area.   In this way it was possible to
gain estimates for the proportions of alpha and gamma motoneurons and
their absolute numbers susbserving each muscle innervated by the fifth
nerve.

In this paper we present the results of a recently finished pilot study
which follows the development of the internal granule cell layer of the rat
cerebellum between days 17 and 27 postnatal.

MATERIALS AND METHODS.

Histological procedures.

2 Porton rats were sacrificed at 17, 20, 24 and 27 days post natal (DPN) respectively.   The animals were anaesthetised with ether and perfused transcardially with 250 mls. of 4% paraformaldehyde, 2.5% glutaraldehyde in 0.1 M phosphate buffer at 80 mm. Hg.   The cerebellum was removed from each animal and post-fixed for one hour at 4 C.   It was then washed and stored in phosphate buffered saline O.1 M.

A sagittal section was made through the vermis in the midline.   A thin slice was taken from the face of one of the blocks and dehydrated through a graded series of alcohols over six hours.   Each slice then had two exposures of thirty minutes in propylene oxide and was then transferred to an equal parts mixture of propylene oxide and TRANSMIT LM (TAAB) for one hour.   Finally the slices were placed in TRANSMIT LM mixture for four hours.   After hardening, 1 micron parasagittal sections were taken using a Jung "Autocut".   Sections were mounted on glass slides and stained with Borax Toluidine Blue and then cover slipped using DPX mounting medium.

Stereological methods.

The number of granule cell profiles per unit area (Q_A) was measured at the superficial and deep portions of the internal granule cell layer (IGL) for each animal respectively, observing an unbiased counting rule (Gundersen, 1977).   This was achieved by measuring twenty fields with a rectangular graticule of side 30 microns in real units at x1,000 under oil immersion with a Nikon Optiphot Research microscope.   The superficial measurements were made just below the Purkinje cell layer and the deep measurements adjacent to the white matter.   This procedure was repeated at the top of adjacent folia until the sample size was attained.   The sample size was decided by performing a pilot study over forty fields and subsequently constructing a graph of the cumulative mean.

The mean diameter of each granule cell profile falling within the test quadrats was measured, using a linear eyepiece graticule.   The diameters for each area for each animal were assembled into a histogram using a class width of 1 micron.   Approximately 500 profiles were measured in each case.   The two dimensional information in each histogram was "unfolded" using the distribution free method of Cruz-Orive (1983), which allows for truncation and overprojection, and an estimate of the mean diameter ($\bar{D}$) for granule cells was obtained.   The higher moments of each unfolded distribution were also calculated and the results are reported below.

The numerical density of granule cells for each level in each animal (N_V) was calculated according to DeHoff and Rhines (1961)

RESULTS

Qualitative examination of the sections revealed the presence of the external granule layer (EGL) in the 17, 20 and 24 DPN specimens, although it was reduced to a patchy unicellular layer in the latter. The external granule cell-layer was absent at 27 DPN.

In the internal granule cell layer (IGL) the presence of degenerating cells was indicated by their pyknotic nuclei, poor cytoplasmic staining and swollen appearance. An example is indicated in Fig. 1. No degenerating cells were detected at 17 DPN. By 20 DPN scanty degenerating cells were evident, mainly in the deeper IGL. At 24 DPN numerous degenerating granule cells were seen at all depths of the IGL, but by 27 DPN they were scarce and confined mainly to the superficial portion of the IGL.

The major stereological parameters are listed in Table 1 and the average thickness, measured perpendicular to the pial surface, of the IGL is indicated in Table 2.

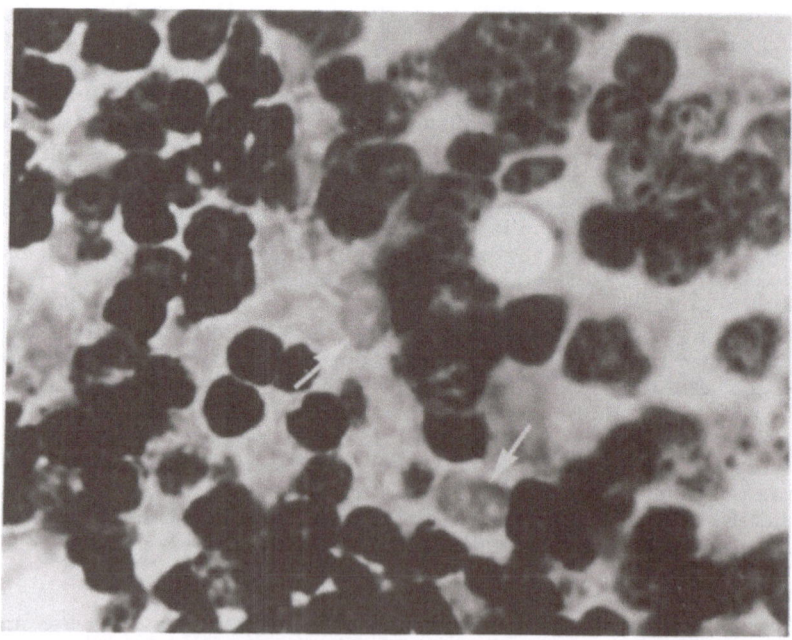

Fig. 1 A photomicrograph of the superficial part of IGL, 24 DPN. Degenerating cells are indicated by the arrows. x5,625.

Table 1   Stereological results for each animal.   Position indicates
              sampling area of IGL.   For symbols see text.

DPN	Animal	Position	$\bar{D}$ μm	$Q_A/\mu m^2$	$N_V/\mu m^3 \times 10^6$
17	1	Sup	5.51	0.0201	3,652
17	2	Sup	5.81	0.0234	4,020
17	1	Deep	5.54	0.0197	3,564
17	2	Deep	5.36	0.0180	3,352
20	1	Sup	4.11	0.0364	8,867
20	2	Sup	3.92	0.0354	9,028
20	1	Deep	3.70	0.0357	9,655
20	2	Deep	3.85	0.0307	7,980
24	1	Sup	4.28	0.0374	8,741
24	2	Sup	4.52	0.0399	8,823
24	1	Deep	4.03	0.0326	8,082
24	2	Deep	4.16	0.0328	7,892
27	1	Sup	4.08	0.0300	7,326
27	2	Sup	4.02	0.0268	6,661
27	1	Deep	3.07	0.0229	7,456
27	2	Deep	3.12	0.0242	7,764

Table 2.   Mean thickness of IGL in microns (S.D.) for each
            developmental age based on 4 measurements in each
            case.

Maturity	17 DPN	20 DPN	24 DPN	27 DPN
Mean (SD) IGL Width	144 (23)	116 (31)	113 (14)	128 (27)

DISCUSSION

It is increasingly evident that the great divide between those who
study structure and those who study function is becoming very blurred.
The physiologist who investigates the firing patterns of a neuron and the
neurochemist who identifies the transmitter substances associated with it
will both in the end want to ask the same questions.   How many similar
cells are there performing the same job and what is their connectivity?
Sadly for the neurobiologist, both these quantities (i.e. number and
connectivity) are only directly obtainable in three dimensions.   We are, of
course, restricted to the dimensionality of the thin slice, which is in
practice two dimensions.   Great advances in the interpretation of two
dimensional data have occurred over the past two decades, under the

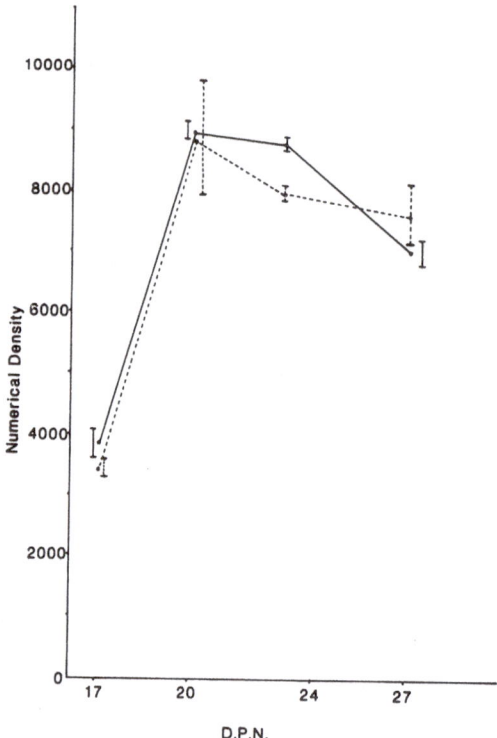

Fig. 2   Variation of granule cell numerical density (N_V) with maturity. Ordinate: number of granule cells within a cube of side 100 microns. Error bars: S.D. Continuous line superficial IGL. Hatched lines: deep IGL.

general heading of stereology. This has happened under the auspices of the International Society for Stereology, which is a group of mathematicians, statisticians, biologists and material scientists who have common interest in the interpretation of 2D data from slices through solids. Advances in stereological technique and theory are regularly published in the Journal of Microscopy. There seems little doubt that the trend of co-operation between morphometrists and neurobiololgists will continue to flourish.

By studying the characteristics of whole populations of neurons we can obtain objective evidence about the patterns of variability of the system under examination, and can gain information of functional significance (Howard and Scales, 1982 and Howard, 1983). The pilot study

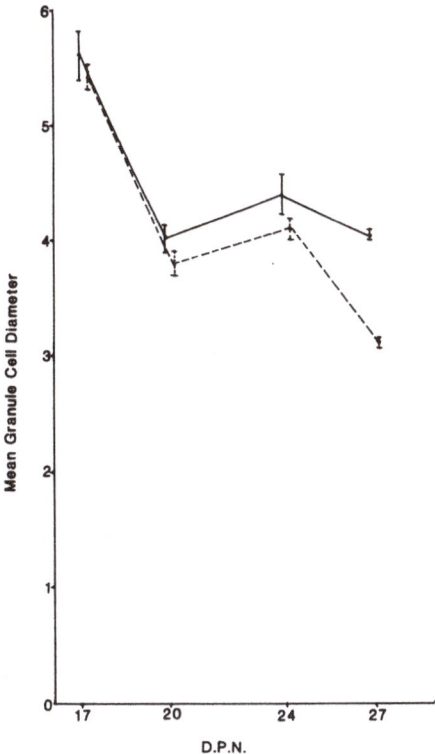

Fig. 3   Variation of granule cell mean diameter with maturity.
Ordinate in micrometers.   Error bars: S.D.   Continuous line:
superficial IGL. Hatched line: deep IGL.

reported above is an attempt to see if there are any gross changes in the
characteristics of such a population during ontogeny and, in particular, at
the time of 'epigenetic' cell death. These preliminary results are part of a
larger and ongoing research project and the authors are at pains to point
out that the sample sizes here are small.   However, they do consider that
the changes shown are of general interest and worth reporting at this stage.

Addison (1911) showed that the external granule cell layer started
involution at about 12 DPN and disappeared at about 22 DPN in the white
rat.   Our findings seem to indicate a dramatic rise in numerical density of
granule cells between 17 and 20 DPN (Fig. 2) and thereafter a gradual fall

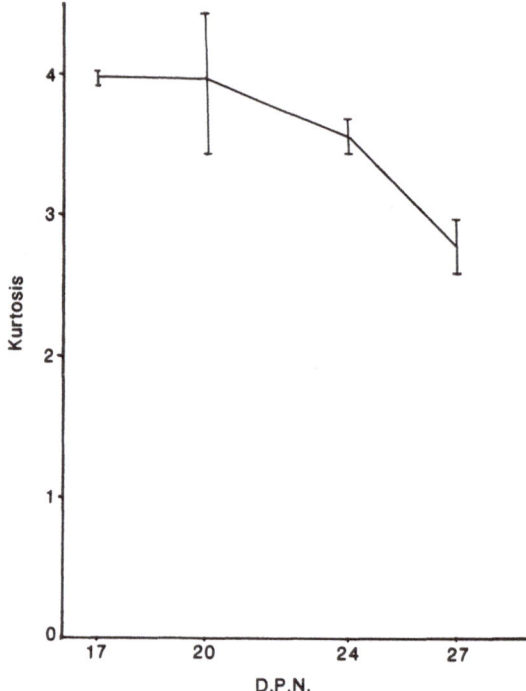

Fig. 4   The mean of the kurtosis estimates from the four distributions at each maturity in Fig. 6 Error bars: S.E. of the four estimates.

to 27 DPN. That could be explained by the fact that the cerebellum may still be increasing in volume or that there is a differential shrinkage during tissue preparation with different maturities. These are factors that we intend to investigate in subsequent studies. However, Addison (1911) showed that there was little change in width of the internal granule cell layer after 24 DPN in the white rat, and our results on the general morphology of the internal and external granule cell layers are in general agreement with his findings. If we accept that there is a real diminution in the total number of granule cells, then this must be due to the fact that epigenetic cell death is occurring.

The results that we have shown indicate a number of other changes associated with the diminution in numerical density. The mean size of granule cells was found to diminish over the time scale (Fig. 3) studied as did the inter-animal variation. Another trend was the gradual decrease in

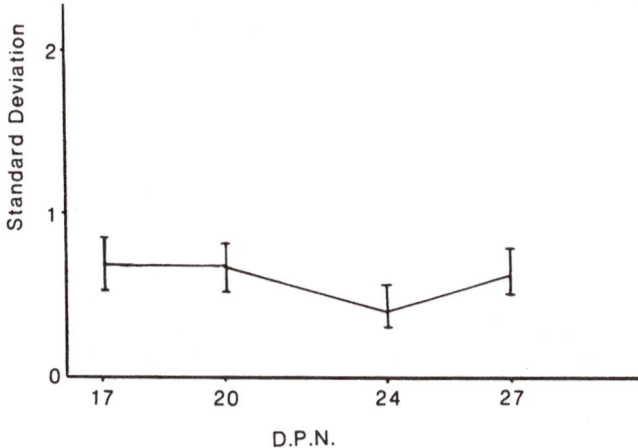

Fig. 5   The mean of the 4 standard deviation estimates from the four
distributions at each maturity in Fig. 6.   Ordinate in microns.
Error bars: S.E. of the four estimates.

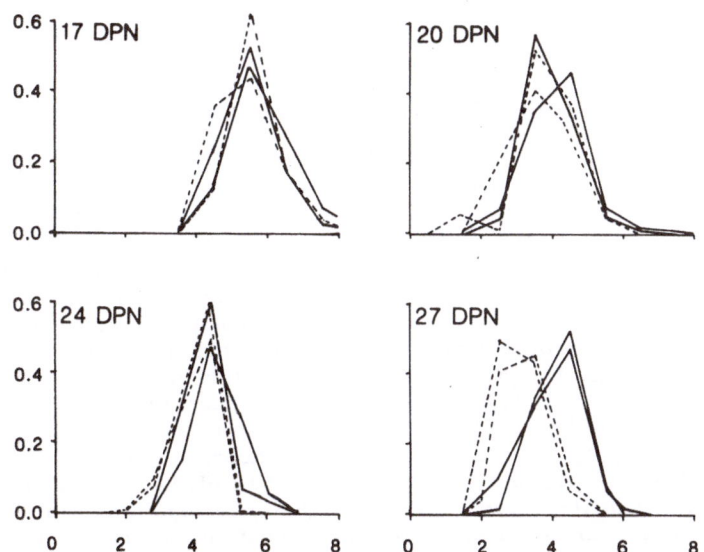

Fig. 6   'Unfolded' (Cruz Orive 1983) size distributions of granule cells
for each animal at each maturity.   Ordinates: Relative
frequency.   Abscissae: Granule cell diameters in microns.
Continuous lines: superficial IGL. Hatched lines: deep IGL.

the average kurtosis calculated from the individual granule cell diameter distributions (Fig. 4) arrived at by the Cruz-Orive (1983) method.   The second (Fig. 5) and third moments of the distributions of granule cell diameters did not appear to alter with increasing maturity.   Because the range of granule cell size is small and, after unfolding, there were often only five classes in the histogram then there is no doubt that the skewness estimate would be very susceptible to the effects of sampling error.

It appears from the results that there are definite changes occurring in the overall pattern of variability of this particular population of neurons.   As the animal matures certain granule cells are disappearing and this has the effect of reducing the mean granule cell diameter and, quite probably, making the distribution of granule cell diameter less 'peaked' and more 'rectangular' in nature. If our qualitative findings on the presence of degenerating granule cells is combined with the quantitative evidence in Fig. 6, we find that at 24 DPN there is massive granule cell death occurring, but that by 27 DPN there is no observable granule cell death in the deeper layers of the IGL.   There is a very obvious dichotomy in mean granule cell size between the superficial and deep parts of the IGL at 27 DPN and it is tempting to theorise that this alteration in the pattern of variability of the deep population is due to selective cell death affecting the larger members of that population.   Examination of Fig. 3 will show that there is a slight rise in mean granule cell diameter between 20 and 24 DPN.   It is possible that this might be accounted for by the increased number of dying granule cells which tend to expand during the process of necrosis.

Sadler and Berry (1983) have described 're-modelling' of the dendritic arbor in Purkinje cells during development.   That may well be an example of very fine tuning during selective stabilisation, which may be preceded by a more gross mechanism which actually removes redundant cells.   Thus it seems possible that the cells that are dying are in some way being selected from among the very large and very small individuals.   Henneman et al (1965) were the first to point out the functional significance of cell size in the spinal motoneurons.   Changeaux and Danchin (1976) put forward the theory of selective stabilisation and discussed possible mechanisms for the specification of neuronal networks.   It is possible that there is a certain optimal size range for granule cell somata within which normal function may occur.   If cells fall outside that size range then there may be a functional incongruity which will make them more susceptible to cell death at the time of epigenetic cell death.   That incongruity could take various forms.   One such possibility is that, if ion channels have a certain density on the surface of the neuron, then their ability to repolarise will become a function of the surface area of the cell and if the latency of firing is of importance in selective stabilisation then that may be a reson why we are observing a 'size clue'.   This line of argument cannot be taken further on the evidence provided by this stereological experiment, although the grounds for it can be made more solid by examining more animals in each group and by controlling for differential shrinkage, as mentioned above. To carry the idea further will require other studies of those cells that are potentially at risk for epigenetic cell death.   Perhaps those cells falling

outside a certain critical size range may be suitable candidates.   It would
be very interesting to know if, by immunofluorescent TIN methods, there
were any differences in the concentration of various transmitter
substances, with size, which might provide very important clues about the
nature of neuronal network specification.   This may be an area in which
the quantitative aspects of TIN research may be fruitfully employed.

   The reduction in the mean diameter of granule cells during ontogeny
is in agreement with the theory of "neuromicrosis" proposed by Howard
(1983).   We are aware that during the time of epigenetic cell death there is
a 'balancing act' going on between the various components of the central
nervous system.   The end result is a finely tuned and cohesive entity, each
part of which relies upon the balanced excitatory, inhibitory and trophic
activities of the other elements with which it comes into contact.   The
effects of disruption are easily seen in conditions such as stroke, when
certain afferents to spinal motoneurons are disinhibited to produce a
spastic paresis.   In the sensory system, denervation injuries such as
"phantom limb", cause the fine balance between the elements of the dorsal
horn of the spinal grey matter to be upset and abnormal firing of neurons
gives rise to the sensation of pain, often of an intractable nature.
'Neuromicrosis' hypothesises that the restricted and often asymmetric
patterns of variability of size in various neuronal elements may well be
evidence that nature uses epigenetic cell death to achieve the delicate
balance that we see in the intact central nervous system.   By dispensing
with over or undersized neurons, it is possible that the sensitivity of any
particular neuronal population to depolarisation or hyperpolarisation can be
finely adjusted.   Clearly the "size clue" that we think we are seeing is
evidence of some aspect of the function of these cells.   If the thinking of
Henneman et al (1965) and of Rall (1964) can be held to be realistic, then it
seems possible that a reduction in the mean cell size of a neuron population
might be accompanied by an overall increase in susceptibility to
depolarisation, i.e. an increased sensitivity.   Admittedly the perikaryon
posesses a small part of the total surface of a neuron.   We may just be
observing a biophysical effect in that larger perikarya may be able to
support larger dendritic trees.   But if that were the case it would not
necessarily invalidate the argument above.   There is another point in
favour of a tendency towards smaller neurons and that is that an increased
packing density brings advantages that have been demonstrated by recent
advances in microchip technology.   The closer the elements of a network
are, the shorter will be their lines of communication and this will enhance
the speed with which they can communicate with one another.

   By presenting the results of this study and mentioning other research
areas in which the characteristics of nerve cell populations can be of
importance, we hope that we will have drawn attention to the power of
stereological techniques and indicated certain areas in which quantitative
morphometrical findings can give strong indications about the function of
the system under review.   It is unlikely to provide the final proof for any
of the theories that we have put forward, but there seems little doubt that
stereology can provide strong clues as to where physiologists and
neurochemists should be looking and, in the case of a developmental

sequence, when. It is impossible to know how much morphometric data has been lost or destroyed because of the injudicious use of desk top calculators! On many occasions scientists who have spent countless hours collecting data then simply waste most of the information by feeding the measurements into the machine and gaining a mean value. If you are studying a bimodal system and you calculate the arithmetic mean of your measurements then you will of course obtain a single answer. However, the simple expedient of plotting the data in a frequency histogram and examining it can demonstrate the presence of multimodal systems. The authors know personally of two experiments where this has come to light on re-examination of raw data. We think that the main message of this paper is that the higher moments of size distributions of the anatomical features that we look at contain important information that we should not discard lightly. Patterns of biological variation are important indicators of the pressures and design constraints that a system is operating under and we ignore them at our own cost.

REFERENCES

Addison, W.H.F., (1911). The development of the Purkinje cells and the cortical layers in the cerebellum of the albino rat. J. Comp. Neurol., 21,459-485.

Changeaux, J.P. and Danchin, A. (1976). Selective stabilisation of developing synapses as a mechanism for the specification of neuronal networks. Nature (Lond.), 264., 705-712.

Cruz-Orive, L.M. (1983). Distribution-free estimation of sphere size distributions from slabs showing over projection and truncation, with a preview of previous methods. J. Microsc., 131, 265-290.

Dahlstrom, A. (1968). Effect of colchicine on transport of amine storage granules in sympathetic nerves of rat. Europ. J. Pharmacol., 5, 111-113.

DeHoff, R.T. and Rhines, F.N. (1961). Determination of the number of particles per unit volume from measurements made on random plane sections. Tr. A.I.M.E., 221, 975.

Fuxe, K., Agnati, L.F., Calza, L., Toni, G., Benfenati, F., Andersson, K., Farabegoli, C., Hokfelt, T. and Zini, I. Transmitter-identified neuron systems at the median eminence level: characterisation by means of quantitative immunocytochemistry and receptor autoradiography. Neuroscience Letters, Suppl. 14, S126.

Gundersen, H.J.G. (1977). Notes on the estimation of the numerical density of arbitrary profiles: the edge effect. J. Microsc., 111, 219-224

Henneman, E., Somjen G. and Carpenter D.O. (1965). Functional significance of cell size in spinal motoneurons. J. Neurophysiol.,28, 560-580.

Howard, C.V. (1981a).   Theoretical and experimental evaluation of size distributions: Their evolution and the phenomenon of "genetic pruning". Stereol. Iugoslav., 3(Suppl. 1), 79-88.

Howard, C.V. (1981b).   On the functional significance of the third moment of size distributions in biological systems.   Stereol. Iugoslav.,3(Suppl. 1), 129-135.

Howard, C.V. (1983).   Neuromicrosis: A process affecting the phylo- and ontogenetic development of the brain.   An hypothesis based on stereological studies of neurone population characteristics.   Ph.D Thesis. University of Liverpool.

Howard, C.V. Allibone, R. and Scales, L.E.   The relationship between Purkinje cell diameter and position within the cerebellar folia of the albino rat: A stereological analysis.   Acta Stereol.2(Suppl. 1), 219-222

Howard, C.V. and Scales, L.E. (1981).   The separate numerical densities of alpha and gamma motoneurons in the spinal cord of the rat.   Sterol. Iugosl., 3(Suppl. 1), 503-510.

Howard, C.V. and Scales, L.E. (1982).   The concept of 'neuromorphotaxis' based on a minimisation principle.   A case for the critical analysis of biological variation.   Acta Sterol., 1, 241-252.

Howard, C.V., Scales, L.E. and Lynch, R.V. (1980).   The numerical densities of alpha and gamma motoneurons in the trigeminal motor nucleus of the rat: A method of determining the separate numerical densities of two mixed populations of anatomically similar cells.   Mikroskopie (Wien) 37, (Suppl.), 229-236.

Lynch, R.V. (1982).   A qualitative and quantitative analysis of the motor nucleus of the trigeminal nerve in the rat: An HRP study.   Ph.D. Thesis. University of Liverpool.

Lynch, R.V. and Howard, C.V. (1983).   The use of Horseradish Peroxidase (HRP) in the stereological analysis of motor nuclei.   Acta Stereol., 1, 253-258.

Mathews, M.A., Cornell, W.J. and Alchediak, T. (1982).   Inhibition of axoplasmic transport in the developing visual system of the rat.   1. Structural changes in the retina and optic nerve with graded doses of intra-ocular colchicine.   Neuroscience., 7, 363-384.

Rall, W. (1964).   Theoretical significance of dendritic trees for neuronal input-output relations. In Neural theory and modelling (R.F. Reiss, ed.). Palo Alto : Stanford University Press. pp. 73-79.

Sadler, M. and Berry, M. (1983).   A morphometric study of the development of Purkinje cell dendritic trees in the mouse using vertex analysis. J. Microsc., 131, 341-354.

Williams, R.G. and Dockray, G.J. (1983).   Distribution of enkephalin-related peptides in rat brain: immunohistochemical studies using antisera to met-enkephalin and met-enkephalin Arg[6] Phe[7]. Neuroscience, 9, 563-586.

LAMINAR SPECIFICITY OF DENDRITIC MORPHOLOGY: EXAMPLES FROM THE GUINEA PIG HIPPOCAMPAL REGION

THEODOR W. BLACKSTAD

Anatomical Institute, University of Oslo, Karl Johans Gate 47, Oslo 0162, Norway

Since the function of neuronal structures everywhere depends in part on spatial, tridimensional features, most neuroanatomical projects have, to some degree, a quantitative dimension. The present note will, essentially, be based on the type of quantitative study consisting of a recording of the tridimensional characteristics of dendrites in Golgi-impregnated specimens. Some aspects of sampling and actual and potential use of data will be illustrated by considering branching pattern and surface morphology in relation to the environment of the dendrites, more specifically the cortical laminae in which they lie. Three different types of cell from the hippocampal region will be mentioned.

As well known, the regular architecture of the hippocampal region has prompted numerous studies of a general, neurobiological nature (structural, electrophysiological, histochemical, etc.) in addition to hodological and behavioural investigations aiming at the elucidation of the role of the region as a whole, which is poorly understood. In briefest outline the most dominant pathways within the region are as follows (cf. Fig.1). Perforant paths from the entorhinal area go to the molecular layer of the fascia dentata (1-3, Fig.1) where synaptic contact is established with the dendrites of the granule cells. The axons of these cells, the mossy fibres, innervate the pyramidal cells of regio inferior (CA3), after having emitted collaterals for cells in the hilus fasciae dentatae (at 5-8, Fig.1). The CA3 cells have heavy axon collaterals activating, in stratum radiatum and oriens, the pyramidal cells of regio superior (CA1). Each septotemporal level of the fascia dentata supplies a specific level of CA3; whatever the nature of the information processed by the fascia dentata may be, it is sent on to the hippocampus proper in a precise spatial order.

PYRAMIDAL CELLS OF CA1

The soma of these cells inhabits stratum pyramidale, and their

55

dendrites form a basal arbor in stratum oriens and an apical arbor
in stratum radiatum and the overlying stratum lacunosum-moleculare.
The dendrites are densely studded with fine spines (gemmules), a
majority having a very thin stalk carrying an expanded portion (a
head). On closer inspection of good Golgi preparations (perfusion-
fixed; epoxy resin embedded; dehydration baths saturated with silver chromate) the author has nevertheless been struck by the existence of a rather dramatic change in dendritic morphology from radiatum to lacunosum-moleculare, in the guinea pig (Fig.2). The amount
of spines decreases to about one fifth and the spines become less
uniform in shape and mostly shorter.

It is not entirely surprising that a difference in dendritic
morphology should exist here, for on the whole totally different
afferent systems supply stratum oriens and radiatum on one hand,
lacunosum-moleculare on the other hand. Moreover, most histochemical methods (for enzymes, transmitters, metals, etc.) demonstrate
an abrupt transition in reactivity from radiatum to lacunosum-moleculare; the synaptological and biochemical milieu offered the dendrites is transformed from one layer to the next. Dendrites that
have reached lacunosum-moleculare do not turn back into radiatum.

When the CA1 pyramidal cell is used as a paradigm in neuronal
modelling, in order to explore by theoretical means biophysical
properties of cortical neurones, it would seem well to recall that
the dendritic mass of these cells has separate divisions with different morphology and possibly different biophysical properties.

How large these divisions are was suggested by measurement, in
two pyramidal cells, of dendritic lengths, after reconstructing the
cells graphically. The cells were selected from the middle of CA1.
In general, in order to secure full information about shape, size,
dendritic length, etc., in impregnated cells it is obviously necessary to measure the cells in their entirety. This was done with the
two cells under consideration. Cell 1 was reconstructed from 11
Golgi sections, cell 2 from 7. The methods were largely those used
by Blackstad et al. (1984). The cell is drawn with a camera lucida
(drawing tube), first in one section, after which dendrites cross-
ing the section surfaces are traced in the neighbouring sections.
The tracking is continued through the series of sections until the
dendritic arbors are fully mapped. Relative heights (vertical levels) of points along the dendrites (called z coordinates) are read
from the microscope and noted on the drawings. With a digitizer the
x and y coordinates are entered into the computer memory and interactively supplemented with the associated z values. Recalculation
of heights according to type of objective lens and immersion medium
is required, if oil immersion has not been used. It is essential in
such investigations that uneven shrinkage be avoided. Therefore,
embedding in epoxy resin which sets uniformly in all directions is
advantageous. Also, it can be cut without compression artifacts,
with a heated steel knife. More automated data sampling than used
here is possible (see other contributions to this book) and in gen-

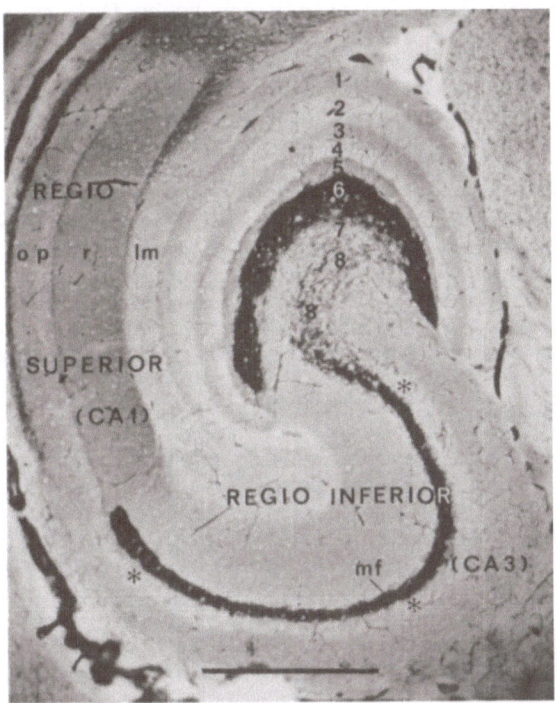

Fig.1. Transverse section of guinea pig hippocampus and area dentata (fascia dentata and hilus fasciae dentatae). Stained with
Timm's sulphide silver method (Haug, 1973). o, stratum oriens; p,
pyramidale; r, radiatum; lm, lacunosum-moleculare; mf, layer of
mossy fibres in CA3; 4, stratum granulosum of fascia dentata. For
other symbols, see text. Bar: 1000 µm.

eral should be aimed at. Notwithstanding, it remains of paramount
importance to invest a maximum of care in the fundamental task of
tracking the dendrites faithfully from section to section. This may
require to scrutinize the sections more than once, change objective lens or illumination, use stereo-microscopy, or study the
slide from the under side.

With the author's computer programs the two CA1 pyramidal
cells were reconstructed for inspection from various angles
(Fig.3). On the whole the mass of dendrites present in stratum
radiatum and oriens is confined within an approximately cylindrical
space about 300 µm in diameter and 600 µm high. (Close to the soma
there are fewer dendrites in this space.) The apical dendrites in

stratum lacunosum-moleculare appear not to be thus restricted but
are seen to meander 3-5 times farther from the axis than the oriens-radiatum branches (Fig.4). Moreover, in these two neighbour
cells these dendrites take different directions (indicating lack of
any consistent anisotropy). One may get the impression that the
dendrites of the CA1 pyramids form two different cell organs, obey-
ing different laws of distribution and perhaps being there to serve
different mechanisms of interneuronal communication at the two levels of the cell: lacunosum-moleculare which is dominated by the
perforant paths (although also receiving other systems), and radi-
atum-oriens. As indicated, the information which the fascia dentata sends to CA3 through the mossy fibre system is further commu-
nicated to CA1. Perhaps the sheet of quasicylindrical postsynaptic
dendritic units the cells in CA1 form in oriens-radiatum is a pattern evolved to allow analysis and use of the differentiated information conveyed through the mossy fibre system.

    As suggested, one way of quantifying the different postsynap-
tic regions on a CA1 pyramidal cell is to determine the length of
the dendrites. This may be of interest if one believes that dif-
ferent·roles are assigned to different regions of the cell. Length
and synaptic density together will reflect amount of synapses. The
calculated total length of the dendrites (in tridimensional space)
was for cell 1 11,130 µm, for cell 2 10,170 µm. Out of this the
basal dendrites formed 36% in cell 1, 33% in cell 2. The apical
dendrites therefore constituted 64% and 67%, respectively. The

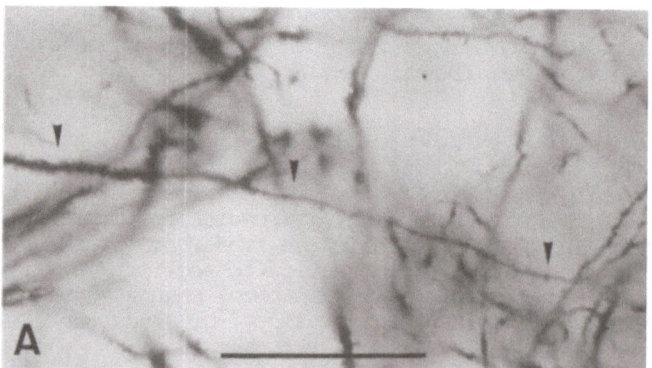

Fig.2. Regio inferior, Golgi impregnation. A: Left half shows
stratum radiatum, right half stratum lacunosum-moleculare. At
arrowheads, pyramidal cell dendrite. Bar: 50 µm. B: Pyramidal cell
dendrites in radiatum. Bar: 10 µm.

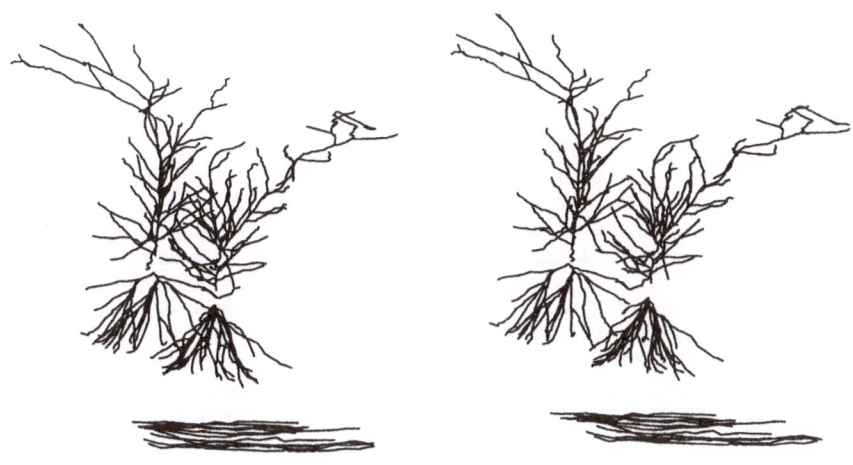

Fig.3. The dendrites of two pyramidal cells from CA1 (together with a part of the underlying ependymal surface) seen from the side, in stereoscopic view. Cell 1 to the right, cell 2 to the left. Computer reconstructions.

proportion present in radiatum was 51% in both cells, in lacunosum-moleculare accordingly 13% in cell 1, 16% in cell 2 (i.e., 20% and 24%, respectively, of all apical dendrites).

If in further studies the proportion of dendrites in the various laminae prove to be as constant as suggested by these two cells, this will represent a further aspect of the laminar specificity of dendritic morphology. - Laminar specificity of CA1 pyramidal cells at the electron microscope level has not yet been systematically examined.

HILUS OF FASCIA DENTATA

The two further examples of neurones to be mentioned are taken from the hilus fasciae dentatae. In this region Amaral (1978) has given a very thorough description of cell types, supplementing Cajal (1911) and Lorente de Nó (1934) very substantially. In the present context Amaral's study has the drawback, however, that it is based on the rat, the hilus of which is not discernibly strati-

fied whatever method is applied. In the guinea pig, like in many
other species, the opposite is true. Below, only a very simplified
description of the hilus will be given, referring to the symbols
used in Figs. 1 and 5. (A provisional terminology selected ad hoc
from Cajal, Lorente de Nó, and earlier authors is used.)

When stratum pyramidale of CA3 reaches the hilus, it continues
as a looser band, CA4 or the first reflected blade (mp, 8). Between
it and the granular layer of the fascia dentata (4) a further cell
layer (f) is present, the layer of fusiform cells or the second
reflected blade. It is bounded by the cell-poorer superficial (5,
op) and deep (7, ip) plexiform layers. In cell-stained preparations the layers (zones) are in no way distinctly delimited. In
contrast, in several kinds of histochemical preparations a stark
pattern of stratification appears. Among these are Timm's sulphide
silver method as modified by Haug (1973) and neuropeptide preparations (Stengaard-Pedersen et al., 1983). The Timm method in most
of the hippocampal region brings out distinct laminae which coin-
cide with specific synaptic zones. In the guinea pig hilus a very
dark band is seen (6) overlapping the second reflected blade without being precisely congruent with it. For lack of a better term

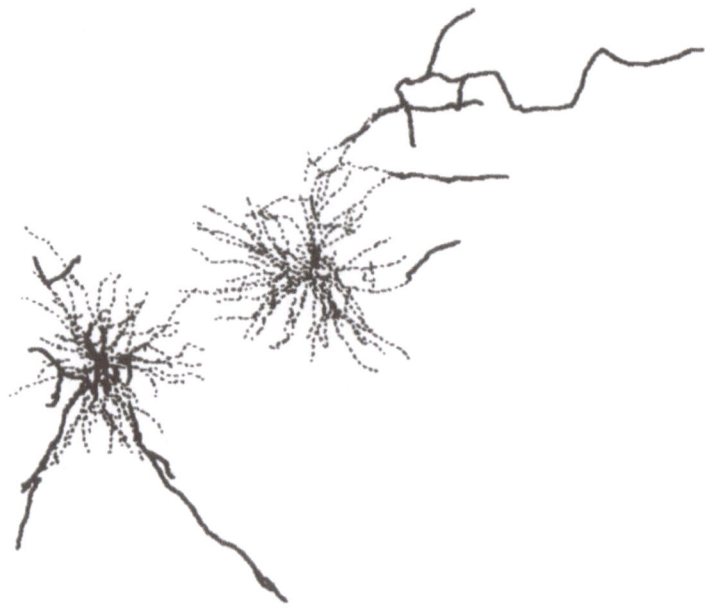

Fig.4. The dendrites of the two pyramidal cells in Fig.3, seen
along the apico-basal axis in stratum oriens-radiatum ("on end").
Dendrites in these strata dotted, those in lacunosum-moleculare
drawn with thick line. Computer reconstruction.

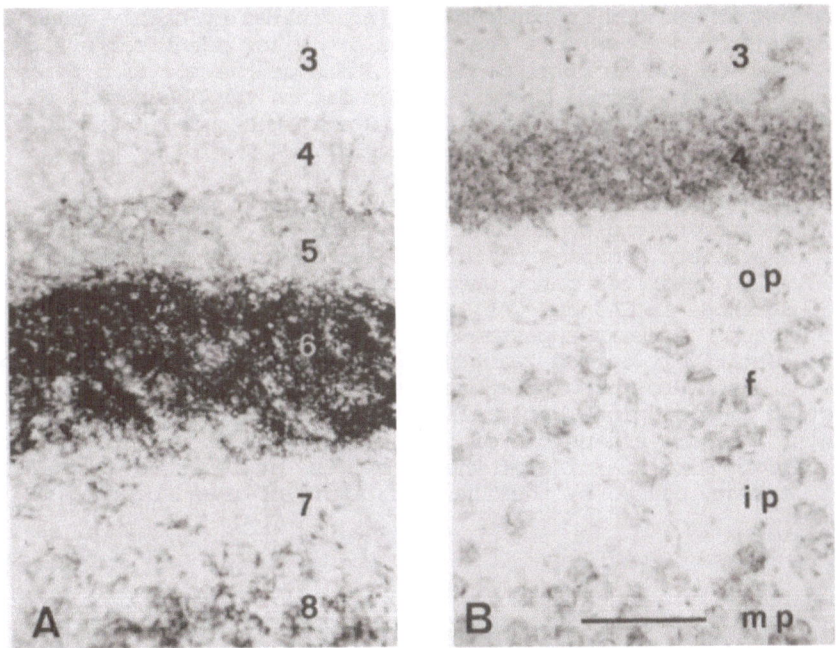

Fig.5. A: Part of area dentata stained and labelled as in Fig.1.
B: Closely adjacent section from the same series, stained with
thionin. op, outer plexiform layer; f, layer of fusiform cells
(second reflected blade); ip, inner plexiform layer; mp, CA4
(first reflected blade) with modified pyramids. Corresponding
laminar levels in A and B are shown in juxtaposition. Bar (for
A and B): 100 μm.

the band will be called the Z zone, as done previously (Blackstad,
1963). - Commissural afferents are absent from the Z zone but
present in both adjacent layers (unpublished observations).

The Mossy Cell

At the level of the Z zone a cell is seen that clearly corresponds to Amaral's mossy cell in the rat. However, as a basis
for the mention of it, it is convenient to briefly recall the synaptology where the mossy fibres contact the CA3 pyramidal cells.

Here, giant boutons en passage are invaginated by highly ramified spines (excrescences). In the guinea pig hilus identical synapses are found on the proximal dendritic stems of the modified pyramids in the first reflected blade. Farther out on the dendrites, both in the deep and superficial plexiform layers, the spines are of the common slender, unramified type (Fig.10).

The mossy cells are multipolar and their dendritic stems have very characteristic synapses sharing some but not all features with the mossy fibre synapses on the CA3 and CA4 cells (Blackstad, 1963). The dendrites carry outgrowths which are ramified but less so than those in CA3 and CA4, and they are less tall (Figs.6 and 7). Because of their appearance in the light microscope they are here tentatively termed "wart spines". In the electron microscope they are seen to invaginate boutons that are larger than those contacting simple spines but smaller than the giant mossy fibre swellings. They are expansions of fine collaterals of the mossy fibres. The strong Timm staining of the Z zone is, at least to a considerable extent, caused by the Timm positivity of these boutons.

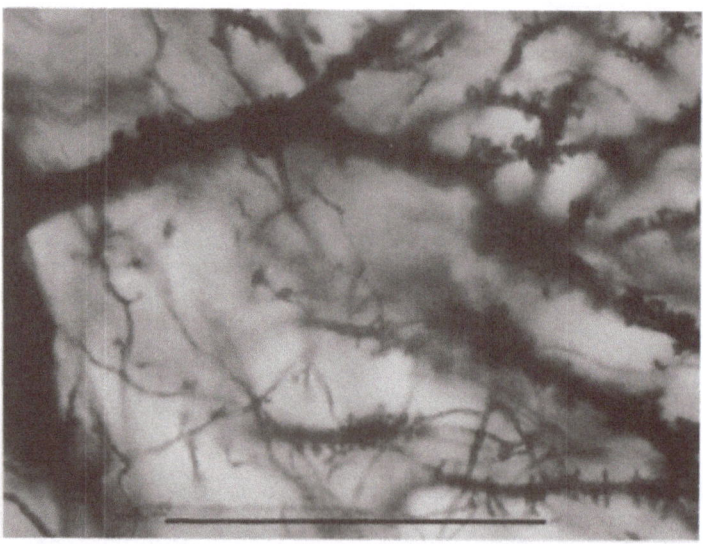

Fig.6. Golgi-impregnated structures from the Z zone (second reflected blade). Ramified dendrite of mossy cell with large "wart spines" in upper half of picture. Smooth initial part of the dendrite to the extreme left. Dendrite of long-spined cell visible at lower right. Bar: 50 μm.

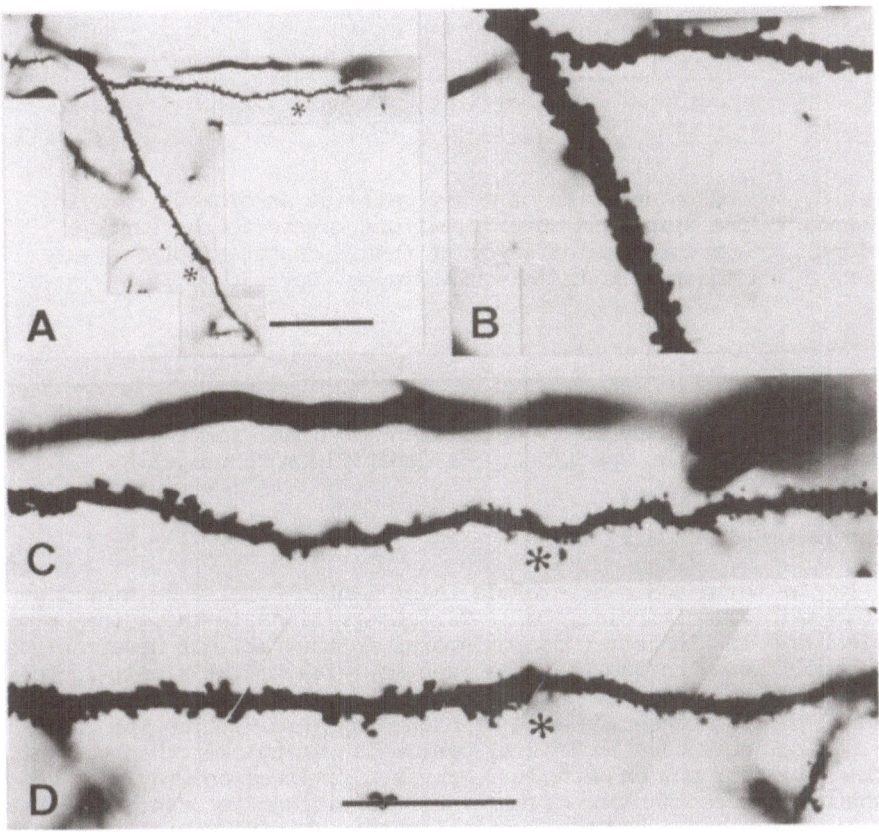

Fig.7. Montage of photomicrographs of mossy cell dendrites. A: a smooth dendritic stem close to the cell body is visible to the upper left. Bar: 50 μm. B, C, and D: Enlarged parts of the same dendrites. At asterisks coarse wart spines are replaced by fine spines. C and D show transition to outer and inner plexiform layers respectively. Bar (for B,C, and D): 20 μm.

The wart spines are present along a certain distance of the dendrites, after which they are replaced by fine spines (Fig.7). Comparison of Timm and Golgi material makes clear that it is at the transition from the Z zone to the adjacent zones that the spine pattern (and thus the synapse type) changes. Some of the dendrites continue up through the fascia dentata (Fig.8), often as far as the pial surface. While passing the granular layer their surface is changed again, the amount of spines being much reduced or negligible. In the molecular layer the surface gets a spine pattern similar to that of the granular cell dendrites. Summing up, we have

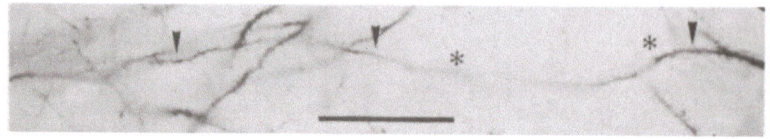

Fig.8. Ascending dendrite of mossy cell (at arrowheads) carrying
abundant fine spines in outer plexiform layer (to the extreme
right), fewer in granular layer of fascia dentata (between asterisks) and again more in molecular layer. Bar: 50 μm.

seen clear-cut differences between dendritic portions of the mossy
cell passing as many as four different laminae, forming another
example of laminar specificity of dendritic morphology.

The Long-Spined Cell

     The third and last example to be mentioned will be Amaral's
long-spined (multipolar) cell. Our observations in the guinea pig
diverge a little from those of Amaral in the rat, but nevertheless
the cell is so unique that the identity with Amaral's cell in the
rat is beyond doubt. The dendrites of the long-spined cell have
very special spines and mode of branching (Fig.9). The spines as
a general rule have an initial part which is thicker than the very
thin stalk of common gemmules. Many are ramified and may even di-
chotomize more than once. There are transitions to branchlets that
ressemble a chain of these special spines (Fig.9, B-D). The mode of
emergence of smaller and larger branches is typical: they arise
from the parent stem or branch very nearly at right angles to it.
During repeated dichotomizations they bend back to a restricted
space instead of growing out into neighbouring layers. Comparison
with Golgi preparations where other kinds of cells are impregnated
simultaneously, and with Timm preparations, proves that the long-
spined cell dendrites are abundant in the Z zone and do not penetrate deeper than it, i.e., into the inner plexiform zone (cf.
Fig.10). They never reach the granular or molecular layers of the
fascia dentata and they at least largely shun the outer plexiform

Fig.9. Photomicrographs of dendrites of long-spined cells. A: Montage of 5 pictures. Spine-free soma and dendritic stem at upper
right. B, C, and D: Other dendrites at three different focal levels. A branchlet (arrowheads) emerges from a thicker branch (aster-
isk). Bar (for B, C, and D): 10 μm. E: Several dendrites (from one
or more cells). Bar (for A and E): 10 μm.

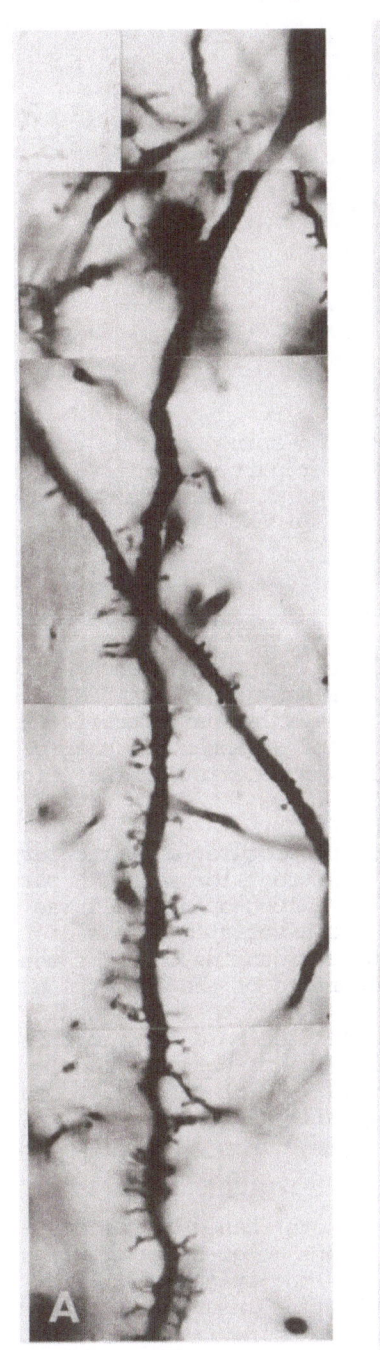

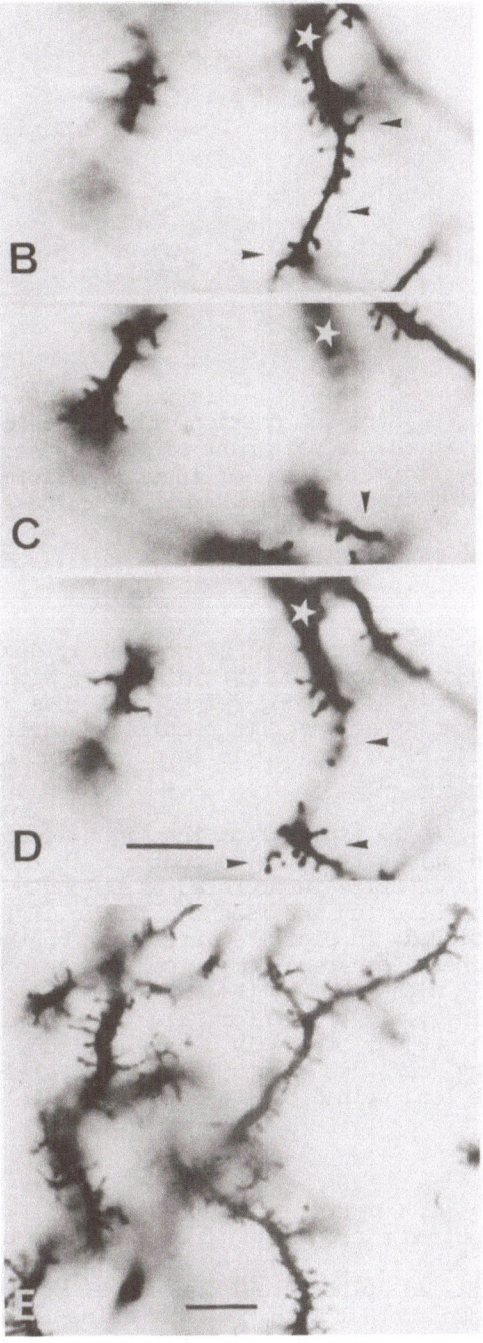

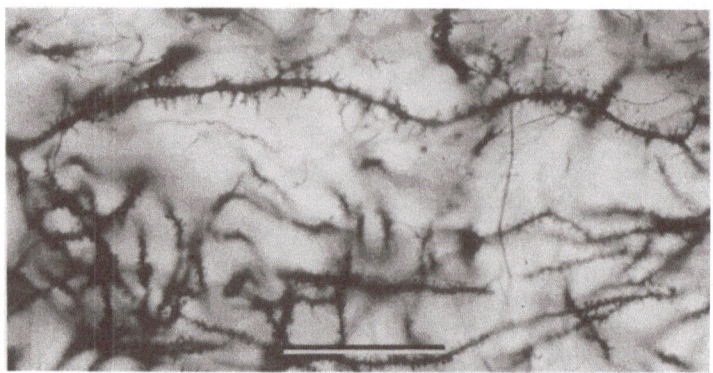

Fig.10. Straight portion of dendrite of long-spined cell is present in the deepest part of the Z zone. Lower part of picture shows superficial portion of inner plexiform layer with numerous dendrites of modified pyramids carrying fine spines. Bar: 50 μm.

layer. It cannot yet be completely excluded that some branches may invade it for a certain distance. With a reservation for this latter point, the long-spined cell thus displays laminar specificity in the sense that its peculiar dendrites and synapses are characteristic of one single layer and not allowed to exist in the neighbouring layers.

As to expect from the light microscope picture, a particular synaptic pattern is encountered in the electron microscope. A part of a dendrite interpreted as belonging to a long-spined cell was reconstructed and is shown in Fig.11A. Two associated spines are included. In Fig.11B-C are shown boutons surrounding one of these spines. Perhaps a comparison of the properties of the long-spined cell with those of the CA1 pyramidal cell, which possesses usual spines, might be informative at a time when the role of spines is the subject of lively debate (Wilson, 1984). For neuronal modelling (refs. in Wilson, l.c., and Pellionisz, 1979) certainly both the long-spined cell and the mossy cell will represent complex

Fig.11. Computer reconstructions (stereo sets) based on serial electron micrographs (from conventional, not Golgi-EM material). A: Dendrite interpreted as belonging to a long-spined cell. Two ramified spines are included. B: Seven boutons in asymmetric synaptic contact with the left spine in A. The spine fits into the central cavity seen in this reconstruction. C: The right reconstruction in B manually shaded.

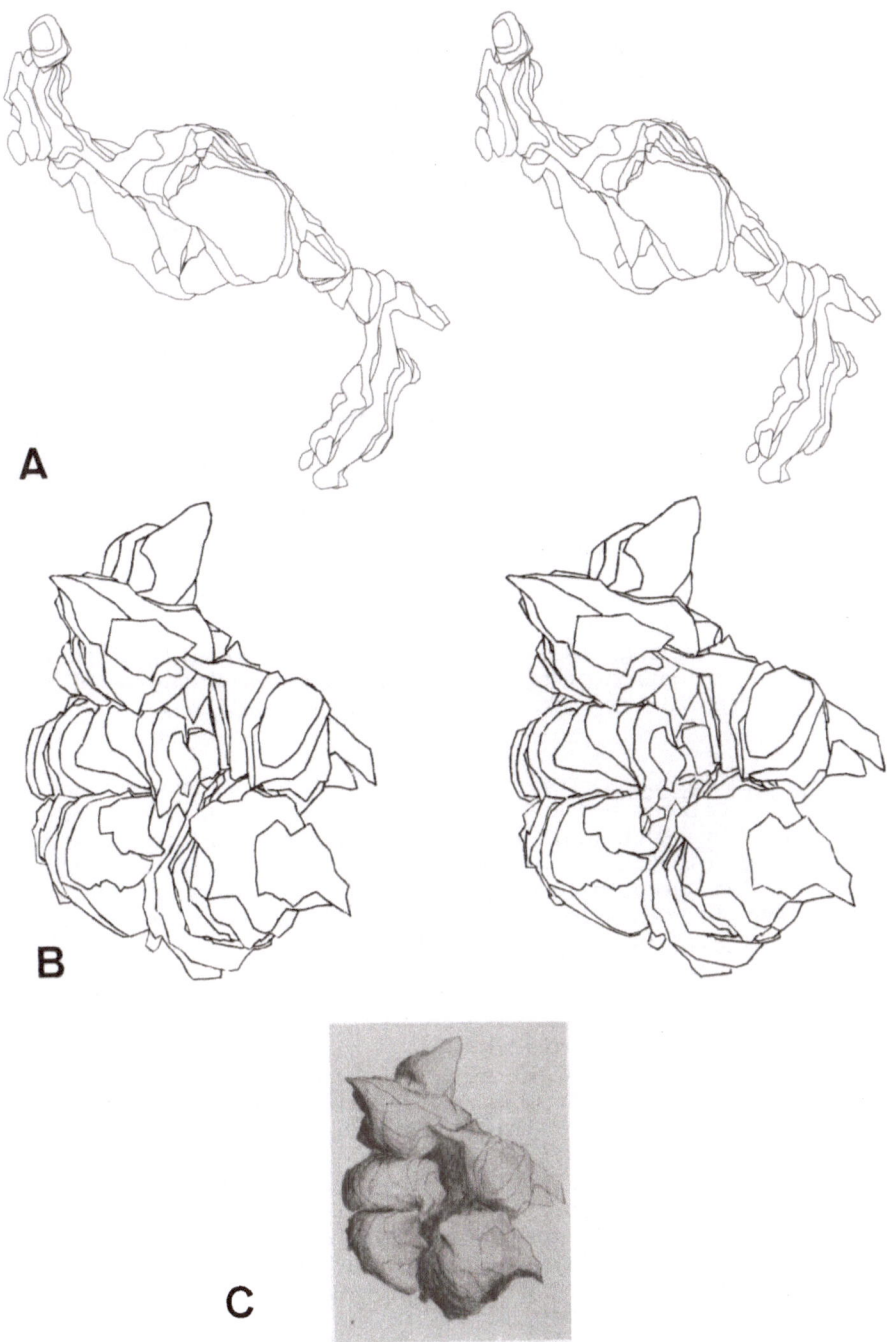

A

B

C

paradigms. But both are very common cells in the hilus and an understanding of the mode of functioning of the fascia dentata and the hippocampus can hardly be achieved without knowledge about the geometry and physiological properties of these cells. In the context of transmitter research stratification, i.e., segregation of structural elements, has long been known to be helpful, as witnessed by studies on retina, cerebellum, hippocampus, and other areas. Morphological specialization of dendrites as they pass such strata is evidence of changing synaptology and invites a search for accompanying biochemical features, be it "conventional" transmitters and neuropeptides or other chemical parameters.

SUMMARY

This note is based on preliminary data from three studies in progress, on pyramidal cells in the hippocampal subfield CA1 and on mossy cells and long-spined cells in the hilus fasciae dentatae. Computer-aided graphic reconstructions of Golgi material and electron micrographs are used. The regions examined are characterized by stratification, each layer having specific histochemical properties, synapse types, and afferent input. By way of the three cell examples it is emphasized that dendrites in stratified regions may have specific surface morphology (mainly spine pattern) and mode of branching in the individual layers. Such local specialization in turn is an indication of specific synaptology which may be of interest in transmitter research and neuronal modelling.

Acknowledgements. For excellent assistance the author is grateful to G.Lothe and E.Risnes (photography), T.Reppen (electronics), K.Stentoft (cell recording), K.F.Urbye (histology), and J.L.Vaaland (electron microscopy). These studies were supported by The Norwegian Research Council for Science and the Humanities, Anders Jahre's Foundation for the Promotion of Science, and Det Videnskapelige Forskningsfond av 1919.

REFERENCES

Amaral, D.G. (1978). A Golgi study of the hilar region of the hippocampus in the rat. J.Comp.Neurol., 182, 851-914.
Blackstad, T.W. (1963). Ultrastructural studies on the hippocampal region. Prog.Brain Res., 3, 122-148.
Blackstad, T.W., Osen, K.K., and Mugnaini, E. (1984). Pyramidal neurones of the dorsal cochlear nucleus: A Golgi and computer reconstruction study in the cat. Neuroscience (in press).
Cajal, Ramón y, S. (1911). Histologie du Système Nerveux de l'Homme et des Vertébrés. Vol.II, Maloine, Paris.
Haug, F.-M.Š. (1973). Heavy metals in the brain. A light microscope study of the rat with Timm's sulphide silver method. Methodological considerations and cytological and regional staining patterns. Adv.

Anat.Embryol. Cell Biol., 47 (Fasc.4), 1-71.
Lorente de Nó, R. (1934). Studies on the structure of the cerebral cortex. II. Continuation of the study of the ammonic system. J. Psychol.Neurol.(Lpz.), 46, 113-177.
Pellionisz, A. (1979). Modeling of neurons and neuronal networks. In The Neurosciences: Fourth Study Program. (eds. F.O. Schmitt and F.G. Worden), pp. 525-546. M.I.T. Press, Cambridge.
Stengaard-Pedersen, K., Fredens, K., and Larsson, L.-I. (1983). Comparative localization of enkephalin and cholecystokinin immuno-reactivities and heavy metals in the hippocampus. Brain Res., 273, 81-96.
Wilson, C.J.(1984). Passive cable properties of dendritic spines and spiny neurons. J. Neuroscience, 4, 281-297.

STUDIES OF LOCAL BLOOD FLOW AND GLUCOSE UTILIZATION IN BRAIN BY COMPUTER ASSISTED AUTORADIOGRAPHY

CHRISTER OWMAN* and NILS H. DIEMER†

*Department of Histology, University of Lund, Sweden
†Department of Neuropathology, University of Copenhagen, Denmark

It has for a long time been assumed that the cerebrovascular bed differs fundamentally from the peripheral circulation in that perivascular nerves are lacking and that the control of flow in the brain is exerted entirely by "chemical" or "metabolic" mediators. With the application of highly sensitive and specific neurohistochemical methods for visualization of aminergic transmitters (Falck, 1962; Falck et al., 1962; Björklund et al., 1972) and, later on, for the demonstration of various neuropeptides (Hökfelt et al., 1980) it has now been established that the cerebrovascular bed is, indeed, extensively supplied with several systems of perivascular nerve fibres (Owman et al., 1984). This has led to the additional concept of "neurogenic" control mechanisms for the brain circulation (Owman and Edvinsson, 1977; Heistad and Marcus, 1982; MacKenzie et al., 1984).

It can be anticipated that the recently devised autoradiographic techniques for precise quantitative and simultaneous measurement of local flow and metabolism in the brain (Reivich et al., 1969; Sokoloff et al., 1977; Sakurada et al., 1978) will provide further help in solving the mechanisms underlying the tight linkage between flow and metabolism, and how "metabolic and "neurogenic" processes are integrated in the physiological regulation of flow in health and disease.

Our understanding of the mechanisms whereby the nervous system influences and regulates cardiovascular performance - via direct nerve connections or through the release of circulating neurotransmitters - does not date further back than the middle of the nineteenth century (see Folkow and Neil, 1971). At this time a number of new techniques were developed for physiological recording of vascular events, and the effects of electrical stimulation of the nerves associated with the heart and blood vessels could be studied. This made it possible to establish the presence of tonically active neurons in the medulla oblongata, adjusted in their activity by the afferent input from the cardiovascular

receptors, regulating the heart and blood vessels by sympathetic
and parasympathetic nerves. The introduction of a great number
of new methods has expanded our knowledge tremendously about
the distribution of autonomic nerves engaged in cardiovascular
mechanisms (Coons, 1942; Falck, 1962; Burnstock and Costa, 1975),
the detailed structure and function of the vascular neuroeffector
system (Bennet, 1972; Bevan et al., 1980), the different types
of receptors mediating the effects (Furchgott, 1972; Ariëns et
al., 1979; Williams and Lefkowitz, 1978), as well as our understanding of the complex operation of the servo control involving
the neural reflexes (Korner, 1979).

Though almost a century has passed since Roy and Sherrington
(1890) first described the powerful effect of carbon dioxide
on the cerebral circulation, the mechanism of action of this
major product of brain metabolism is still poorly understood.
It is still not known to what extent carbon dioxide or other
possible mediators under physiological conditions couple the
blood flow in the brain to its metabolic demands. An important
synthesis of how the brain circulation operates was achieved
by the nitrous oxide technique of Kety and Schmidt (1948), and
the method of Lassen and Ingvar (1961, 1972) for measuring the
blood flow regionally has contributed much to the elucidation
of adjustments and impairments of the cerebral blood flow in
health and disease.

Even if earlier methods for measurements of average blood
flow and metabolism throughout brain have contributed much to
our understanding of normal and pathological mechanisms in the
cerebral circulation, they were not able to disclose detailed
phenomena related to the enormously complex circuitry of the
brain and its structural and functional heterogeneity. Later
techniques for analyzing regional changes in blood flow suffered
from the disadvantage of measuring "cylinders" of brain tissue
(with the xenon technique) containing a mixture of areas with
different function, or only regions at or very close to the brain
surface (with the krypton technique). The disadvantage has been
compensated in studies with the intracarotid ^{133}Xe method by
increasing the number of detectors, each with only a small colli-
mator apperture (Sveinsdottir et al., 1977). In this way, a dynamic system for mapping of human cerebral functions has been
achieved (Lassen et al., 1977).

However, there are serious problems in applying the inert
gas technique to laboratory animals. Most species (except the
primates) have a great number of anastomoses between the cerebral
circulation and the extracranial tissues. Such anastomoses exist
on both the arterial and the venous sides. The presence of these
anastomoses has invalidated measurements of blood flow by inert
gas techniques in laboratory animals such as dogs and cats. Even
in primates the extracranial tissues should be removed. Measurement of cerebral blood flow in the rat by an external scintilla-

tion detector, placed over the animal's head, will result in substantial contamination, even after selective injection of the tracer into the internal carotid artery and attempts to ligate the collaterals.

The problem of a mixture of functionally different areas in measurements of flow even in discrete regions holds true also for the tissue sampling approach applied to animal experiments and utilizing such tracers as radioactive antipyrine (Reivich et al. 1969) and ethanol (Eklöf et al., 1974), or microsphere techniques based on the bolus fractionation principle (Sapirstein, 1962). With these methods it has not been possible to measure metabolic parameters in the same regions as those defined for the flow measurements, which is evidently of great disadvantage considering the close coupling between flow and metabolism in the brain.

The introduction by Sokoloff et al. (1977) of the ^{14}C-deoxyglucose technique meant a revolution in this field of neurochemistry in that one fundamental aspect of cerebral metabolism – glucose utilization – could be measured with great precision and accuracy in discrete areas of the animal brain. In the normal brain, energy is formed mainly when glucose is oxidized to carbon dioxide and water (Siesjö, 1978). Energy is expended in transporting ions and other compounds across membranes and in synthesizing cell constituents. It is stored in the form of high--energy phosphate groups which are readily available whenever needed. Thus, under physiological conditions, the oxidation of glucose is tightly coupled to ATP formation. Since the stores of ATP as well as phosphocreatine suffice for only about 20 seconds' uninterrupted utilization, a continuous process of oxidative metabolism is essential for a normal function of the brain.

When the glucose molecule enters the cell, its subsequent metabolism requires that it is first phosphorylated to yield glucose-6-phosphate. Deoxyglucose is transported between plasma and brain (and vice versa) by the same carrier as glucose and it is also phosporylated by hexokinase. However, the further metabolism of deoxyglucose is impeded by the fact that it is a poor substrate for the enzymes, glucosephosphate isomerase and glucose-6-phosphate dehydrogenase. Since the deoxyglucose-6--phosphatase activity is low in brain tissue, the sugar is essentially trapped in the cells as deoxyglucose-6-phospate, since phosphorylated hexoses do not easily pass cell membranes. The amount trapped at any given time after its introduction into the blood equals the integral of the rate of phosphorylation by hexokinase during that interval of time. A quantitative relationship thus exists between the phosphorylation rate of glucose and the amount of deoxyglucose-6-phosphate formed. This depends on the relative concentrations of the sugars and on their relative affinities for the carrier mechanisms and for hexokinase.

General Equation for Measurement of Reaction Rates with Tracers:

$$\text{Rate of Reaction} = \frac{\text{Labeled Product Formed in Interval of Time, O to T}}{\begin{bmatrix}\text{Isotope Effect}\\\text{Correction Factor}\end{bmatrix}\begin{bmatrix}\text{Integrated Specific Activity}\\\text{of Precursor}\end{bmatrix}}$$

Operational Equation of $[^{14}C]$ Deoxyglucose Method:

Labeled Product Formed in Interval of Time, O to T

$$R_i = \frac{\overbrace{C_i^*(T)}^{\substack{\text{Total }^{14}\text{C in Tissue}\\\text{at Time, T}}} - \overbrace{k_1^* e^{-(k_2^*+k_3^*)T}\int_0^T C_p^* e^{(k_2^*+k_3^*)t}\,dt}^{^{14}\text{C in Precursor Remaining in Tissue at Time, T}}}{\underbrace{\left[\frac{\lambda\cdot V_m^*\cdot K_m}{\Phi\cdot V_m\cdot K_m^*}\right]}_{\substack{\text{"Isotope Effect"}\\\text{Correction}\\\text{Factor}}}\left[\underbrace{\int_0^T\left(\frac{C_p^*}{C_p}\right)dt}_{\substack{\text{Integrated Plasma}\\\text{Specific Activity}}} - \underbrace{e^{-(k_2^*+k_3^*)T}\int_0^T\left(\frac{C_p^*}{C_p}\right)e^{(k_2^*+k_3^*)t}\,dt}_{\substack{\text{Correction for Lag in Tissue}\\\text{Equilibration with Plasma}}}\right]}$$

Integrated Precursor Specific Activity in Tissue

FIG. 1. Operational equation of radioactive deoxyglucose method and its functional anatomy. T represents the time at the termination of the experimental period; λ equals the ratio of the distribution space of deoxyglucose in the tissue to that of glucose; Φ equals the fraction of glucose which, once phosphorylated, continues down the glycolytic pathway; and K_m and V_m and K_m and V_m represent the familiar Michaelis—Menten kinetic constants of hexokinase for deoxyglucose and glucose, respectively. The correction factor for the isotope effect is conveniently termed the "lumped constant". (From Sokoloff, 1981).

Since the latter are known, it is possible to determine glucose utilization rates at steady state by measuring the arterial history of the specific activity in plasma of deoxyglucose, the plasma glucose concentration, and the deoxyglucose activity in the brain tissue.

For calculations of glucose utilization rates with the equation (Fig. 1) described by Sokoloff et al. (1977), certain conditions must be fulfilled. (a) The plasma glucose concentrations should be relatively constant during the experiment. (b) The deoxyglucose should occur only in tracer concentrations, (c) homogenous tissue compartments for deoxyglucose and glucose should exchange directly with plasma. (d) Since measurements of ^{14}C-deoxyglucose activity in tissue are made by autoradiography, it is not possible to distinguish between the free and phosphorylated forms. Hence, a sufficient period should pass after the injection of ^{14}C-deoxyglucose to allow most of the free form to disappear by phosphorylation or transport. The lumped constant and the rate constants of the operational equation have been determined for several animal species, including rat.

On the basis of the autoradiographic approach for the measurement of local glucose utilization in the animal brain, an autoradiographic method was also introduced for determination of local cerebral blood flow (Sakurada et al., 1978). Like the previously mentioned tissue sampling technique for regional measurements it is based on the Fick principle, according to which the time rate for the change of the amount of a biologically inert diffusible tracer substance within a tissue is equal to the difference between the rate at which the substance is brought to the tissue in the arterial blood and removed from its venous blood. This can be expressed mathematically according to the following equation (Kety, 1960):

$$ Ci(T) = \lambda k \int_0^T C_a(t) e^{-k\,(T-t)} dt $$

where $Ci(T)$ = the concentration of tracer substance in the tissue at time T; λ = the tissue-blood partition coefficient for the tracer; k = the rate of blood flow per unit weight of tissue divided by the partition coefficient for that tissue; and C_a = the arterial concentration of tracer substance. Thus, to calculate the blood flow to a given homogenous region of the brain one must know: (a) the tissue-blood partition coefficient of the tracer substance for that region of the brain, (b) the concentration of the tracer substance in this region at a time, T (usually 1 min), after the start of the infusion, and (c) the time course of change of arterial concentration of tracer material. ^{14}C-iodoantipyrine was used as tracer, since this was found to have a higher oil-to-water partition coefficient than antipyrine.

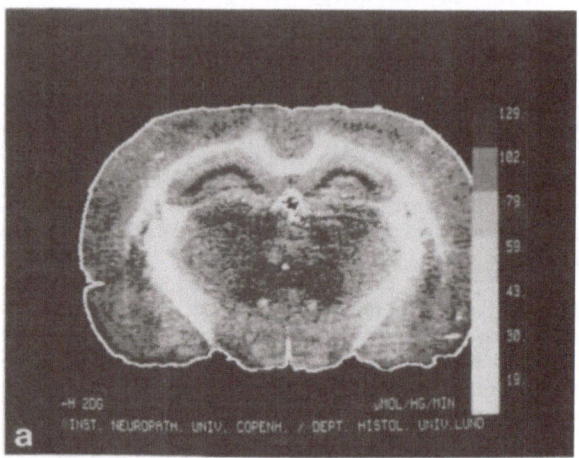

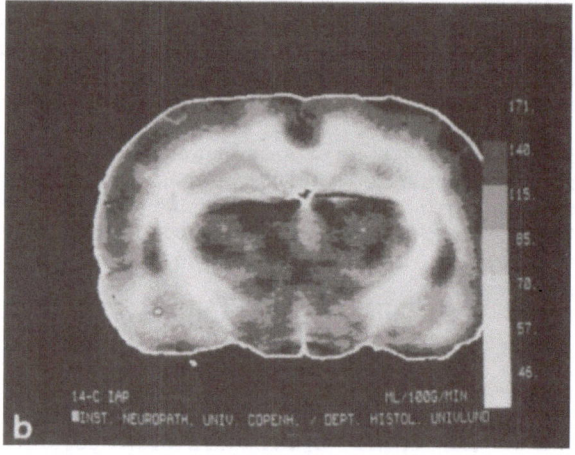

FIG. 2. Simultaneous demonstration of local cerebral blood flow and glucose utilization in normal rat brain. Autoradiograms obtained from two adjacent cryostat sections and digitized in 8 grey levels. (a) ^{3}H-deoxyglucose autoradiogram following washing of section with DMP for complete removal of ^{14}C-iodoantipyrine. Metabolism expressed in μmol/100g/min. (b) ^{14}C-iodoantipyrine autoradiogram in which flow is expressed in ml/100g/min. Note close correlation between rates of flow and metabolism in different brain areas. (Unpublished data).

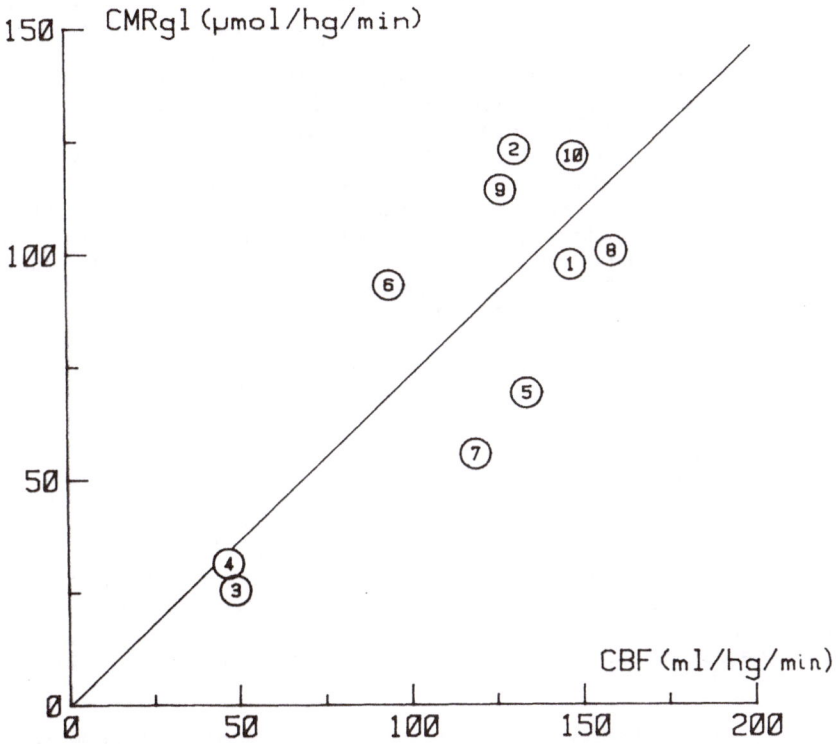

FIG. 3. Correlation between local blood flow and glucose metabolism in normal rat brain. Values for 10 brain areas were determined by means of the dual-tracer, double-label technique described in the text and quantified by computer assisted autoradiography, as illustrated in Fig. 2. (Unpublished data).

Various attempts have been made to measure local cerebral blood flow and glucose metabolism simultaneously in the same circumscribed areas by double radionuclide autoradiographic techniques (Blasberg et al., 1981; Diemer and Rosenörn, 1981; Lear et al., 1981; Mies et al., 1981; Furlow et al., 1983). In the method devised by Diemer and Rosenörn (1981) a double autoradiogram is made from two neighbour sections after extracting one of the isotopes (^{14}C-iodoantipyrine) with a solvent (Fig. 2). Rats are injected i.v. with 500-1500 µ Ci kg of ^{3}H-deoxyglucose, and during the circulation period frequent samples of blood are taken for the determination of the plasma deoxyglucose integral as described by Sokoloff et al. (1977). After the last deoxyglucose sample, the arterial catheter is connected to a constant velocity withdrawal pump for mechanical integration of the tracer concentration (Gjedde et al., 1980) and simultaneously 50 µCi/kg body weight of ^{14}C-iodoantipyrine (Sakurada et al., 1978) is injected as an i.v. bolus. After 20 seconds the animal is decapitated and the brain frozen. Twenty µm thick serial cryostat sections were used for autoradiography.

Experiments with extraction of the iodoantipyrine with different solvents have given various degrees of isotope retention in the sections (Diemer and Rosenörn, 1981). It was found that 2,2-dimethoxypropane (DMP) removes all iodoantipyrine. DMP reacts chemically with water, forming methanol and acetone, and this procedure did not influence the content or distribution of deoxyglucose (and -phosphate) in the sections as measured by microdensitometry of the deoxyglucose autoradiograms on ^{3}H sensitive film (LKB Ultrofilm).

The autoradiograms (Fig. 2) were analyzed with computer-assisted microdensitometry using the Leitz TAS PLUS equipment. The ^{3}H autoradiograms from the combined flow/glucose metabolism experiments have the same high resolution (Fig. 2a) as described for single tracer experiments (Sokoloff et al., 1977; Faraco-Cantin et al., 1980). Although ^{3}H is present in the sections for blood flow autoradiography, the irradiation from tritium is too weak to reach the emulsion on the X-ray film due to the thickness of the protective (gelatine) layer, and these autoradiograms thus reflect only the ^{14}C label.

The close relationship between local glucose utilization and blood flow in the same area of the rat's brain could be confirmed in the double-label experiments (Fig. 3)

The mechanical integration of ^{14}C-iodoantipyrine concentration in arterial blood in the 20-second experiments for determination of local cerebral blood flow was based on the following basic equation relating integrated in- and efflux rates of butanol in the brain (Gjedde et al., 1980):

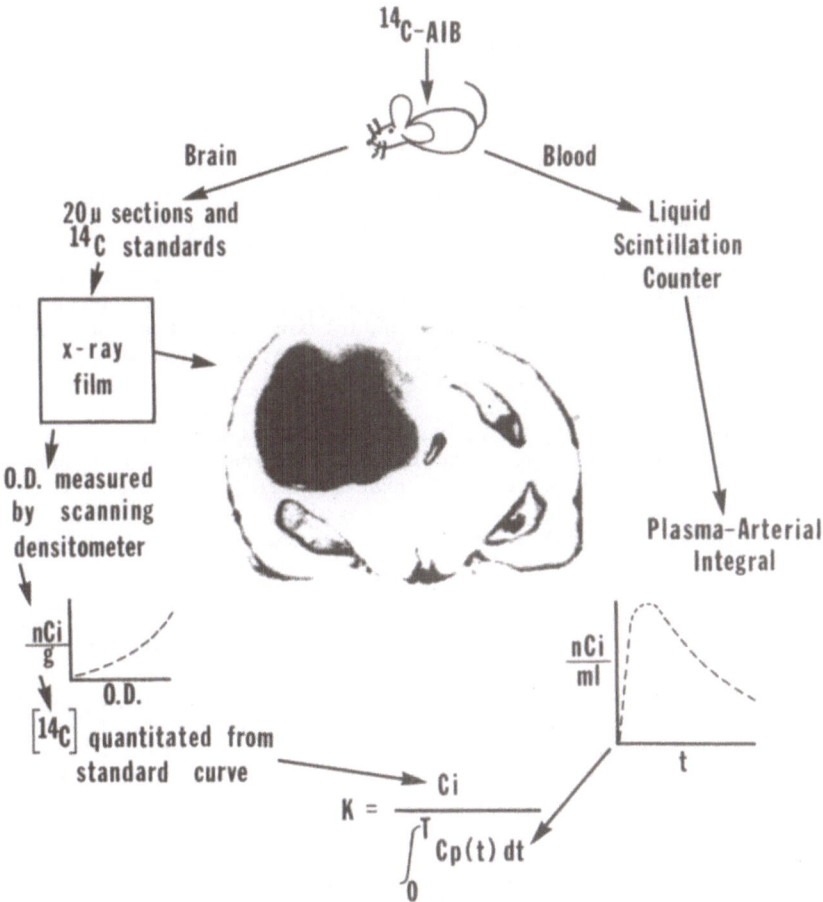

FIG. 4. Schematic outline of a quantitative autoradiographic experiment to determine the blood-to-tissue transfer constant, K, for α-aminoisobutyric acid. The plasma samples are analyzed as shown on the right. The brain sections are analyzed as shown on the left. An example of an X-ray image of a brain section containing a permeable avian sarcoma virus-induced tumor is shown in the center. (From Blasberg et al., 1981).

$$f^{bl} = \frac{C_{br}(T)}{E(T) \displaystyle\int_0^T C_a^{bl}(t)\,dt}$$

in which f^{bl} is the blood flow rate per unit weight of a sample of brain, $C_{br}(T)$ the indicator content per unit weight of brain tissue, $E(T)$ the net extraction fraction of indicator in the time from introduction of the indicator into the circulation (t=0) to the termination of the experiment (t=T), and $C_a^{bl}(t)$ the arterial blood concentration of the indicator at the time t. The usefulness of the equation depends entirely on the knowledge of $E(T)$. The assumptions underlying the equation, its development, and methods of estimation of $E(T)$ have been presented in the study by Gjedde et al. (1980).

The use of the equation requires that two criteria are fulfilled: (a) that the time integrals of internal carotid and femoral arterial concentrations of tracers from injection to arrest of cortical circulation are equal, and (b) that efflux of tracer from brain is negligible in the 20 seconds and that, therefore, an extraction fraction (E) equal to 1 may be used as a good approximation. With regard to the first criterion, Gjedde et al. (1980) showed that the integral was independent of the site of sampling from 3 to 30 seconds after injection of the tracer. The second criterion was confirmed by Gjedde et al. (1980) for butanol and by Sakurada et al. (1978) for iodoantipyrine. They compared cerebral blood flow values calculated with these indicators with values measured directly on the basis of arterial and venous concentrations or by means of inert gases of known, essentially infinite, permeability in the blood-brain barrier.

Since the effect of circulating vasoactive neurotransmitters on cerebral blood flow is highly dependent on the state of the blood-brain barrier, which efficiently prevent these substances from reaching beyond the vascular endothelial lining (Owman and Hardebo, 1982) it is of considerable value to detect barrier lesions with precise autoradiography. This can be achieved with the inert α-aminoisobutyric acid as radioactive tracer (Blasberg et al., 1983). This non-metabolized amino acid does not cross normal brain capillaries. Once having entered the extracellular fluid of the brain in case of barrier break-down, it is avidly taken up, concentrated, and trapped within brain cells by the small neutral amino acid transport mechanisms in the cell membrane. Since backflux of α-aminoisobutyric acid to blood during the experimental period is thus negligible, it can be conveniently used for determination of the unidirectional blood-to-brain transfer constant (k) as shown in Fig. 4.

These autoradiographic approaches, together with positron emission tomography (Raichle, 1979) which is really nothing

TABLE I. Examples of different types of approaches for quantitative structural and functional mapping of various parameters in the brain based on the principles developed by Falck and co-workers for histo-chemical studies and Sokoloff and co-workers for autoradiographic studies.

Parameter studied	Model tracer or reagent	Technique	Level of resolution	Reference
Blood flow	^{14}C-iodoantipyrine	Autoradiography	Local	Sakurada et al. 1978
Glucose utilization	^{14}C-deoxyglucose	Autoradiography	Local	Sokoloff et al. 1977
BBB permeability	^{14}C-α-aminoisobutyric acid	Autoradiography	Local	Blasberg et al. 1981
Protein synthesis	^{14}C-isoleucine	Autoradiography	Local	Bodsch et al. 1984
Potassium	Sodium cobaltinitrite	Staining	Regional	Mies et al. 1984
pH	Umbelliferone	Fluorescence	Regional	Csiba et al. 1983
ATP	Luciferase	Bioluminescence	Regional	Paschen et al. 1981
NADH	Luciferase	Bioluminescence	Regional	Paschen et al. 1981
Amine transmitters	Formaldehyde	Fluorescence	Cellular	Björklund et al. 1972
Transmitter enzymes	Antibodies	Fluorescence	Cellular	Hökfelt et al. 1979
Peptide transmitters	Antibodies	Fluorescence	Cellular	Larsson 1983
Receptors	Ligands	Autoradiography	Local	Kuhar 1983

but in vivo autoradiography of the human brain, have formed the
basis for an extensive development of brain imaging techniques.
By these techniques brain function, including local protein synthesis (Bodsch et al., 1984), and its dependency on an uninterrupted supply of energy sources - oxygen and substrate - by the
blood stream can be studied both in animals (for references,
see Blasberg et al., 1981; Lear et al., 1984) and in man (for
references, see Baron et al. 1984; Heiss et al., 1984; Kanno
et al., 1984) with a high degree of precision and dynamics.

CONCLUDING REMARKS

    The double isotope technique for precise quantitative measure-
ment of local cerebral blood flow and metabolism offers a tre-
mendous potential for functional mapping of the brain. The autoradiographic procedure provides a resolution level in the order
of 10 - 40 μm, allowing for quantitative measurements within
very circumscribed anatomical areas of the brain. The technique
for processing the tissue is well suitable also for simultaneous
histochemical visualization of aminergic and peptidergic nuclear
complexes and fiber tracts in the brain, forming a structural
and dynamic correlate at the cellular level to the local rates
of flow and metabolism. Numerous new additional methods are currently being developed for regional assessment of several other
parameters, taking advantage of the same principal processing
protocol as for the mentioned autoradiographic and histochemical
methods. In this way, as compiled in Table I, it is now possible
to map a whole array of structural and functional aspects of
the animal brain at the cellular, local, or regional level of
resolution, either in one and the same or in adjacent histological sections of the brain.

REFERENCES

Ariëns, E.J., Beld, A.J., Rodrigues de Miranda, J.F. and Simonis,
A.M.(1979). The pharmacon-receptor-effector concept. In The Recep-
tors, Vol. I: General Principles and Procedures. (ed. R.D.
O'Brien). Plenum Press, New York-London, pp. 33-91.

Baron, C., Rougemont, D., Soussaline, F., Bustany, P., Crouzel,
C., Bousser, M.G. and Comar, D. (1984). Local interrelationsships
of cerebral oxygen consumption and glucose utilization in normal
subjects and in ischemic stroke patients: A positron tomography
study. J. Cerebral Blood Flow Metab. 4, 140-149.

Bennet, M.R. (1972). Autonomic Neuromuscular Transmission. University Press, Cambridge.

Bevan, J.A., Bevan, R.D. and Duckles, S.P. (1980). Adrenergic
regulation of vascular smooth muscle. Handbook of Physiology,
Section 2, Volume II, 515-566.

Björklund, A., Falck, B. and Owman, Ch. (1972). Fluorescence microscopic and microspectrofluorometric techniques for the cellular localization and characterization of biogenic amines. In Methods of Investigative and Diagnostic Endocrinology Vol. I (ed. S.A. Berson), The Thyroid and Biogenic Amines. (eds. J.E. Rall and I.J. Kopin). North Holland, Amsterdam, pp. 318-368.

Blasberg, R.G., Fenstermacher, J.D. and Patlak, C.S. (1983). Transport of α-aminoisobutyric acid across brain capillary and cellular membranes. J. Cerebral Blood Flow Metab. $\underline{3}$, 8-32.

Blasberg, R.G., Groothuis, D. and Molnar, P. (1981). Application of quantitative autoradiographic measurements in experimental brain tumor models. Seminars in Neurology, $\underline{1}$, 203-221.

Bodsch, W., Takahashi, K., Ophoff, B.G and Hossmann, K.-A. (1984). Local rates of cerebral protein synthesis in the gerbil and monkey brain. In Methods for Measurement of Cerebral Blood Flow and Metabolism. (eds. A. Hartmann and S. Hoyer). Springer Press, Heidelberg and New York. In press.

Burnstock, G. and Costa, M. (1975). Adrenergic Neurons. Their Organization, Function and Development in the Peripheral Nervous System. Chapman and Hall, London.

Coons, A.H., Creech, H.J., Jones, R.N. and Berliner, E. (1942). The demonstration of pneumococcal antigen in tissues by the use of fluorescent antibody. J. Immunol. $\underline{45}$, 159-170.

Csiba, L., Paschen, W. and Hossmann, K.-A. (1983). A topographic quantitative method for measuring brain tissue pH under physiological and pathophysiological conditions. Brain Res. $\underline{289}$, 334-337.

Diemer, N.H. and Rosenörn, J. (1981). Determination of local cerebral blood flow and glucose metabolism or transfer by means of a double autoradiographic method. J. Cerebral Blood Flow Metab. $\underline{1}$, suppl. 1, S72-S73.

Edvinsson, L. and MacKenzie, E.T. (1977). Amine mechanisms in the cerebral circulation. Pharmacol. Rev. $\underline{28}$, 275-348.

Eklöf, B., Lassen, N.A., Nilsson, L., Norberg, K., Siesjö, B.K. and Torlöf, P. (1974). Regional blood flow in the rat measured by the tissue sampling technique; a critical evaluation using four indicators C^{14}-antipyrine, C^{14}-ethanol, H^{3}-water and xenon^{133}. Acta Physiol. Scand. $\underline{91}$, 1-10.

Falck, B. (1962). Observations on the possibilities of the cellular localization of monoamines by a fluorescence method. Acta Physiol. Scand. $\underline{56}$, suppl 197, 1-25.

Falck, B., Hillarp, N.-Å., Thieme, G. and Torp, A. (1962). Fluorescence of catecholamines and related compounds condensed with formaldehyde. J. Histochem. Cytochem. 10, 348-354.

Faraco-Cantin, F., Courville, J. and Lund, J.P. (1980). Methods for ^{3}H-2-D-deoxyglucose autoradiography on film and fine-grain emulsions. Stain Technol. 55, 247-252.

Folkow, B. and Neil, E. (1971). Circulation. Oxford University Press, London.

Furchgott, R.F. (1972). The classification of adrenoceptors (adrenergic receptors). An evaluation from the standpoint of receptor theory. Handbook of Experimental Pharmacology 33, 283-335.

Furlow Jr, T.W., Martin, R.M. and Harrison, L.E. (1983). Simultaneous measurements of local glucose utilization and blood flow in the rat brain: An autoradiographic method using two tracers labeled with carbon-14. J. Cerebral Blood Flow Metab. 3, 62-66.

Gjedde, A, Hansen, A.J. and Siemkowicz, E. (1980). Rapid simultaneous determination of regional blood flow and blood-brain glucose transfer in brain of rat. Acta Physiol. Scand. 108, 321-330.

Heiss, W.-D. Pawlik, G., Herholz, K., Wagner, R., Göldner, H. and Wienhard, K. (1984). Regional kinetic constants and cerebral metabolic rate for glucose in normal human volunteers determined by dynamic positron emission tomography of ^{18}F-2-fluoro-2-deoxy-D--glucose. J. Cerebral Blood Flow Metab. 4, 212-223.

Heistad, D.D. and Marcus, M.L. (eds.) (1982). Cerebral Blood Flow. Effects of Nerves and Neurotransmitters. Elsevier/North-Holland, Amsterdam.

Hökfelt, T., Fuxe, K., Goldstein, M., Johansson, O., Ljungdahl, Å., Lundberg, J.M. and Schultzberg, M. (1979). Immunocytochemical studies on catecholamine cell systems with aspects on relations to putative peptide transmitters. In Catecholamines: Basic and Clinical Frontiers. (eds. E. Usdin, I.J. Kopin and J. Barchas). Pergamon Press, New York, pp. 1007-1019.

Hökfelt, T., Johansson, O., Ljungdahl, Å, Lundberg, J.M. and Schultzberg, M. (1980). Peptidergic neurones. Nature, 284, 515-521.

Kanno, I., Lammertsma, A.A., Heather, J.D., Gibbs, J.M., Rhodes, C.G., Clark, J.C. and Jones, T. (1984). Measurement of cerebral blood flow using bolus inhalation of C^{15}O$_2$ and positron emission tomography: Description of the method and its comparison with the C^{15}O$_2$ continuous inhalation method. J. Cerebral Blood Flow Metab. 4, 224-234.

Kety, S.S. (1960). Blood-tissue exchange methods. Theory of blood-tissue exchange and its application to measurement of blood flow. In Methods in Medical Research, vol. 8 (ed. H.D. Bruner). Year Book Publishers, Chicago, pp. 223-227.

Kety, S.S. and Schmidt, C.F. (1948). The nitrous oxide method for the quantitative determination of cerebral blood flow in man: theory, procedure and normal values. J. Clin. Invest. 27, 476-483.

Korner, P.I. (1979). Central nervous control of autonomic cardiovascular function. Handbook of Physiology, Section 2, Volume I, 691-739.

Kuhar, M.J. (1983). Autoradiographic localization of drug and neurotransmitter receptors. In Handbook of Chemical Neuroanatomy, vol. 1: Methods in Chemical Neuroanatomy (eds. A. Björklund and T. Hökfelt). Elsevier, Amsterdam, pp. 398-415.

Lassen, N.A. and Ingvar, D.H. (1961). Blood flow of the cerebral cortex determined by radioactive krypton. Experientia, 17, 42-45.

Larsson, L.-I. (1983). Methods for immunocytochemistry of neurohormonal peptides. In Handbook of Chemical Neuroanatomy, vol. 1: Methods in Chemical Neuroanatomy (eds. A. Björklund and T. Hökfelt). Elsevier, Amsterdam, pp. 147-209.

Lassen, N.A. and Ingvar. D.H. (1972). Radioisotopic assessment of regional cerebral blood flow. Progr. nucl. Med. 1, 376-409.

Lassen, N.A., Roland, P.E., Larsen, B., Melamed, E. and Soh, K. (1977). Mapping of human cerebral function: A study of the regional cerebral blood flow pattern during rest, its reproducibility and the activations seen during basic sensory and motor functions. Acta Neurol. Scand. Suppl. 64, 56, 262-263.

Lear, J.L., Ackermann, R., Kameyama, M., Carson, R. and Phelps, M. (1984). Multiple-radionuclide autoradiography in evaluation of cerebral function. J. Cerebral Blood Flow Metab. 4, 264-269.

Lear, J.L., Jones. S.C., Greenberg, J.H., Fedora, T.J. and Reivich, M. (1981). Use of ^{123}I and ^{14}C in a double radionuclide autoradiographic technique for simultaneous measurement of LCBF nad LCMR$_{gl}$. Theory and method. Stroke, 12, 589-597.

MacKenzie, E.T., Seylaz, J. and Bès, A. (eds.) (1984). Neurotransmitters and the Cerebral Circulation. Raven Press, New York.

Mies., Kloiber, O., Drewes, L.R. and Hossmann, K.-A. (1984). Cerebral blood flow and regional potassium distribution during focal ischemia of gerbil brain. Ann. Neurol. In press.

Mies, G., Niebuhr, I. and Hossmann, K.-A. (1981). Simultaneous measurement of blood flow and glucose metabolism by autoradiographic techniques. Stroke, 12, 581-588.

Owman, Ch., Andersson, J., Hanko, J. and Hardebo, J.E. (1984). Neurotransmitter amines and peptides in the cerebrovascular bed. In Neurotransmitters and the Cerebral Circulation (eds. E.T. MacKenzie, J. Seylaz and A. Bès). Raven Press, New York, pp. 11-38.

Owman, Ch. and Edvinsson, L. (eds.) (1977). Neurogenic Control of the Brain Circulation. Pergamon Press, Oxford.

Owman. Ch. and Hardebo, J.E. Mechanisms of cerebral vasodilatation: Amines, peptides and the blood-brain barrier. In Vasodilatation. (eds. P.M. Vanhoutte and I. Leusen). Raven Press, New York, pp. 159-179.

Paschen, W., Niebuhr, I. and Hossmann, K.-A. (1981). A bioluminescence method for the demonstration of regional glucose distribution in brain slices. J. Neurochem. 36, 513-517.

Raichle, M.E. (1979). Quantitative in vivo autoradiography with positron emission tomography. Brain Res. Rev. 1, 47-68.

Reivich, M., Jehle, J., Sokoloff, L. and Kety, S.S. (1969). Measurement of regional cerebral blood flow with antipyrine-^{14}C in awake cats. J. appl. Physiol. 27, 296-300.

Roy, C.S. and Sherrington, C.S. (1980). The regulation of the blood supply of the brain. J. Physiol. 11, 85-108.

Sakurada, O., Kennedy, C., Jehle, J., Brown, J.D., Carbin, C.L. and Sokoloff, L. (1978). Measurement of local cerebral blood flow with ^{14}C-iodoantipyrine. Amer. J. Physiol. 234, H59-H66.

Sapirstein, L.A. (1962). Measurement of the cephalic and cerebral blood flow fractions of the cardiac output in man. J. Clin. Invest. 41, 1429-1435.

Siesjö, B.K. (1978). Brain Energy Metabolism. John Wiley, Chichester.

Sokoloff, L. (1981). Localization of functional activity in the central nervous system by measurement of glucose utilization with radioactive deoxyglucose. J. Cerebral Blood Flow Metab. 1, 7-36.

Sokoloff, L., Reivich, M., Kennedy, C., Des Rosiers, M.H., Patlak, C.S., Pettigrew, K.D., Sadurada, O. and Shinohara, M. (1977). The ^{14}C-deoxyglucose method for the measurement of local cerebral glucose utilization: theory, procedure, and normal values in the conscious and anaesthetized albino rat. J. Neurochem. 28, 897-916.

Sveinsdottir, E., Larsen, B., Rommer, P. and Lassen, N.A. (1977). A multidetector scintillation camera with 254 channels. J. Nucl. Med. 18, 168-174.

Williams, L.T. and Lefkowitz, R.L. (1978). Receptor Binding Studies in Adrenergic Pharmacology. Raven Press, New York.

Session II

MORPHOMETRY AND MICRODENSITOMETRY OF TRANSMITTER–IDENTIFIED NEURONS

Chairman: G. WOODWARD and G. JONSSON

MORPHOMETRICAL AND MICRODENSITOMETRICAL STUDIES ON MONOAMINERGIC AND PEPTIDERGIC NEURONS IN THE AGING BRAIN

LUIGI F. AGNATI[1], KJELL FUXE[2], LAURA CALZA[1],
LUCIANA GIARDINO[1], ISABELLA ZINI[1], GINO TOFFANO[3],
MENEK GOLDSTEIN[4], PAOLO MARRAMA[1],
JAN–ÅKE GUSTAFSSON[5], ZHAO–YING YU[1,5],
A. CLAUDIO CUELLO[6], LARS TERENIUS[7],
RUDOLF LANG[8] and DETLEV GANTEN[8]

[1]Departments of Human Physiology and Endocrinology, University of Modena, Modena, Italy
[2]Department of Histology, Karolinska Institutet, Stockholm, Sweden
[3]Fidia Research Laboratories, Abone Terme, Italy
[4]Department of Psychiatry, New York University Medical Center, New York, USA
[5]Department of Medical Nutrition, Huddinge Hospital, Huddinge, Sweden
[6]Department of Pharmacology and Human Anatomy, University of Oxford, Oxford, UK
[7]Department of Pharmacology, Biomedical Center, Uppsala, Sweden
[8]Department of Pharmacology, University of Heidelberg, Heidelberg, German Federal Republic

INTRODUCTION

In a recent study we have characterized aging processes in transmitter-identified neurons demonstrated by immunocyto-chemistry or by radioreceptor autoradiography using computer assisted morphometry and microdensitomotry (Agnati et al. 1984). The results indicated the existence of heterogeneities in the degenerative patterns taking place in transmitter-identified nerve cells and receptors in relation to aging. It was i.a. dis-covered that the meso-striatal and meso-accumbens dopamine (DA) neurons underwent degeneration in the aging brain both pre- and postsynaptically, while the DA nerve terminal networks within the tuberculum olfactorium were resistant to the aging processes. Furthermore a marked and widespread disappearance of the μ- and Δ-type of opiate receptors were found in the aged brain, while the benzodiazepine binding was enhanced in several brain areas of the aging brain compared with the 3 month old rat brain. Thus, it is possible that in aging also supersensitivity development can take place in certain types of receptors such as the benzodiaze-pine receptors, which possibly represent co-transmitter binding sites in GABA synapses (see Guidotti et al. 1983). Instead, the GABA receptor binding sites are reduced in number in the aged brain, suggesting that some compensatory changes in aging may take place within the co-transmitter binding sites.

In the present paper we have continued these studies on the aged brain by analyzing the 5-HT Substance P costoring nerve cells of the raphe nuclei of the medulla oblongata, and the nor-

adrenaline (NA) nerve cells of the locus coeruleus and their
innervation by phenylethanolamine-N-methyltransferase (PNMT)
immunoreactive and neuropeptide Y (NPY) immunoreactive nerve
terminal networks as well as their contents of glucocorticoid
receptor immunoreactivity present predominantly in the nuclei.
Finally, a detailed analysis has been performed on the subnuclei
of the paraventricular hypothalamic nucleus focusing on oxytocin,
vasopressin and neurophysin I and II immunooreactive nerve cells
of this nucleus.

MATERIAL AND METHODS

     3 month and 24 month old male Sprague-Dawley rats have been
used and kept under standardized light and dark conditions
(lights on at 6 a.m. and off at 8 p.m.). Analysis of the 5HT
Substance P costoring nerve cell bodies of the raphe nuclei of
the medulla oblongata, of the TH-positive cells of the locus
coeruleus and its NPY and PNMT innervation and of the neurophysin
I-vasopressin and neurophysin II-oxytocin positive nerve cell
bodies of the paraventricular hypothalamic nucleus was performed
in 3 month and 24 month old male Sprague-Dawley rats. The gravity
centers of the 5-HT-Substance P immunoreactive nerve cell bodies
and the density maps obtained from the locus coeruleus and the
paraventricular hypothalamic nucleus were obtained by the use of
a semiautomatic image analyzer (MOP-AMO2) coupled to an Apple-II
computer. The microdensitometrical analysis as well as the determination of field area was made in an IBAS Image Analyzer
(Zeiss-Kontron). The results obtained in 24 month old rat were
usually expressed in per cent of the corresponding mean values
found in the 3 month old animals (control value). The variability
was usually expressed as s.e.m. The statistical analysis was
performed by means of Student´s t-test. In the analysis of the
5-HT Substance P costoring nerve cell bodies of the raphe nuclei
of the medulla oblongata the occlusion method was used to analyze
possible changes in the entity of coexistence of 5-HT and Substance P during the aging process (see Agnati et al. 1982, Fuxe
et al. this symposium).
Abbreviations used in figures:
SNC - substantia nigra, pars compacta; SNR - substantia nigra,
pars reticulata; CC - crus cerebri; LM - lemniscus medialis; FL
- fasciculus longitudinalis; PVPO - paraventricular nucleus, pars
posterior; PVL - paraventricular nucleus, pars lateralis; PVM -
paraventricular nucleus, pars medialis; DC - paraventricular
nucleus, pars dorsalis; LC - locus coeruleus; DPR - nucleus parabrachialis, pars dorsalis; VPR - nucleus parabrachialis, pars
ventralis; CGPn - substantia grisea centralis pontis; Me5 -
nucleus mesencephalicus nervi trigemini; me5 - radix nervi trigemini; scp - pedunculus cerebellaris superior.

RESULTS

Studies on the Substance P/5-HT costoring nerve cell bodies of the medulla oblongata during aging

An example of 5-HT and SP-positive cells of the nuc. raphe magnus of the medulla oblongata in adult and old rats, visualized by means of the PAP technique, is given in fig. 1 A-D. A marked disappearance of SP immunoreactive nerve cells is noted in the old rat. The morphometrical analysis is summarized in fig. 2 and fig. 3. The location of the x-value and the y-value of the gravity centers in the young and the old rats is shown for the 5-HT and Substance P immunoreactive nerve cell bodies in the nucleus raphe pallidus, nucleus raphe obscurus and nucleus raphe magnus (B_1, B_2 and B_3 cell group, respectively). As seen the main change in the 5-HT immunoreactive nerve cell population of the B_1, B_2 and B_3 groups in relation to aging is a shift of the y-value of the gravity center towards a more ventral position. (fig. 2). In the case of the Substance P immunoreactive nerve cell body populations the aging induced changes in the location on the gravity center is considerably more complex and varies in the B_1, B_2 and B_3 regions. Thus, the y-value of the gravity center in the nucleus raphe pallidus is shifted to a more ventral position, while in the B_3 area it is shifted to a more dorsal position (see fig. 3). Also in the B_2 region the gravity center is shifted into a more dorsal position in the old rat compared with the adult animal. Thus, as evaluated by means of this overall parameter, the 5-HT and Substance P immunoreactive nerve cell body populations present in the raphe of the medulla oblongata are differentially altered by the aging process. Furthermore, there is a 45% reduction in the number of SP-immunoreactive profiles in the old rat, while the 5-HT immunoreactive profiles are only reduced by about 15%. The results obtained in the analysis of coexistence in the old rat are summarized in fig. 4. Using the occlusion method the number of Substance P and 5-HT immunoreactive nerve cell bodies were determined at various rostrocaudal levels of the medulla oblongata. Since the occlusion method was used to evaluate coexistence, the sections were also incubated, at each rostrocaudal level, with both antibodies simultaneously and the number of immunoreactive profiles was determined. It could be demonstrated that during the aging process the entity of coexistence expressed in per cent of the mean number of 5-HT nerve cell bodies was reduced from 55 % in the 3 month old rat to 31 % in the 24 month old rat. Thus, it is possible that the synthesis and/or utilization of the comodulator Substance P is more markedly affected during aging than the synthesis of the transmitter 5-HT.

L.F. Agnati, et al.

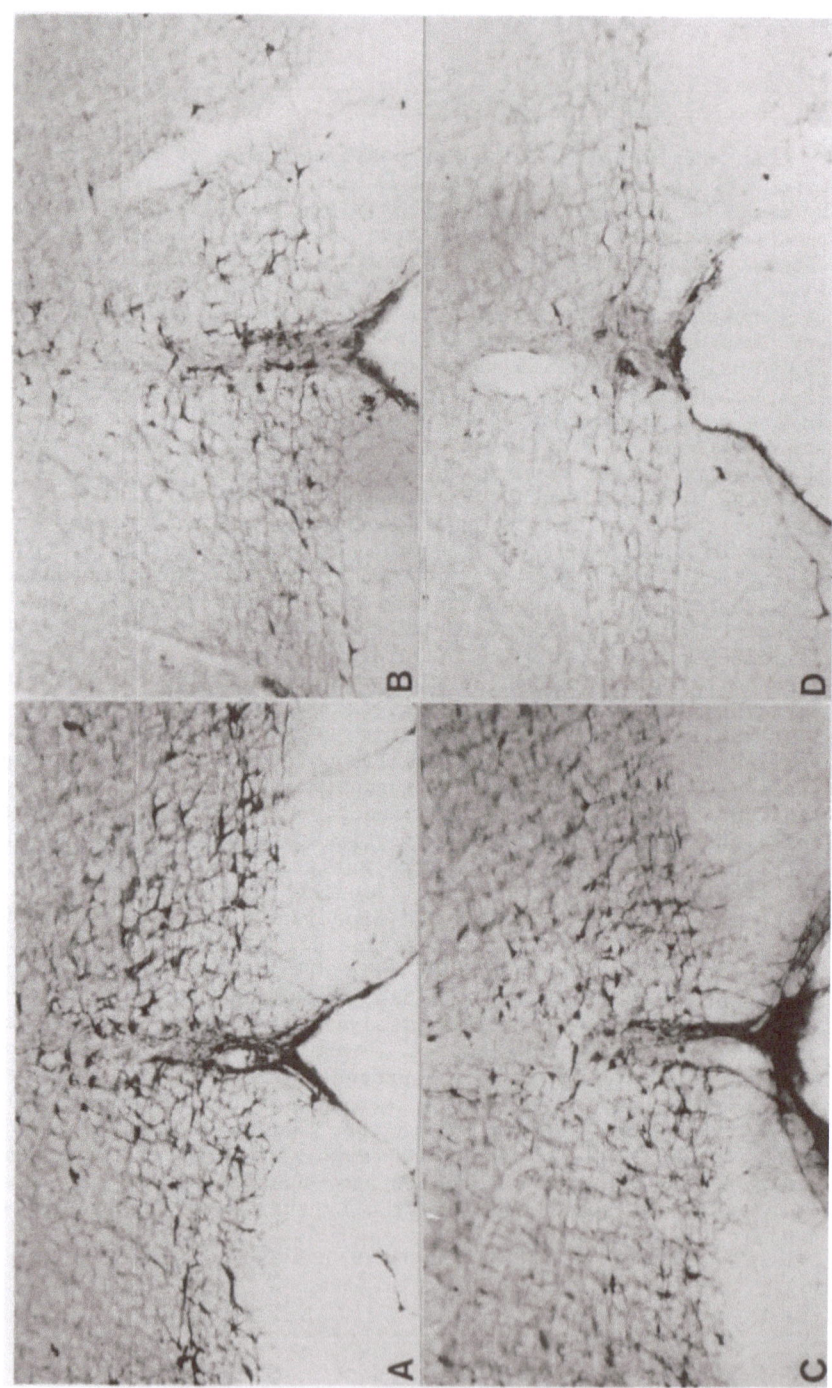

5HT-IMMUNOREACTIVE CELL BODIES

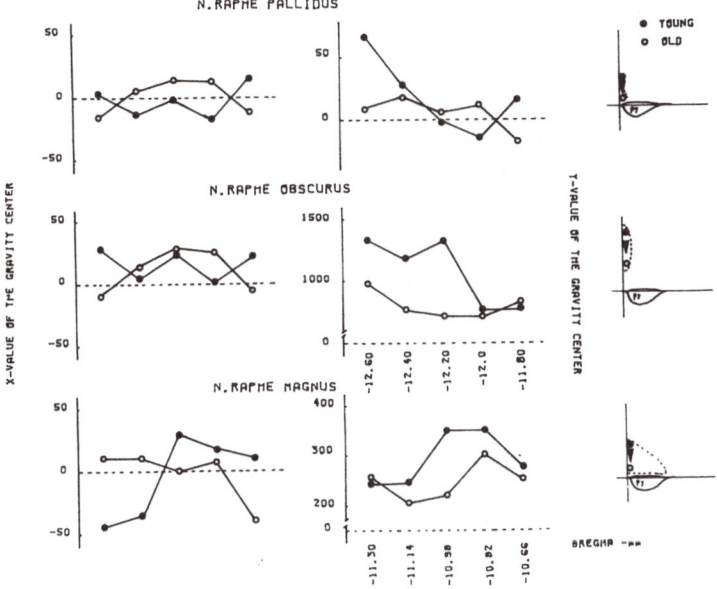

Fig. 2. Location of the gravity centers of the 5-HT nerve cell bodies of the various raphe nuclei of the medulla oblongata of the 3 month and 24 month old male rat. The x and y values of the coordinates of the gravity centers are reported at various rostro-caudal levels of the medulla oblongata. (Paxinos and Watson 1982). In the right part of the figure is given a schematical illustration of the main changes in the location of the gravity centers during aging as illustrated in the frontal plane.

Opposite:

Fig. 1 A, B, C and D. 5-HT and Substance P-like immunoreactivity is shown in frontal sections within nerve cell bodies of the nucleus raphe magnus and surrounding reticular areas in the 3 and 24 month old male rat. Colchicine treatment (75 μg/10μl; i.c., 18 h earlier). PAP technique. Serial 50 μm thick vibratom sections were used. Primary antisera were diluted 1:800. The rostrocaudal level was according to the Paxinos and Watson atlas -13.3 mm (in relation to bregma). 100 X. For further details, see text.

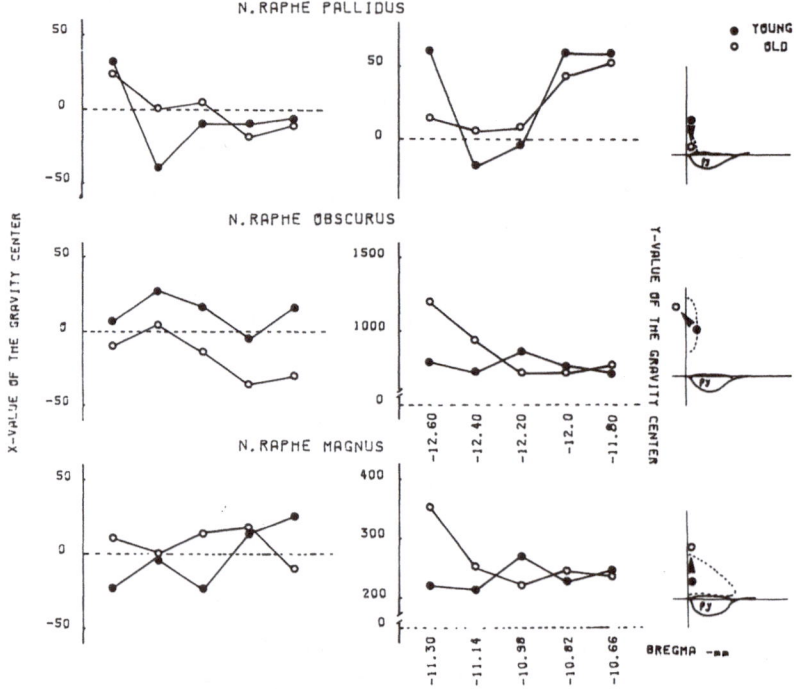

Fig. 3. The position of the x and y coordinates of the gravity centers for the Substance P immunoreactive nerve cells in the raphe nuclei at various rostrocaudal levels of the medulla oblongata of the 3 and 24 month old male rat is shown. For further details, see text of fig. 2.

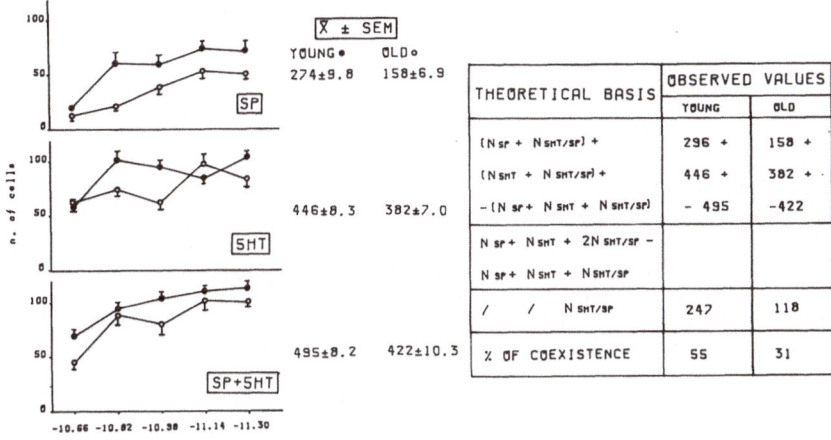

Fig. 4. In the left part of the figure the number of 5-HT immunoreactive, Substance P (SP) immunoreactive and 5HT plus SP immunoreactive profiles are given at various rostrocaudal levels of the
rat medulla oblongata in the 3 month and 24 month old male rat.
In the central part of the figure the total number of cells found
after incubation with the SP antiserum alone, 5-HT antiserum
alone and the two SP and 5-HT antisera are reported (n = 3; means
± s.e.m.). In the right part of the figure the occlusion method
has been applied on the number of 5-HT, SP and 5-HT plus SP
immunoreactive nerve cell bodies has been applied. In this way an
overall evaluation of coexistence of Substance P and 5-HT has
been obtained in the 24 month and 3 month male rat. The number of
SP and 5-HT costoring cells has been expressed in per cent of the
number of 5-HT immunoreactive nerve cell bodies.

Studies on tyrosine hydroxylase immunoreactive nerve cell bodies of the locus coeruleus and its innervation by NPY and PNMT immunoreactive nerve terminals during aging

In the microdensitometrical analysis of the locus coeruleus
semiquantitation of the immunoreactivity of tyrosine hydroxylase
and NPY present in cell bodies was made by gray tone histograms
(fig. 5 and 6). As shown in the figures it is possible to obtain
the specific gray value by means of the differences between the
mean of the gray tone histogram of the specific immunoreactivity
and the mean of the gray tone histogram for the unspecific staining. In this way it can be seen that the mean difference of gray
values ($\Delta \bar{X}$) is similar for tyrosine hydroxylase immunoreactivity
in the locus coeruleus of the old and 3 month old male rat, while
it is substantially reduced for NPY immunoreactivity in the 24
month old rat compared with the 3 month old rat. The results

obtained with this procedure are summarized in fig. 7, showing a
significant reduction of the gray value of the NPY immunoreactive
nerve cell bodies of the locus coeruleus in the old rat. Five
different rostrocaudal levels of the locus coeruleus were considered in this analysis. As seen in fig. 7 also the field area
was determined with the help of the IBAS Image Analyzer. The
field area of the NPY immunoreactive cell bodies within the locus
coeruleus of the old rat is shown to be significantly reduced
compared with the corresponding field area in the 3 month old
rat. However, following colchicine treatment this reduction is no
longer present, showing that only a reduction in the synthesis of
the NPY comodulator had taken place. The specific field area
could be determined by erasing all gray tone values above the
mean gray tone value for the background minus 2 standard deviations. By means of this procedure it was possible to select in an
objective way only the gray tones darker than the background. As
seen in fig. 8, also the specific field area of the PNMT immunoreactive nerve terminals has been analyzed in the old rat and
compared with the changes in the NPY immunoreactive field area
(cell bodies + terminals (major component)). It is shown that
also the PNMT immunoreactive area is reduced in the old rat when
compared with the 3 month old rat, although this reduction is
substantially less than that found for the NPY immunoreactive
area. It should be noticed that the diameters of these types of
transmitter identified nerve terminals are not significantly
changed during the aging process. Again these results underline
the possibility that during the aging process the synthesis of

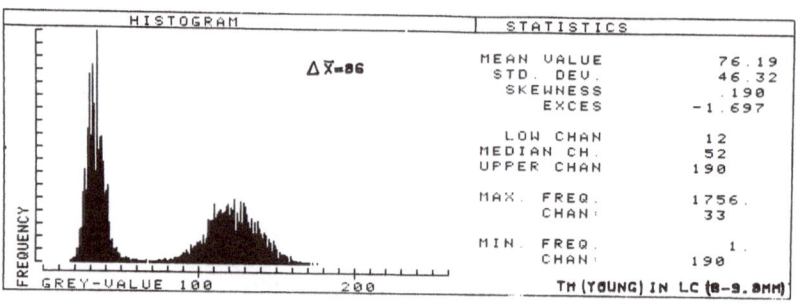

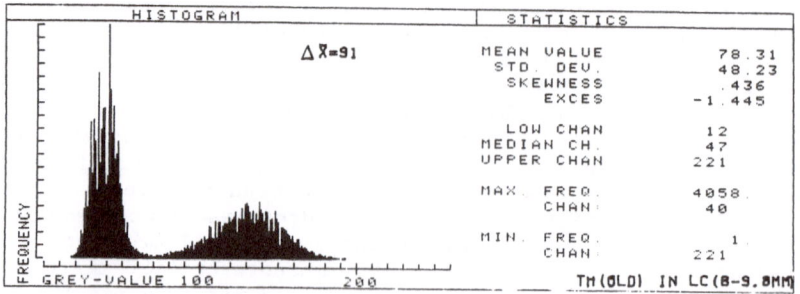

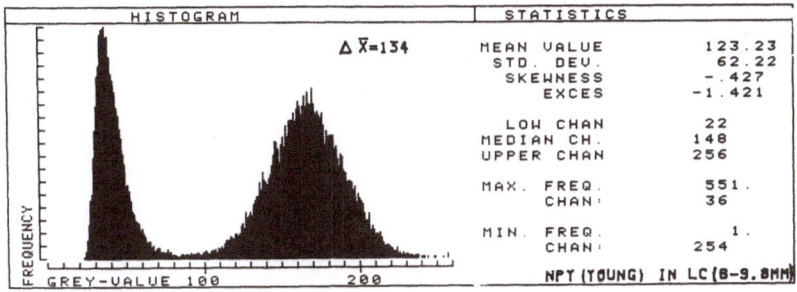

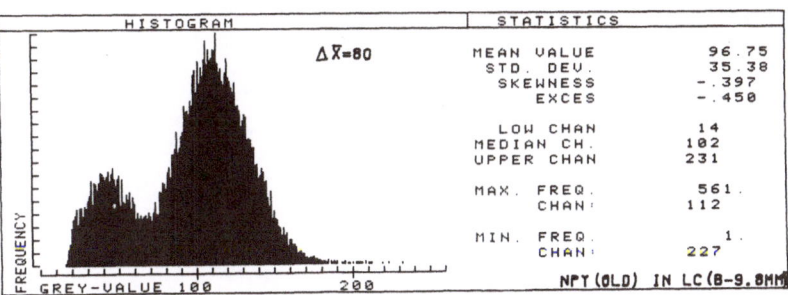

Fig. 5 and 6. Gray tone histograms showing a specific (modal value on the left) and unspecific (modal value on the right) immunoreactivity for TH (fig. 5) and NPY immunoreactivity (fig. 6) in the 3 month and 24 month old male rat. The differences between the two mean values observed in the gray tone histogram for each antiserum has been reported as a measurement of the specific immunoreactivity within the locus coeruleus of the 3 month and 24 month old male rat.

the comodulator, in this case NPY, may be preferentially reduced in the adrenaline nerve cells compared with, in this case, the synthesis of the enzyme phenylethanolamine-N-methyltransferase (PNMT). NPY is to a very large degree costored in the PNMT immunoreactive nerve terminals of the locus coeruleus (see Fuxe et al. this symposium). The NPY innervation of the locus coeruleus in the old and 3 month old rat is illustrated in fig. 9A, B. as well as the loss of NPY immunoreactive nerve cell bodies in the locus coeruleus of the old rat. Furthermore, the tyrosine hydroxylase immunoreactive nerve cell bodies and the dendrites in the locus coeruleus and subcoeruleus in the old and 3 month old rat are illustrated in fig. 10 A, B, C and D. In this figure if is also illustrated (see panel D) that the dentritic tree of the TH immunoreactive cells appears to be reduced in the old animal compared with 3 month old rat.

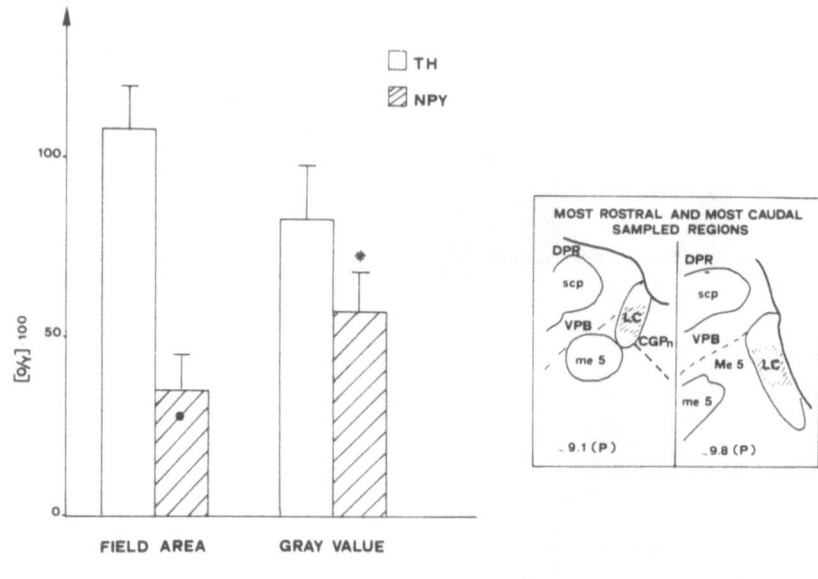

● in colchicine treated animals [%] 100 = 95±15

Fig. 7. Tyrosine hydroxylase and NPY immunoreactive nerve cell bodies within the locus coeruleus of the 3 and 24 month old male rat has been studied by means of the IBAS Image Analyzer. PAP technique. The antisera were diluted 1:8000. The field area and the gray value (specific immunoreactivity, see figs. 5 and 6) are shown. Means ± s.e.m. Student´s t-test. The statistical analysis has been carried out by Student´s t-test; * = p < 0.05. It is noted that in colchicine treated animals (75 μg/rat; i.c.) the percentage ratio of the field area for the NPY immunoreactive nerve cell bodies in the locus coeruleus is close to 100 per cent.

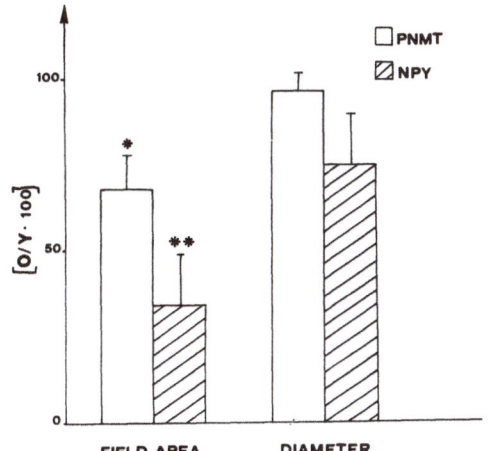
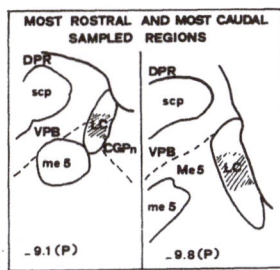

Fig. 8. PNMT and NPY immunoreactive nerve terminals within the locus coeruleus of the 3 month and 24 month old male rat. PAP technique. The PNMT antiserum was diluted 1:1500 and the NPY antiserum 1:750. The field area and the diameter of the respective nerve terminals have been evaluated by means of the IBAS Image Analyzer. Means ± s.e.m. (n= 4) are shown for the ratios of the values obtained in the old and adult animals. The statistical analysis has been carried out by Student´s paired t-test. Student's t-test * = p < 0.05, ** = p < 0.01.

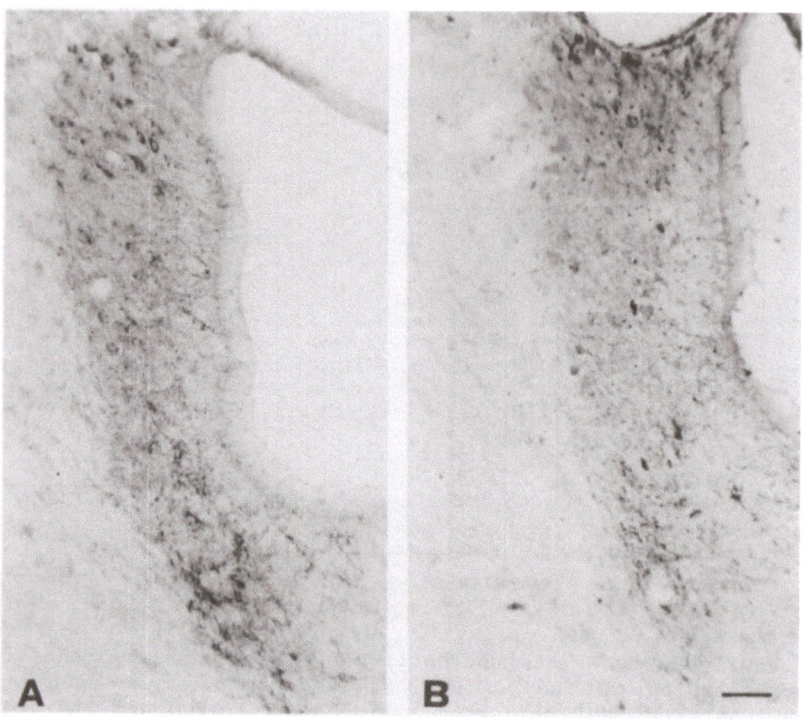

Fig. 9 A and B. NPY immunoreactive nerve terminals and cell
bodies are illustrated in the LC of the adult (A) and 24 month
old male rat (B). The rostrocaudal level is - 0.8 mm posterior to
bregma (Paxinos and Watson 1982). The primary antiserum was
diluted 1:750. The bar present a length of 50 μm. Please note the
loss in the number of NPY immunoreactive nerve cell bodies in the
old rat.

Fig. 10 A, B, C and D. Tyrosine hydroxylase immunoreactive nerve
cell bodies and their dendrites are shown within the locus
coeruleus and subcoeruleus of the 3 and 24 month old male rat. (A
and C is taken from the 3 month old male rat and B and D is taken
from the 24 moth old male rat). PAP technique. Primary antiserum
was diluted 1:800. The bar represents a length of 50 μm. A and B
have been photographed at a rostrocaudal level of -9.8 mm posterior to bregma (Paxinos and Watson 1982); C and D have been photographed at the rostrocaudal level -8.2 mm behind bregma (Paxinos
and Watson 1982).

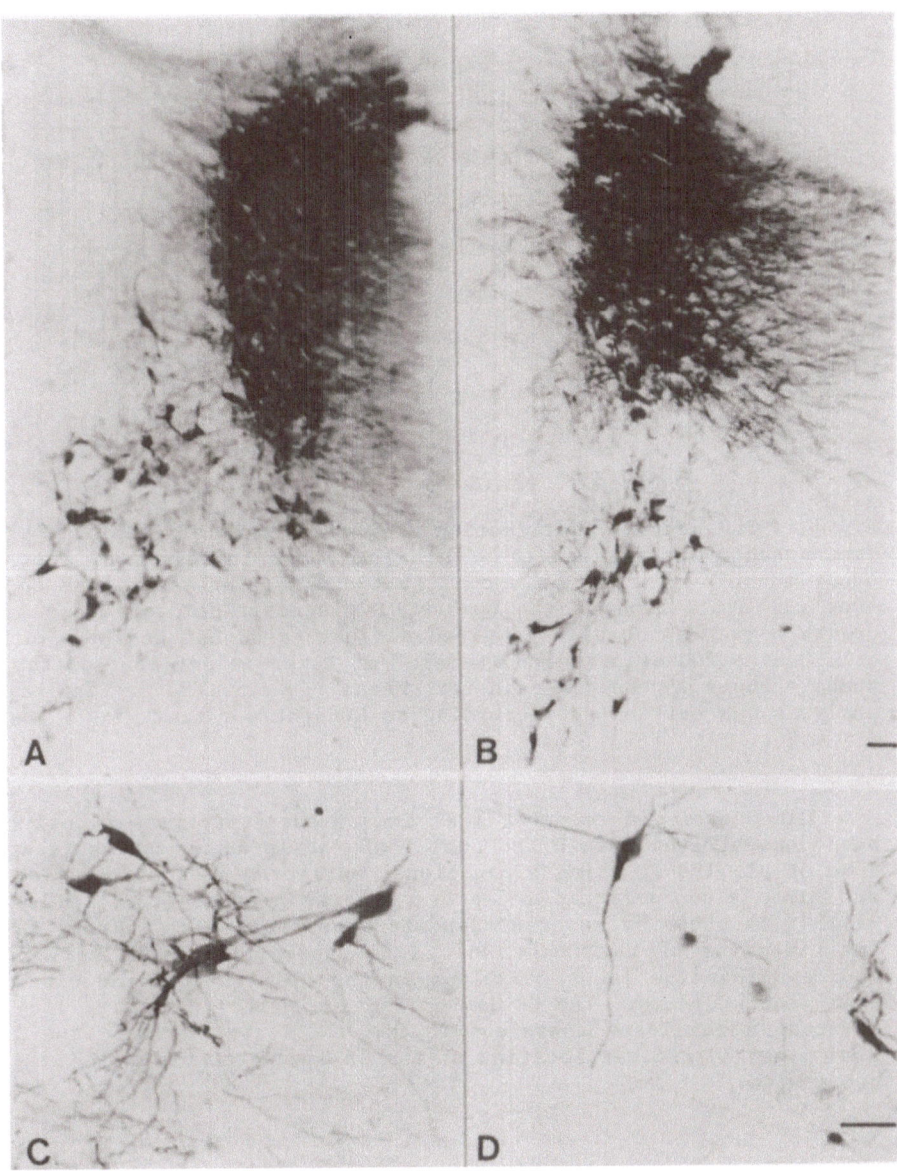

Fig. 10 A, B, C and D

104   L.F. Agnati, et al.

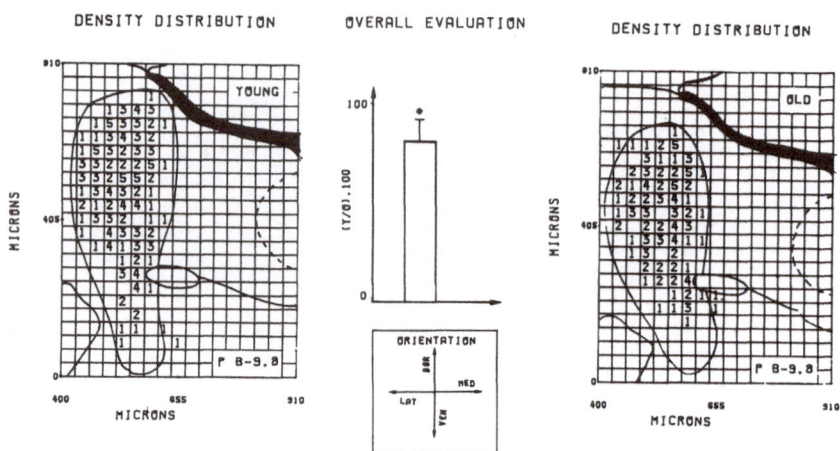

Fig. 11. The density distribution of glucocorticoid receptor (GR)
immunoreactive nerve bodies within the locus coeruleus of the
adult and old male rat is shown. In the middle of the figure an
over all evaluation of the number of glucocorticoid immunoreactive nerve cell bodies is given and expressed as the ratio of
cell bodies found in the adult male rat taken in per cent of the
numbers found in the old male rat. Means ± s.e.m. The statistical
analysis was carried out according to Student´s t-test. * =
p < 0.05.

        Glucocorticoid receptor (GR) immunoreactivity has recently
been demonstrated within the nuclei of the locus coeruleus (see
Fuxe et al. 1984), using a monoclonal antiserum against purified
rat liver glucocorticoid receptors. In the present study (fig.
11), it is shown by means of density maps that there is only a
small decline of the number of GR immunoreactive nerve cell
bodies within the locus coeruleus in the old rat compared with
the 3 month old rat. The GR immunoreactive nerve cells are illustrated in the locus coeruleus in fig. 12 A and B, where the
predominantly nuclear location of the GR immunoreactivity is well
documented.

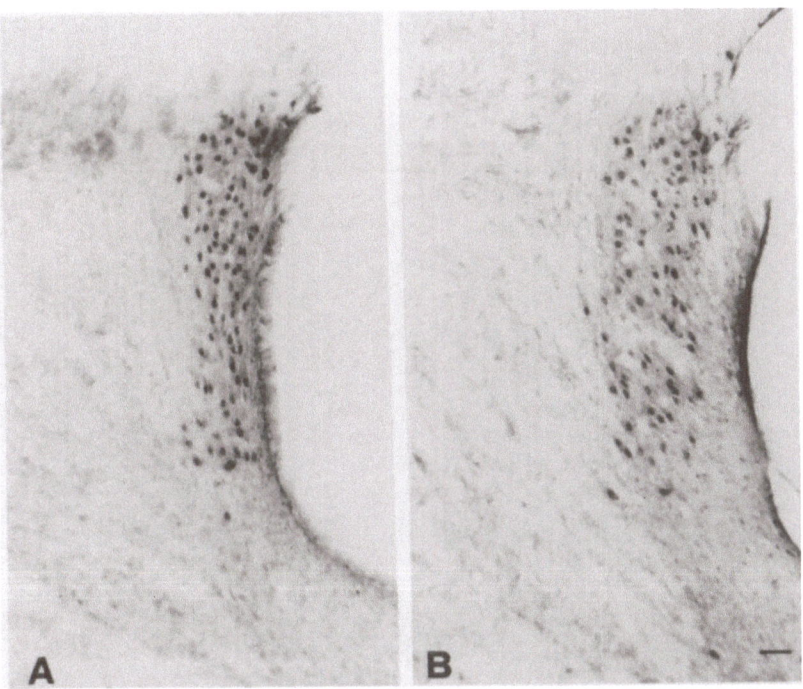

Fig. 12. Illustration of the GR immunoreactive nerve cell bodies within the locus coeruleus of the adult (A) and old (B) male rat. PAP technique. Primary antiserum was diluted 1:750. Rostrocaudal level was according to Paxinos and Watson (1982) -9.8 mm posterior to bregma. The bar represent the length of 50 μm. It is shown that the GR immunoreactivity is predominantly located within the nuclei of the nerve cells of the locus coeruleus.

Studies on neurophysin I-vasopressin, neurophysin II-oxytocin immunoreactive neurons in subnuclei in paraventricular hypothalamic nucleus during aging

The oxytocin immunoreactive nerve cell bodies are illustrated in fig. 13 and the corresponding density maps are shown in fig. 14. It is demonstrated that the overall profile number is not clearly reduced during the aging process. However, when analysing the subnuclei it can be demonstrated that during the aging process there is a preferential disappearance of oxytocin immunoreactive nerve cell bodies within the dorsal parvocellular component of the paraventricular hypothalamic nucleus present at the posterior level. In fig. 15 density maps of oxytocin immunoreactive nerve cell bodies are also shown within the posterior subnucleus of the paraventricular hypothalamic nucleus which

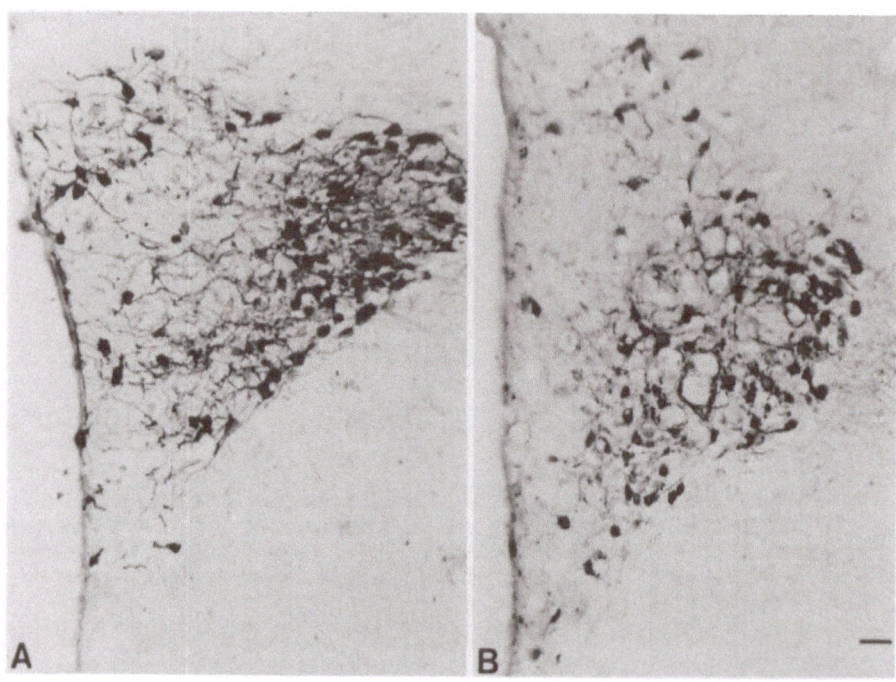

Fig. 13. Oxytocin immunoreactive nerve cell bodies are
illustrated in the adult (A) and old (B) male rat within the
subnuclei of the paraventricular hypothalamic nucleus (see fig.
14). The rostrocaudal level was according to Paxinos and Watson
(1982) B - 1.6 mm behind bregma. The ventricle is shown to the
left in the figure. PAP technique. The primary antiserum was
diluted 1:2000. The bar represents a length of 50 µm.

represents the caudal end of this nucleus. This nucleus is a
parvocellular component of the paraventricular nucleus having a
medial and lateral component. Also in these studies it can be
shown that there is a reduction in the number of immunoreactive
profiles during the aging process. These results are of particular interest, since they illustrate a high degree of heterogeneity in the degeneration processes within the oxytocin positive nerve cell population present in this nucleus. It should be
emphasized that this posterior component of the paraventricular
hypothalamic nucleus mainly sends pathways into the brain stem
and spinal cord (Swanson et al. 1983). Thus, it seems as if these
oxytocin projections may be preferentially degenerated during the
aging process. In fig. 16 the morphometrical and microdensitometrical characteristics of the subnuclei of the paraventricular
hypothalamic nucleus have been evaluated with regard to oxytocin,

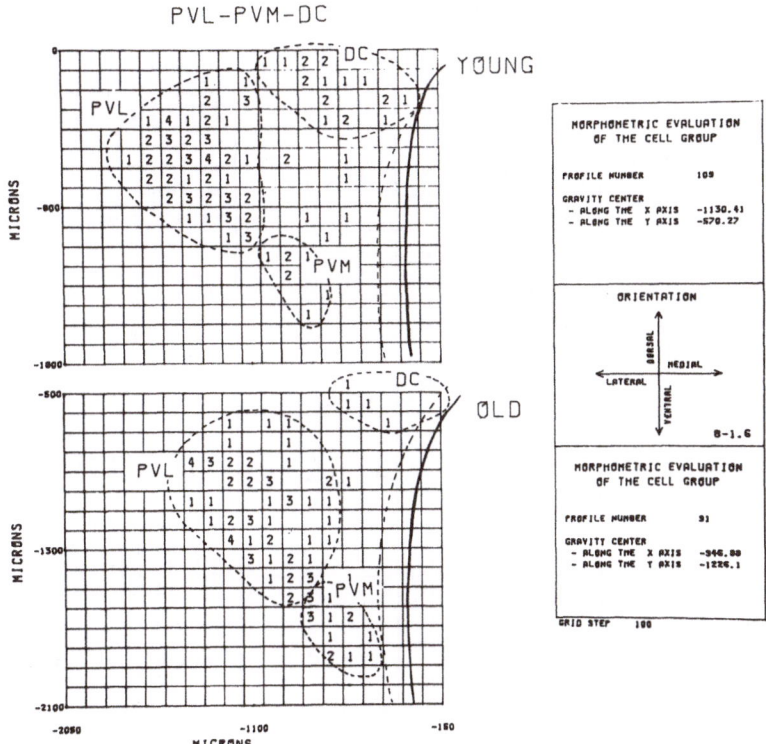

Fig. 14. The density distribution of oxytocin immunoreactive neurons is shown in the adult and old male rat within the subnuclei of the paraventricular hypothalamic nucleus (PVL = paraventricular nucleus, pars lateralis, PBM = paraventricular nucleus, pars medialis, DC = paraventricular nucleus pars, dorsalis). Colchicine treatment (75 µg/rat; i.v.t.; 18 h earlier). The morphometrical evaluation is shown to the right in the figure. Note the preferential disappearance of oxytocin immunoreactive nerve cells within the DC. The ventricular border is indicated as a black thick line.

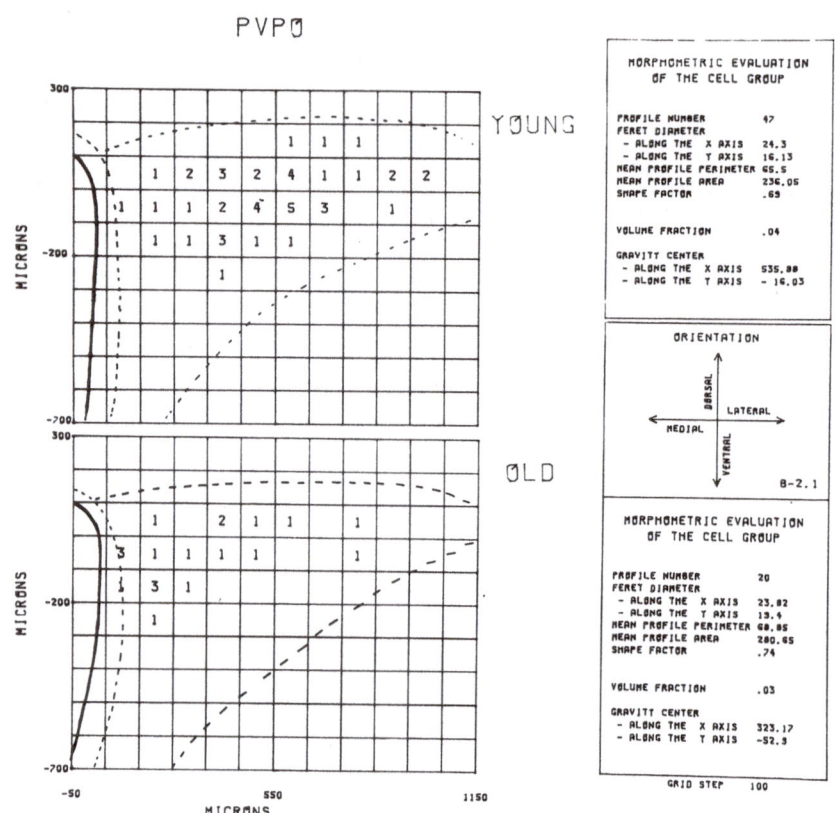

Fig. 15. Density distribution of oxytocin immunoreactive nerve
cell bodies in the adult (upper panel) and old (lower panel) male
rat within the posterior part of the paraventricular hypothalamic
nucleus (PVPO). Colchicine treatment (75 μg/rat; i.v.t.; 18 h
earlier). The morphometrical evaluation is shown to the right.
Note the disappearance of oxytocin immunoreactive nerve cell
bodies within PVPO of the old animal. The ventricular border is
indicated as a black thick line.

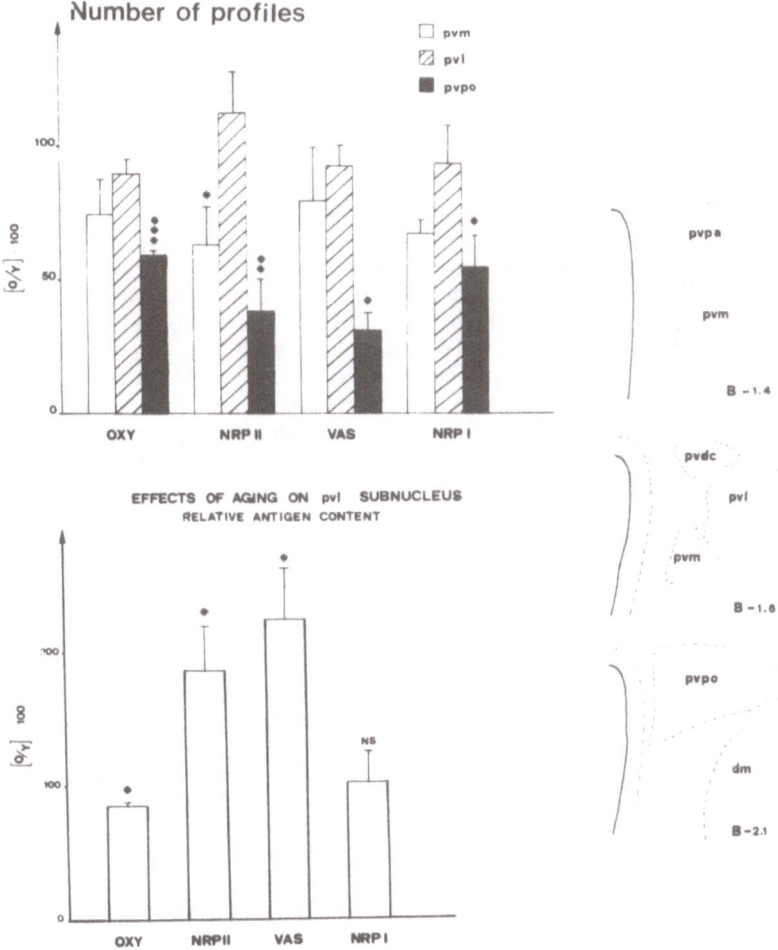

Fig. 16. Effects of aging on the number of profiles and of the relative antigen contents within the paraventricular hypothalamic nucleus as evaluated at different rostrocaudal levels (B -1.4, B -1.6, B - 2.1 mm; according to Paxinos and Watson 1982). Oxytocin, vasopressin and neurophysin I and II (NRP) immunoreactive nerve cell bodies have been evaluated. Means ± s.e.m. n= 3 are shown for the ratio of profiles and antigen contents found in old and adult animals taken in per cent of the values obtained in adult rats. PAP technique. Serial vibratom sections were made (40 mm thick). The primary antisera were diluted 1:2000. The animals had 24 hrs previously been treated with colchicine (75 μg/25μl/rat). This dose had been administered i.v.t. Student´s t-test has been carried out. * p = < 0.05, ** p = < 0.01, *** p = < 0.001.

vasopressin, neurophysin I and neurophysin II immunoreactivities.
It is shown that with regard to the number of immunoreactive
profiles no changes can be observed within the medial and lateral
subnucleus of the paraventricular hypothalamic nucleus in agreement with the findings reported above. Instead within the posterior subnucleus of the paraventricular hypothalamic nucleus there
is a large reduction in the number of vasopressin oxytocin neurophysin I and neurophysin II immunoreactive nerve cell bodies
(Fig. 16) again underlining the high vulnerability of vasopressin
and oxytocin immunoreactive nerve cells in this subnucleus to the
aging processes. On the other hand, it must be underlined as also
seen in fig. 16 that the analysis of gray tone values do not
indicate any substantial reduction in the immunoreactivity in the
remaining oxytocin, neurophysin I and II and vasopressin immunoreactive nerve cells within the lateral subnucleus of the paraventricular hypothalamic nucleus. Instead, a substantial increase
was observed for vasopressin and neurophysin II immunoreactivity
within the magnocellular component of the paraventricular hypothalamic nucleus during the aging process. Thus, the results
indicate that a compensatory activation of the synthesis of
carrierproteins and of active neuropeptides can take place during
the aging process in selected subnuclei.

GENERAL DISCUSSION

     The present results again underline the view that it is
possible by means of computer-based morphometry and microdensitometry to reveal marked heterogeneities in the responses of
transmitter identified neurons to the aging process. This is
particularly evident in analysing the subnuclei of the nucleus
paraventricularis hypothalami with regard to the number of oxytocin and vasopressin immunoreactive cell bodies. Thus, a marked
and selective disappearance of oxytocin immunoreactive nerve
cells and vasopressin immunoreactive nerve cells could only be
observed within the posterior part of the paraventricular hypothalamic nucleus. Another feature of the degeneration process
during aging is exemplified in the study of the 5HT and SP
positive cells of the medulla oblongata. Infact, a reduction in
the entity of coexistence in the 5-HT and Substance P immunoreactive in the nerve cell bodies has been demonstrated, indicating a preferential reduction in the synthesis and/or utilization of Substance P, i.e. of the 5-HT comodulator, during
aging. In line with these findings are the results obtained in
the analysis of the NPY innervation as well as of NPY positive
cell bodies of locus coeruleus, again indicating the preferential
disappearance of the comodulator NPY in the adrenaline nerve
terminal networks and in the NA nerve cell bodies of this nucleus. Instead the number of tyrosine hydroxylase and glucocorticoid receptor immunoreactive nerve cell bodies of the locus
coeruleus was not clearly affected by the aging process. These
results therefore open up the possibility that the neuropeptide

comodulators in nerve cell bodies and nerve terminals, at least of the monoamine neurons, may be particularly vulnerable to the aging processes which may lead to deficiencies in the comodulator transmission lines. Thus, it seems possible that in the human brain during aging there is a deficiency in the comodulator release and receptor activity. Therefore, when considering development of drugs for geriatric patients, possible treatments should also involve drugs, which can activate preferentially the comodulator transmission line.

ACKNOWLEDGEMENT

This work has been supported by a grant (MH25504) from the NIH, by a grant from L. Osterman's Stiftelse, by a grant (04X-715) from the Swedish Medical Research Council, by a CNR grant and by a grant from the Wellcome Trust Foundation and by a grant from the British Medical Research Council. For excellent technical assistance we are grateful to Mrs. Beth Andbjer, Mrs. Ulla-Britt Finnman, Miss Katarina Nilsson, Mrs Siv Nilsson, Mrs Birgitta Nyberg, Miss Barbro Tinner, Mr Giuseppe Maucinelli and Mr Leonida Sabatini. For excellent secreterial assistance we are grateful to Mrs Ulla-Britt Wedin, Mrs Anne Edgren and Mrs Roberta Presato.

REFERENCES

Agnati, L.F., Fuxe, K., Benfenati, F., Toffano, G., Cimino, M., Battistini, N., Calza, L. and Merlo Pich, E. (1984). Computer assisted morphometry and microdensitometry of transmitter identified neurons with special reference to the mesostriatal dopamine pathway. III. Studies on aging processes. Acta Physiol Scand., Suppl. 532, 45-61.

Agnati, L.F., Fuxe, K., Locatelli, V., Benfenati, F., Zini, I., Panerai, A.E., El Etreby, M.F. and Hökfelt, T. (1982). Neuro-anatomical methods for the quantitative evaluation of coexistence of transmitters in nerve cells. Analysis of the ACTH- and beta-endorphin immunoreactive nerve cell bodies of the mediobasal hypothalamus of the rat. J Neurosci Methods 5, 203-214.

Fuxe, K., Yu, Z.-Y., Agnati, L.F., Härfstrand, A., Wikström, A.-C., Okret, S., Granholm, L., Vale, W., Goldstein, M. and Gustafsson, J.-Å. (1984). Mapping out of glucocorticoid receptor (GR) immunoreactive neurons in the rat brain using a monoclonal antibody against rat liver GR. Neurosci Lett., Suppl 18, 106.

Fuxe, K., Agnati, L.F., Zoli, M., Härfstrand, A., Grimaldi, R., Bernardi, R., Tucci, F. and Goldstein, M. (1984). Development of quantitative methods for the evaluation of the entity of coexistence of neuroactive substances in nerve terminal populations in discrete areas of the central nervous system: Evidence for hormonal regulation of cotransmission. In: Quantitative neuroanatomy in transmitter research. (eds. L.F. Agnati and K. Fuxe), MacMillan Press, in press.

Guidotti, A., Saiani, L., Wise, B.C., Costa, A. (1983). Cotransmitters: Pharmacological Implications. J. Neural Transmission, Suppl. 18, 213-225.

Paxinos, G. and Watson, C. (1982). The rat brain in stereotaxic coordinates. Academic Press.

Swanson, L.W., Sawchenko, J., Vale, W.W. (1983). Organization of ovine corticotropin-releasing factor immunoreactive cells and fibers in the rat brain: An immunohistochemical study. Neuroendocrinology 36, 165-186.

THREE–DIMENSIONAL COMPUTER RECONSTRUCTIONS OF CATECHOLAMINERGIC NEURONAL POPULATIONS IN MAN

DWIGHT C. GERMAN[1,2], BRANDY S. WALKER[1],
KATHY McDERMOTT[1], WADE K. SMITH[3],
DANIEL S. SCHLUSSELBERG[3] and DONALD J. WOODWARD[3]

Departments of Physiology[1], Psychiatry[2] and Cell Biology[3], University of Texas Health Science Center, Dallas, Texas 75235, USA

INTRODUCTION

Catecholamine-containing neurons have been found to subserve numerous functions (see reviews by Moore and Bloom, 1978, 1979; Mason, 1981). For example, dopamine (DA) and norepinephrine (NE) containing cells in the hypothalamus regulate pituitary hormone output. The midbrain DA neurons of the substantia nigra (nucleus A9) which innervate the neostriatum, play a role in motor control and cell loss in this nucleus is pathognomonic for Parkinson's disease. The ventral tegmental area DA neurons (nucleus A10), which innervate limbic and specific cortical regions (frontal, cingulate and entorhinal cortex) have been implicated in emotional regulation. Alteration of the receptors of these neurons is thought to be the mode of antipsychotic action for neuroleptic drugs. The locus coeruleus (LC) NE neurons, which diffusely innervate the entire cerebral cortex and spinal cord regions, have been hypothesized to play a role in arousal, attention and anxiety.

The focus of the present paper is on the midbrain DA neurons and the LC-NE neurons in the human brain. These neurons were first shown to contain catecholamines in the rat (Dahlstrom and Fuxe, 1964; Ungerstedt, 1971), and subsequent work (Nobin and Bjorklund, 1973) revealed that these neurons were also catecholaminergic in the human brain. Both of these neuronal groups have been reported to decrease in number with aging (McGeer et al., 1977; Mann et al., 1983) and with certain diseases. For example, LC neurons are markedly reduced in number in brains from Parkinsonian (Mann and Yates, 1983) and Alzheimer's disease patients (Tomlinson et al., 1981). Likewise, midbrain DA cell number is reduced in Parkinsonian patients (McGeer et al., 1977). Because of the large number of cells within these nuclei, a computer graphics system is very useful in characterizing the

cellular topographies of these nuclei and mapping the changes
which occur due to aging and disease.

METHODS

Cell Recognition

    Human brains are immersion fixed in 10% neutral buffered
formalin. The brain is suspended by the basilar artery so that it
is fixed in a relatively undistorted shape. After a weeks time,
the midbrain is dissected out. Fifty μm thick frozen sections are
cut, perpendicular to the long axis of the brainstem, on a sliding
microtome. Sections are stained for Nissl substance, neuromelanin
pigment (Graham, 1979; Saper and Petito, 1982) and tyrosine
hydroxylase (Pearson et al., 1983) (see Figure 1).

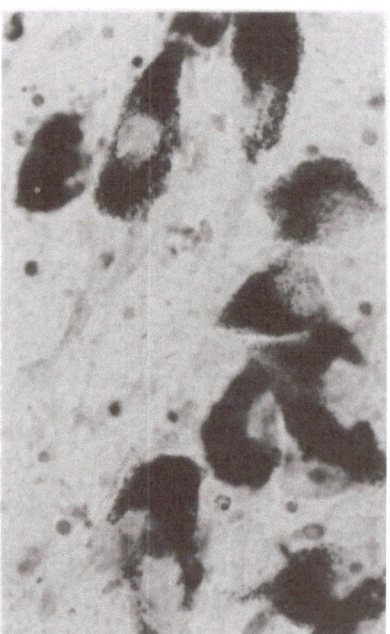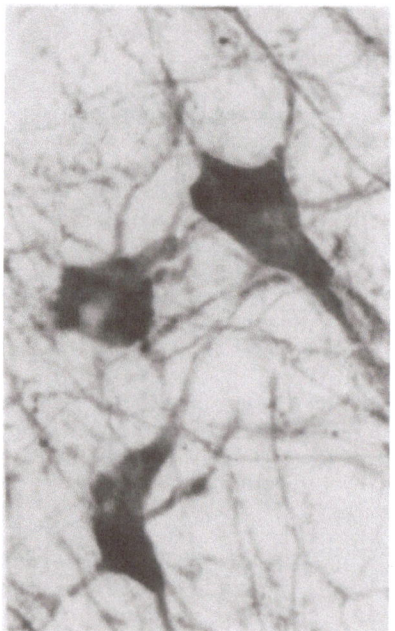

Figure 1 Catecholamine-containing neurons in the human brainstem. Left panel: neuromelanin pigment within substantia nigra
neurons (Nissl stain). Notice that 2 to 3 somas are often in very
close proximity. Right panel: tyrosine hydroxylase immunoperoxidase staining of substantia nigra neurons.

Computer Reconstruction

    Only a brief description of the computer reconstruction
methods will be presented as these methods have been published

previously (Schlusselberg et al., 1982; German et al., 1983).
Brain sections are mounted on glass slides and placed in a
photographic projector. The whole section is projected onto the
surface of a digitizing tablet. An outline of the section is
digitized into the computer along with coordinate zero points used
for section-to-section alignment.

The glass slide is transferred to a Leitz microscope with a
stepper motor driven stage. The zero points are again designated
but now at higher magnification. The previous information input
at low power is merged with information at high power by coding
data with respect to a biological coordinate system relative to
the zero points.

The data acquisition takes place via a "video lucida" system,
analogous to the "camera lucida" system used to normally draw
objects seen in the microscope field. A microscope drawing tube
is aimed at a video display so that vectors which appear on the
display are optically mixed with the image seen in the microscope.
The user then sees a cursor and vectors overlaid on the image
viewed through the eyepiece. The stepper motors are given
commands to move to locations within the biological coordinate
system which contain the catecholamine-containing cells. The
operator, looking at the tissue section in the microscope, sees a
rectangular box around the critical area. A button-cursor system
on a digitizing tablet is programmed to control a small bright
cursor seen through the eyepiece. The cursor is successively
positioned on top of the cells and a button is pressed to record
cell locations. A small triangle appears on top of each cell when
its location has been recorded. When all cells within the
rectangle have been recorded, the stepper motor moves to a new
location and a new adjacent rectangle is drawn in the field of
view along with the viewable data in the neighboring rectangle.
The computer scans a grid system designed to include all cells for
a given brain section.

Every n[th] section in a series is plotted in this way.
Software has been written to allow the construction of two-
dimensional plots of individual sections as well as groups of
sections to be plotted as a reconstructed solid body
representation. The data files can be scanned in a number of ways
to obtain total cell counts, or cell counts within any designated
regions (i.e., within columns, boxes, etc.). The combination of
flexible counting, two and three-dimensional graphic data
manipulation, and visualization offers a primary tool for
appreciating regional variation in cellular densities.

RESULTS

Midbrain DA Neuronal Distribution

The midbrain DA neurons occupy a region approximately 12 x 24

mm (length X width).   Rostrally the cells are found at the level
of the caudal mammillary bodies.   Caudally the nucleus A8 cells
(in the lateral reticular formation) are at the level of the
rostral pons.   The DA neurons are primarily found within the
extent of the interpeduncular fossa region.

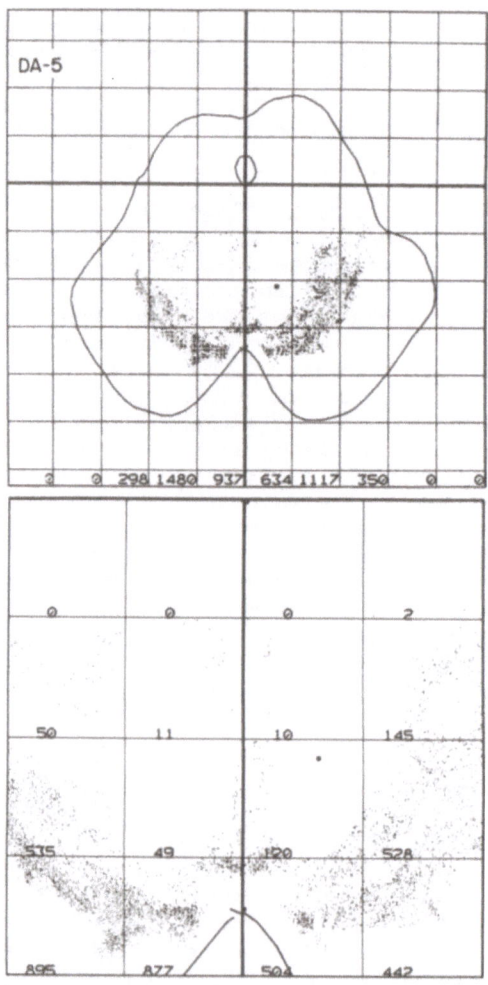

Figure 2   Section through the midbrain DA cell region.   Upper
panel: each DA neuron is illustrated as a dot.   The grid bars are
4 mm apart.   The number of cells within a column appears at the
bottom of each column.   Notice that the cells appear to be
clustered into subnuclear regions.   Lower panel: enlargement of
the ventral tegmental area, from the section above, showing the
number of DA cells within 4 mm square regions.

The DA neurons are clustered into various subnuclear
regions. Figure 2 shows an illustration of a printout of the
distribution of neuromelanin-containing midbrain DA neurons in one
brain section. In the close-up of this printout the subregions
are prominent, as previously observed (Hassler, 1937; Olszewski
and Baxter, 1954). Furthermore, the cell bodies are often in
close proximity to one another, forming groups of 3-4 somas (see
Figure 1, Nissl stain). In the rat, these neurons appear to be
connected via gap junctions (Grace and Bunney, 1983).

The midbrain region can be reconstructed with the computer by
overlaying a number of sections, from rostral to caudal. Figure 3
shows such a reconstruction and illustrates various prominent
midbrain structures. Using depth shading and certain lighting
algorithms, a three-dimensional reconstruction is possible.
Figure 4 shows the distribution of DA neurons within the
midbrain.

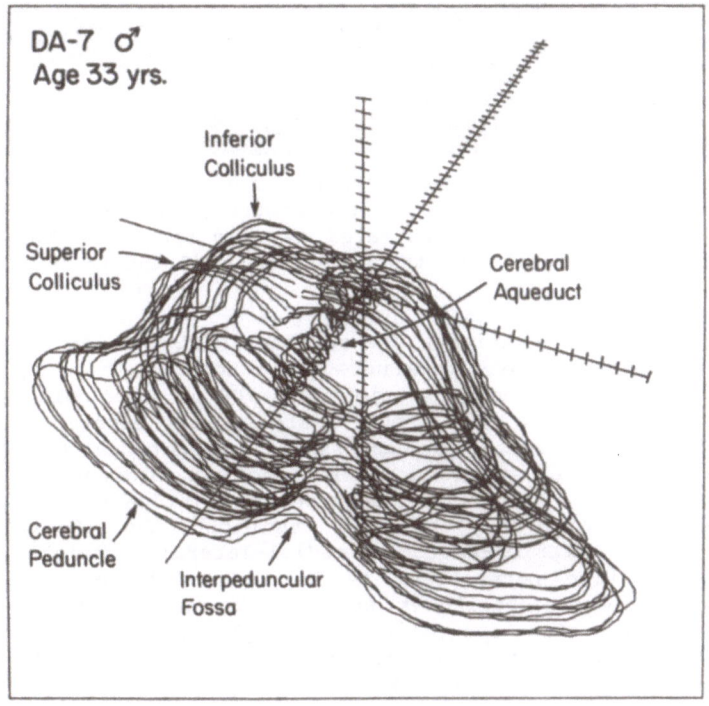

Figure 3   Computer reconstruction of the midbrain. By superimposing serial sections through the midbrain, various structures
became visible in three-dimensions.

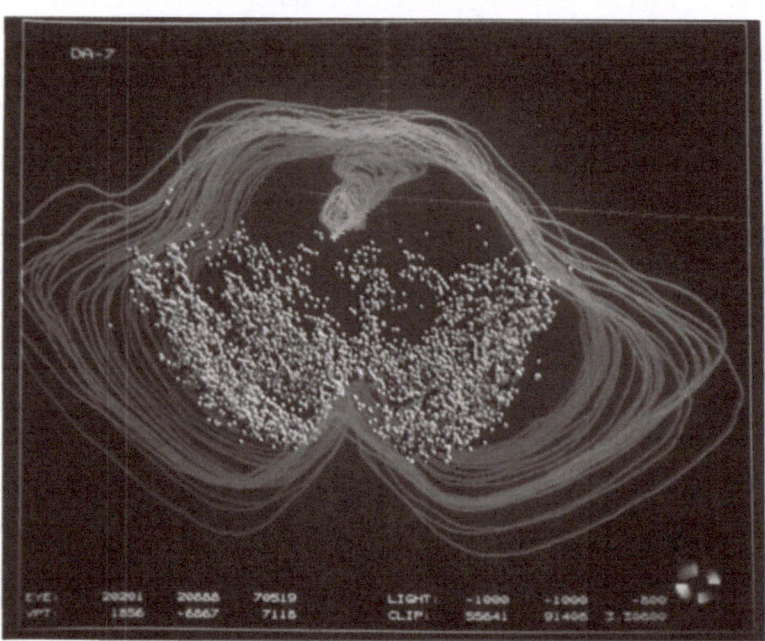

<u>Figure 4</u> Three-dimensional computer reconstruction of the mid-
brain DA neurons. The view is from rostral, in the foreground, to
caudal. DA cells are illustrated as gold spheres. The size of
the sphere is larger than the actual size of the DA cell in this
reconstruction.

        In order to quantitate the regional distribution of neurons,
cell density topography maps were constructed. This was done by
first dividing each brain section into 100 μm wide columns across
the section. Next a cell number histogram was constructed at the
top of each section. By superimposing sections (12 sections
spaced 1 mm apart) and then interconnecting adjacent cell number
histograms, a cell density topography map ("mountain range") can
be created. The higher the peak, the higher the cell density at
that specific anterior-posterior/medial-lateral coordinate. Where
the eye-position perspective is located, with respect to this
"mountain range", determines which peaks are visible. In our
standard view, the eye position is placed directly over the
"mountain range." By color coding peak height, a cell density
topography map is created. Figure 5 illustrates DA cell density
topography maps from two brains. One is from a 58 year old white
male "normal", and the other is from a 67 year old white male who
had end-stage Parkinson's disease. At the time of death, the
patient had been bed-ridden from severe rigidity. Notice the
marked reduction in DA cell number in the parkinsonian brain
compared to the non-parkinsonian control.

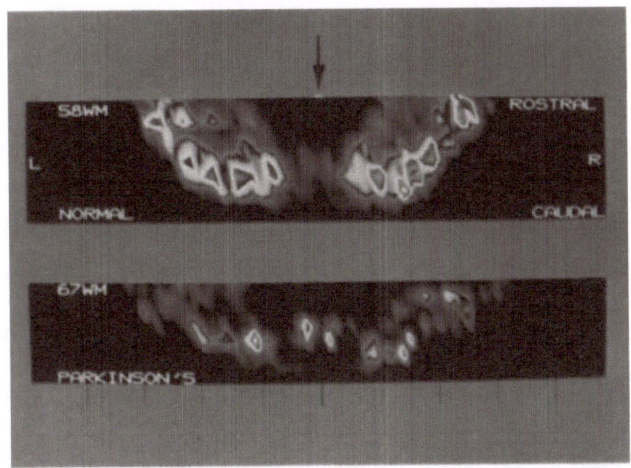

Figure 5 Cell density topography maps of the DA neurons. Upper panel: regional DA cell density in a 58 year old white male. The midline is indicated by the arrow. Right (R) and left (L) halves of the brain are indicated. The grid lines are 4 mm apart. The regions with the highest cell densities are illustrated in white, the next highest in blue, followed by yellow, and finally, the lowest cell density in orange. Lower panel: regional DA cell density in a 67 year old white male who had Parkinson's disease. Notice the marked reduction in cell densities.

Locus Coeruleus NE Neuronal Distribution

The LC NE neurons occupy a region approximately 13 x 2 mm (length X width) per side. The cells are located in the rostral half of the pons and are situated lateral to the cerebral aqueduct (rostrally) and fourth ventricle (caudally).

The LC neurons are a relatively homogeneous population of cells. Figure 6 illustrates computer printouts of the distribution of neuromelanin-containing LC neurons in six rostral (panel A) to caudal (panel F) sections. At the most caudal section the subcoeruleus (Olszewski and Baxter, 1954) neurons can be seen extending ventral-laterally from the LC.

The three-dimensional computer reconstruction of LC neurons is illustrated in Figure 7. The outline of the brain is in green and the LC neurons are represented as blue spheres. Notice that as the cerebral aqueduct opens up into the fourth ventricle, the LC neurons are displaced laterally.

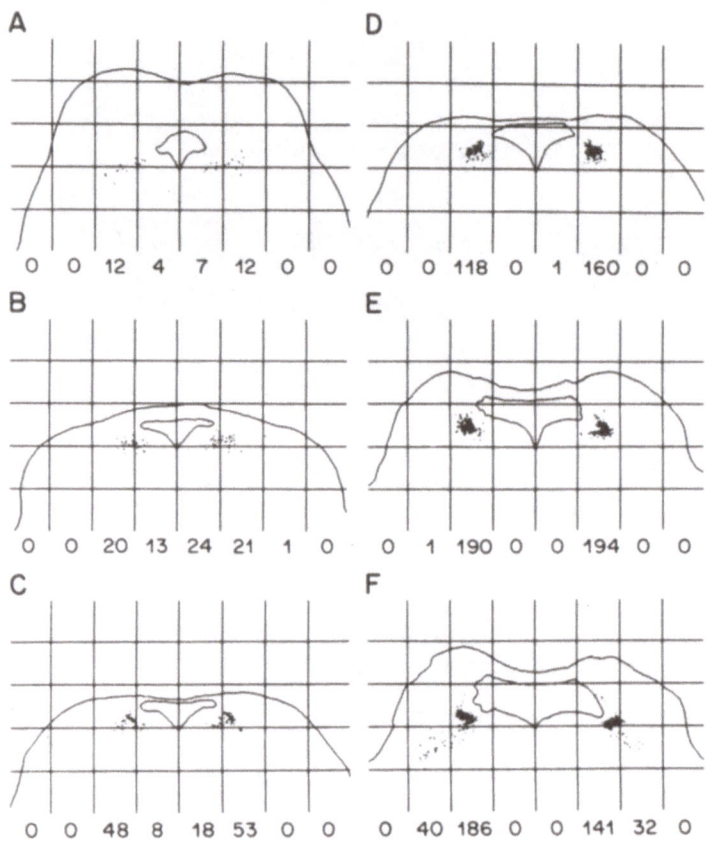

Figure 6   Sections through the locus coeruleus NE cell region. Sections are 2.4 mm apart and extend from rostral (A) to caudal (F). Each NE neuron is illustrated by a dot. The number of cells within a 3 mm column appears at the bottom of each column. Nucleus subcoeruleus is seen extending ventrolaterally from the LC in Panel F.

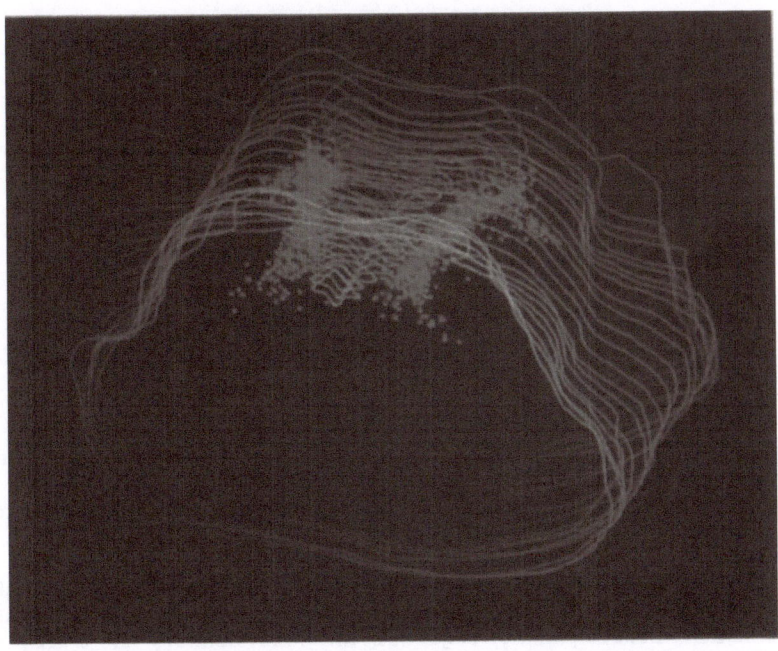

<u>Figure 7</u> Three-dimensional computer reconstruction of the locus coeruleus NE neurons. The view is from rostral, in the foreground, to caudal. NE neurons are illustrated as blue spheres. The size of the sphere is larger than the actual size of the NE neurons in this reconstruction.

LC cell density topography maps enable quantitative comparisons to be made. Figure 8 illustrates the distribution of LC neurons from two brains. The top map came from a 5 year old white female and the bottom map from a 104 year old white female. The differences in cell densities are striking. Using four categories of cell density, the older brain was completely devoid of the two highest categories of cell density.

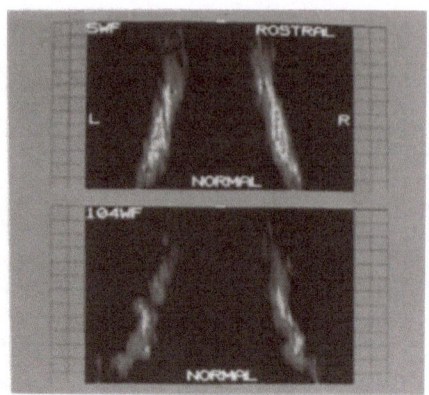

Figure 8 Cell density topography maps of the locus coeruleus NE
neurons.   Upper panel: regional NE cell density in a 5 year old
white female.   The midline extends vertically down the center of
the map.   Right (R) and left (L) halves of the brain are
indicated.   The grid lines are 1 mm apart.   The regions with the
highest cell densities are illustrated in white, the next highest
in blue, followed by yellow, and finally, the lowest cell density
in orange.   Lower panel: regional NE cell density in a 104 year
old white female.   Notice the marked reduction in cell densities.

DISCUSSION

     Quantitation of catecholamine cell number within a biological
coordinate system will help answer many questions.   For example,
what is the pattern of midbrain DA and LC NE cell loss which
occurs in man with aging?   Is there a relationship between the
midbrain DA cell density distribution and the specific
parkinsonian symptoms that the patient exhibited?   Do
schizophrenics, which have been reported to have increased levels
of DA (Mackay et al., 1982), have more than the normal number of
midbrain DA neurons?   DA levels (Glick et al., 1982), the number
of DA receptors (German et al., 1982) and NE levels (Oke et al.,
1978) are unequal on the two sides of specific human brain
regions.   Are these differences due to differences in the number
of cell bodies on the two sides of the brain?

     Before such questions can be answered, a number of technical
problems must be overcome.   The human brain contains hundreds of
thousands of catecholamine neurons.   Manual counting of these
large nuclei is tedius and time-consuming.   We have developed an
automated cell counting program which should greatly speed data
input into the computer (Schlusselberg et al., 1982).   The
algorithm counts objects in a 512 X 512 pixel array in about one-
third of a second, allowing program values and microscope lighting

to be adjusted interactively to achieve the optimal accuracy of cell recognition. After selection of these parameters, the user can add or delete cells, and then signal the computer to store the list of cell descriptions obtained from a single field scan. It is also important to calculate the minimal number of brain sections which should be counted in order to accurately represent the three-dimensional cell distribution. "Standard brains" need to be calculated so that comparisons between different brains (within a particular species) can be made considering that brains have variation in size (especially in the human brain). The accuracy with which neuromelanin marks all catecholamine neurons must be determined by comparing adjacent tyrosine hydroxylase and neuromelanin-stained sections.

In summary, the computerized reconstruction of neuronal distributions represents an important tool which can be used to quantitate normal age- and disease-related changes in cell numbers and to correlate changes in regional cell densities with various behavioral changes.

ACKNOWLEDGEMENTS

The authors wish to thank Laura Boynton for excellent secretarial assistance. Research supported by the Biological Humanics Foundation, the Dallas Area Parkinsonism Society and grants NS-20030, DA-2338, AA-0390.

REFERENCES

Dahlstrom, A. and Fuxe, K. (1964). Evidence for the existence of monoamine containing neurons in the central nervous system. I. Demonstration of monoamines in the cell bodies of brainstem neurons. Acta Physiol. Scand., 62, (Suppl. 232), 1-55.

German, D.C., Schlusselberg, D.S., McMillen, B.A., McDermott, K., Smith, W.K. and Woodward, D.J. (1982). Asymmetries in human brain dopamine receptor binding: relationship to midbrain dopamine cell number. Neurosci. Abstracts, 8, 114.

German, D.C., Schlusselberg, D.S. and Woodward, D.J. (1983). Three-dimensional computer reconstruction of midbrain dopaminergic neuronal populations: from mouse to man. J. Neural Trans., 57, 243-254.

Glick, S.D., Ross, D.A. and Hough, L.B. (1982). Lateral asymmetry of neurotransmitters in human brain. Brain Res., 234, 53-63.

Grace, A.A. and Bunney, B.S. (1983). Intracellular and extracellular electrophysiology of nigral dopaminergic neurons-3. Evidence for electrotonic coupling. Neurosci., 10, 333-348.

Graham, D.G. (1979). On the origin and significance of neuromelanin. Arch. Path. Lab. Med., 103, 359-362.

Hassler, R. (1937). Zur normalanatomie der substantia nigra. J. Fur Psychologie und Neurologie, 48, 1-55.

Mackay, A.V.P., Iversen, L.L., Rossor, M., Spokes, E., Bird, E., Arregui, A., Creese, I. and Snyder, S.H. (1982). Increased brain dopamine and dopamine receptors in schizophrenia. Arch. Gen. Psychiat., 39, 991-997.

Mann, D.M.A. and Yates, P.O. (1983). Pathological basis for neurotransmitter changes in Parkinson's disease. Neuropath. & Applied Neurobiol., 9, 3-19.

Mann, D.M.A., Yates, P.O. and Hawkes, J. (1983). The pathology of the human locus ceruleus. Clin. Neuropathol., 2, 1-7.

Mason, S.T. (1981). Noradrenaline in the brain: progress in theories of behavioural function. Prog. in Neurol., 16, 263-303.

McGeer, P.L., McGeer, E.G. and Suzuki, J.S. (1977). Aging and extrapyramidal function. Arch. Neurol., 34, 33-35.

Moore, R.Y. and Bloom, F.E. (1978). Central catecholamine neuron systems: anatomy and physiology of the dopamine systems. Ann. Rev. Neurosci., 1, 129-169.

Moore, R.Y. and Bloom, F.E. (1979). Central catecholamine neuron systems: anatomy and physiology of the norepinephrine and epinephrine systems. Ann. Rev. Neurosci., 2, 113-168.

Nobin, A. and Bjorklund, A. (1973). Topography of monoamine neurone systems in the human brain as revealed in foetuses. Acta Physiol. Scand., (Suppl. 388), 1-40.

Oke, A., Keller, R., Mefford, I. and Adams, R.N. (1978). Lateralization of norepinephrine in human thalamus. Science, 200, 1411-1413.

Olszewski, J. and Baxter, D. (1954). Cytoarchitecture of the human brain stem. J.B. Lippincott Company, Philadelphia.

Pearson, J., Goldstein, M., Markey, K. and Brandeis, L. (1983). Human brainstem catecholamine neuronal anatomy as indicated by immunocytochemistry with antibodies to tyrosine hydroxylase. Neurosci., 8, 3-32.

Saper, C.B. and Petito, C.K. (1982). Correspondence of melanin-pigmented neurons in human brain with A1-A14 catecholamine cell groups. Brain, 105, 87-101.

Schlusselberg, D.S., Smith, W.K., Lewis, M.H., Culter, B.G. and Woodward, D.J. (1982). A general system for computer based acquisition, analysis and display of medical image data. Proc. ACM, 18-25.

Tomlinson, B.E., Irving, D. and Blessed, G. (1981). Cell loss in the locus coeruleus in senile dementia of Alzheimer type. J. Neurolog. Sci., 49, 419-428.

Ungerstedt, U. (1971). Stereotaxic mapping of the monoamine pathways in the rat brain. Acta Physiol. Scand., (Suppl. 367), 1-48.

TOPOGRAPHIC DISTRIBUTION OF CATECHOLAMINERGIC NEURONS IN THE RAT MEDULLA OBLONGATA USING QUANTITATIVE THREE–DIMENSIONAL RECONSTRUCTION

M. KALIA[1], D.J. WOODWARD[2], W.K. SMITH[2], K. FUXE[3], T. HÖKFELT[3] and M. GOLDSTEIN[4]

[1]Department of Pharmacology, Thomas Jefferson University, Philadelphia, Pennsylvania, USA
[2]Department of Cell Biology, University of Texas Health Science Center at Dallas, Dallas, Texas, USA
[3]Department of Histology, Karolinska Institutet, Stockholm, Sweden
[4]Department of Psychiatry, New York University Medical Center, New York, USA

In 1964 Dahlstrom and Fuxe (3), using the formaldehyde histofluorescence technique, first described a variety of monoamine containing neurons in the brain. A number of studies on the catecholamine (CA) neurons followed, (2,4-8,12-19,,24,) resulting ·in the demonstration of the existence of the following populations of CA neurons: adrenaline containing neurons – C cell groups (Hokfelt, Fuxe et al, 1974), noradrenaline containing neurons – A cell groups (Dahlstrom and Fuxe, 1964) and dopamine containing neurons. Historically, the question of monoaminergic neurons located in the brain stem has been addressed using fluorescence histochemistry and immunofluorescence techniques (1,2,3,18,19,30).

One major drawback of using techniques involving fluorescence microscopy is that cytoarchitectonic distinctions are not readily discernible under the fluorescence microscope. The immunoperoxidase method of Sternberger (29) permits light microscopic visualization of immunoreactive nerve cell bodies, nerve fibers and preterminal processes. This allows subsequent staining of the material with Nissl stain to determine the cytoarchitectonic boundaries of the region of the CNS being studied. We have recently developed a method of overlay drawing of immunoreactive regions of the brain (21) and have defined detailed cytoarchitectonic subdivisions of the dorsal medullary nuclei – the nucleus of the tractus solitarius (nTS), the dorsal motor nucleus of the vagus (dmnX), area postrema (ap) and adjacent regions of the dorsal medulla. These regions have been previously defined as the A2 and C2 cell groups and have been shown to contain noradrenaline and adrenaline neurons respectively (1,3,12,13,14,15,17,18,19,30).

127

In the present study we have combined four techniques: 1)
bright-field immunoflourescence , 2) overlay drawings of
photomontages of the A1, A2, C1 and C2 cell groups, 3) Nissl
staining and 4) three-dimensional computer reconstruction.
The goal was to answer the following questions regarding
these catecholaminergic cell groups: 1. What is the
relationship of these cell groups to the
cytoarchitectonically and functionally distinct nuclear
regions in the medulla (20,21)? 2. What is the precise
rostro-caudal and medio-lateral extent of the individual cell
groups? 3. Do the adrenaline-containing, rostrally located C
cell groups and the noradrenaline-containing, caudally
located A cell groups overlap or do they represent a
continuous column of cells? 4. What is the spatial
organization of these catecholaminergic cell groups within
the cytoarchitectonic boundaries of the brain stem?

Methods

Immunocytochemical procedure

(i). Perfusion protocol The rats were anesthetized with
sodium pentobarbital (Nembutal, 27-50 mg/kg intraperitoneally
(i.p.), given 5,000 Units of Heparin i.p. and perfused
through the ascending aorta with the following solutions: (1)
50-100 ml oxygenated tyrode solution at 4 degrees centigrade
to which 1 ml of 1% sodium nitrite solution had been added;
(2) 500 ml of 4% paraformaldehyde solution in 0.1 M sodium
phosphate buffer pH 7.4 for 27 minutes. The brainstem and
upper cervical spinal cord were removed and immersed in the
fixative solution for 90 minutes. The brain stem was blocked
in the Horsley-Clarke coronal plane with the aid of a
stereotaxic blocking device. The brain stem was then rinsed
throughly in 0.1 M phosphate buffer at 4 degrees centigrade
and cut as soon after the perfusion as possible.

(ii). Vibratome sectioning Serial sections through the
entire block were cut and counted in order to have reliable
information of the location of sections in the rostrocaudal
axis with reference to the obex. A map was kept of this
tissue, and the areas selected for sectioning were marked and
numbered. The cutting chamber was filled with 0.1M phosphate
buffer at 4 degrees centigrade. The serial sections were cut
into 0.1M phosphate buffer at 4 degrees centigrade (Brain
Research Laboratories) and quickly transferred to reaction
chambers for incubation with the primary antiserum.

(iii). Immunocytochemical procedure (16,29) Sections were

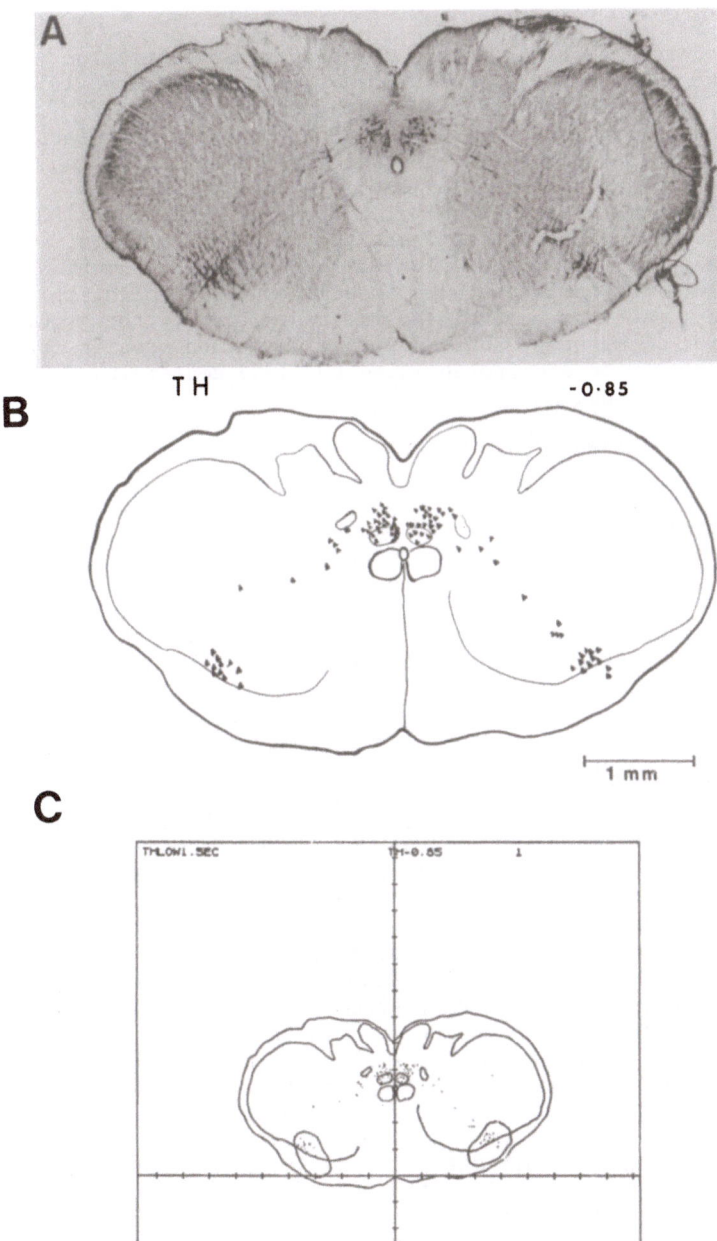

Figure 1: A. Low-power photomicrograph of the caudal medulla at a level -0.85 mm caudal to obex showing TH immunoreactive cells. B. Overlay drawing of section A showing the TH cells and major nuclear boundaries. C. Computer reconstruction of same section.

incubated free floating in the diluted primary antiserum
(tyrosine hydroxylase - TH X 750; dopamine beta hydroxylase -
DBH X 750; phenyethanolamine N-methyltransferase - PNMT X
1500) containing 0.3 % Triton X 100 on a shaker table for
24-36 hours at 4 degrees centigrade. Following incubation
with the primary antisera the sections were washed in
Tris-saline containing 1% normal goat serum (two rinses 10
minutes each). The sections were then treated with goat
antirabbit IgG preadsorbed with rat 1:20 dilution (American
Qualex) at room temperature on a shaker table. Following
this step, sections were rinsed in Tris-saline (without goat
serum) for 10 minutes each. Sections were then treated for
7-10 minutes with diaminobenzidine (0.013%) containing 0.003%
hydrogen peroxide.

Computer Reconstruction

   Histological sections through the brainstem were studied
as follows: First, the sections containing monoamine
immunoreactive neurons were photographed at 2X magnification
with an Olympus Vanox Photomicroscope using bright field
illumination with an open field aperture and condenser
setting to enhance the contrast. The total image of the
section was photographed within one frame and high contrast
prints were made. From the photographic prints overlay
drawings of the location of neuronal profiles were carefully
marked by triangles and the major nuclear boundaries were
marked by straight lines. The number and location of cells
 was monitored by continuous checking of the section under
the microscope during the drawing procedure (Figures 1 and
2). The overlay drawing was made to fit the surface of a
digitizing tablet. The drawing was laid on the digitizing
tablet and an origin and axis were determined as follows. In
sections caudal to the obex, the lowest point of the central
canal was taken as the origin (a)and the highest point of the
area postrema on the surface of the medulla vertically above
it in the region of the area postrema was taken as the second
coordinate point (b) to define an axis on the midline. The
magnification of the section was also defined. The provision
of two coordinate points enabled definition of a axis on the
midline through which alignment adjustments could be made and
through which the sections could be rotated in various planes
during the computer data manipulation and processing. This
procedure allowed the information to be input rapidly at low
power magnification. An outline of the section as well as
outlinesof major landmarks, particularly fiber tracts and
nuclei,were input as separate segments in sequential order in
an anticlockwise direction throughout the series. In
addition, the cell groups were input into the computer

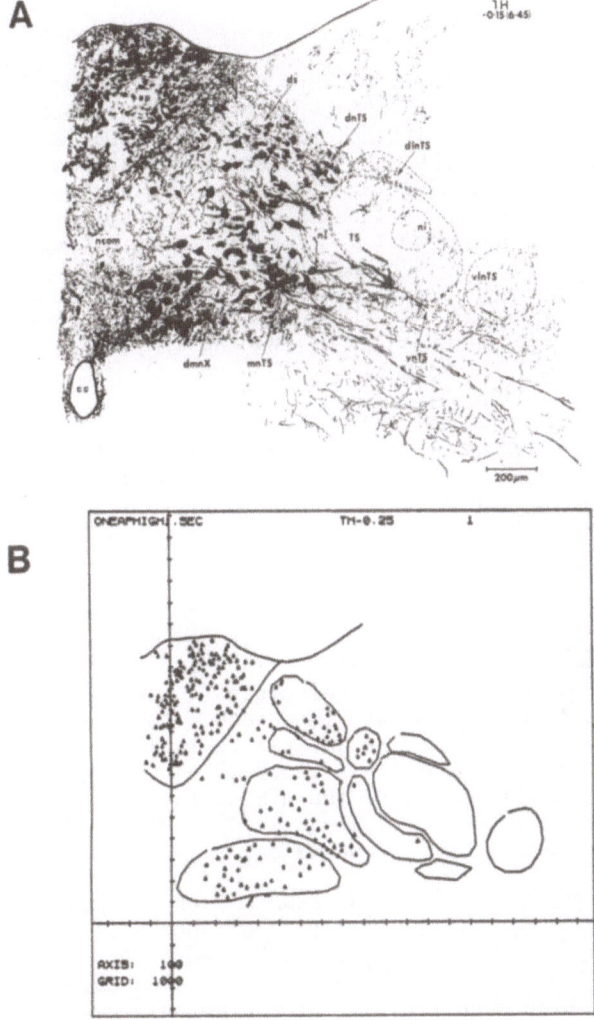

Figure 2: A. Line drawing of TH immunoreactive neurons, fibers and immunoreactive processes in the dorsal medulla at a level -0.15 mm caudal to obex. This drawing was made from an over-lay of a photomontage. B. Computer plot of the same section. Cell bodies and key nuclear boundaries are plotted. Grid marks are 100 μm.

through the digitizing tablet as separate cell·groups within
   individual subnuclei so that a subsequent analysis of a
single cell group could be performed. This information was
stored in the computer at the Department of Cell Biology,
University of Texas Health Sciences Center at Dallas. Every
section stained with the same antibody was drawn sequentially
and the rostrocaudal position of the section was marked. The
graphic system software allowed the two–dimensional plots of
individual sections as well as groups of sections to be
plotted as a serial reconstruction. The data files could be
scanned in a number of ways and permitted the data
manipulation and visualization in a variety of planes for
appreciating the location of cell bodies in relation to one
another (10, 11, 22, 25–28).

RESULTS

   The main purpose in this type of analysis of
monoaminergic neurons in the medulla was to visualize the
spatial location and distribution of different groups of
dopamine, epinephrine, and norepinephrine containing cells.
The use of computer assisted three dimensional reconstruction
of the different neuronal cell groups has enabled us to
clearly visualize the distinction between the noradrenergic
and the adrenergic cell populations in the medulla as is
evident from Figures 1 through 6. The distribution of these
cells from the medial lateral direction could be plotted in
terms of a histogram, and more importantly the relationship of
the cells in relation to the lateral reticular nucleus (LRt)
and nucleus paragigantocellularis (PGi) on the ventral side
and the various subnuclei of the nTS on the dorsal aspect of
the medulla could be easily visualized by viewing these
sections in the high power analysis.

   The following levels were examined at low power: PNMT
–0.8, –0.7, –0.5, 0.1, 1.6, 1.85, 2.05; DBH –0.75, –0.3,
–0.25, 0.05, 1.31, 1.5, 1.9; TH –0.85, –0.6, –0.4, –0.15,
0.05, 0.3, 1.3, 1.75, 2.2, 2.8. We have also compared
drawings at high magnification of the dorsal medulla which
includes the nucleus of the tractus solitarius (nTS) at the
following rostrocaudal levels: 1) levels ranging from 0.1 to
0.2 (TH 0.1, DBH 0.15, PNMT 0.2); and at levels ranging from
–0.25 to –0.15 (TH –0.25, DBH –0.2 and PNMT –0.15).

   Analysis of dopamine containing neurons : These cells
show positive immunoreactivity only with TH antibody, and are
coded green in this series of computer reconstructions. By
means of the three–dimensional computer reconstruction we
have visualized three regions of the dorsal medulla
containing dopamine (D) neurons : 1) the medial and dorsal

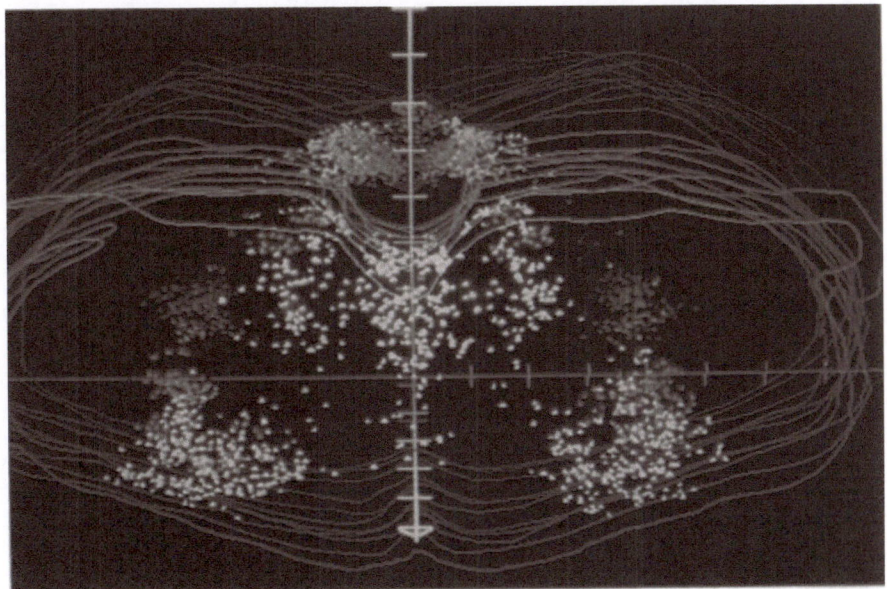

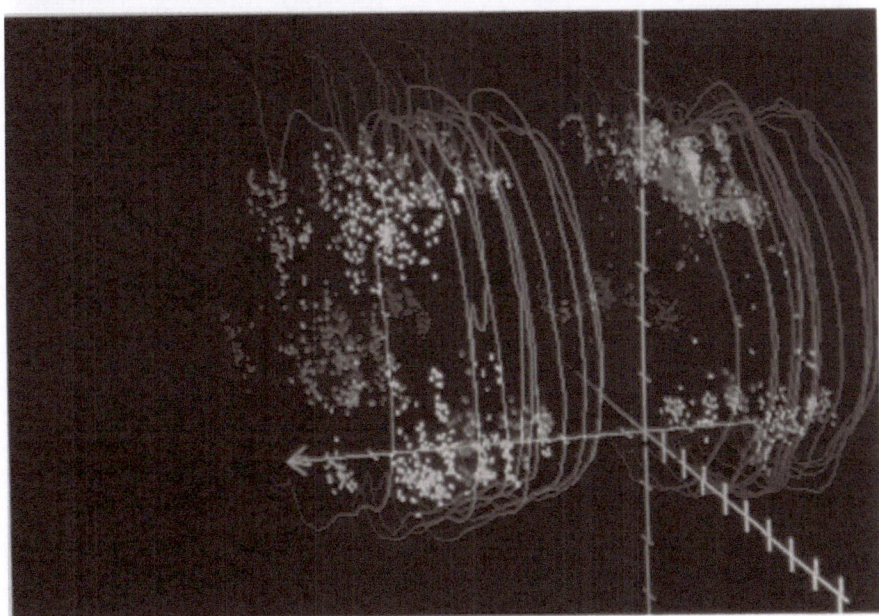

Figure 3: A. Three-dimensional reconstruction of D, NA and A neuronal distribution in the medulla. TH immunoreactive cells are green, DBH cells are blue and PNMT cells are white. View is from the rostral side at a 30 degree tilt. B. 30 degree lateral tilt of same cell groups. Rostral is left, grid=500 μm.

part of the area postrema (ap), 2) the dorsal motor nucleus
of the vagus (dmnX), and 3) the periventricular region (PVR).
These neurons are located at levels caudal to the obex and
are part the previously considered noradrenergic (NA) A2 cell
group (Dahlstrom and Fuxe, '64) (Figure 3A and 6). The
rostrocaudal extent of the dopamine containing neurons can be
seen in Figure 3 where the dopamine neurons in the ap are
most clearly visible. In Figure 6 all three groups of
dopamine containing neurons (staining with TH only) can be
seen most clearly. There appears to be no overlap between
the dopamine containing neurons and the noradrenaline cells.
The population of dopamine containing neurons can be seen in
the dorsal medial part of the area postrema (ap) in Figure 6.
Notice the mixture of TH, DBH and PNMT containing neurons in
the other subnuclei of the nTS.

   Analysis of noradrenaline neurons in the A2 and A1 cell
groups : These cells stain only with TH and DBH antibody and
not with PNMT antibody and are coded with blue for DBH and
green for TH in these series of computer reconstructions.
Figures 3 and 6 shows the differential distribution of TH,
DBH and PNMT containing neurons. The analysis of these
section at a glance shows the following features: 1. area
postrema contains dopamine neurons in the dorsal and medial
part - Note that this population appears green and is missing
in the blue and the white cell population (see Figure 6). 2.
The second population of cells in the area postrema consists
of adrenaline containing cells which are labelled with TH,
DBH and PNMT shown here as green, blue and white cells.
3.The dorsal motor nucleus of the vagus is heavily filled
with dopamine containing cells shown here in green (Figure
6). Note the absence of blue and white cells in this figure.
4. The medial nTS appears to contain primarily noradrenaline
neurons. These neurons are stained both with the green and
blue and are not stained with the white label. 5. The dorsal
parasolitarius region (dPSR) which consists of the region
just dorsal to the medial nTS contains primarily adrenaline
containing cells since they are labelled with both white and
blue and green stain. This also appears to be a mixed group
with adrenaline and noradrenaline containing cells. The
dorsal strip region (ds) which is stained green, blue and
white appears to be predominantly an adrenaline containing
cell group since the most dense population of cells is
stained with white . The dorsal nucleus of the tractus
solitarius (dnTS) also appears to contain primarily
adrenaline containing neurons. The intermediate nucleus (nI)
located between the TS and the rest of the nTS complex also
appears to contain adrenaline neurons although there is
definitely some degree of mixing. The TS itself contains an

occassional adrenaline neuron.

In summary , at the level of the obex in the A2 cell
group there are a mixture of all three CA neurons. The
adrenaline neurons are located in the ds and dnTS, the
noradrenaline neurons are located in the mnTS and ncom and
the dopaminergic neurons are located in the dorso-medial
ap, dmnX and PVR. The A2 cell group extends from -2.7 to +0.3
mm and has dimensions of 0.4 X 3 mm. The Al cell group a
homogeneous population of NA cell extending from -2.5 to +0.2
mm with dimensions of 1.3-2.7 mm. (see Figure 3 - cells
staining blue only).

Analysis of adrenaline containing neurons These neurons
stain with all three antibodies TH, DBH and PNMT. However,
since neither D or NA neurons stain with PNMT, the PNMT
antibody is considered specific for A neurons (12-15). These
neurons are coded white in this series. Figures 4 and 5 show
the adrenaline cell groups and serve as an example of how
three dimensional computer reconstruction can aid in the
visualization of a new cell group. Figure 4 shows a caudal
and rostral view of the adrenaline neurons. A new population
of adrenaline neurons in the ds and dnTS which is quite
separate from trhe classical A2 group can be seen in the
foreground in Figure 4A. Figure 4B shows the classical C2 and
C1 cell groups. Figure 5 shows these same adrenaline groups
viewed from a dorsal aspect with a 70 degree tilt. The
dimensions of the Cl cell group are 1.5 X 1.5 mm extending
from +1 to +2.5. The dimensions of the C2 cell group are 2.5
X 3 mm and extend from +1 to +3.

DISCUSSION

Use of three-dimensional computer reconstruction has
enabled us to view the catecholaminergic cell groups in the
medulla oblongata in a very distinct and special manner. The
ability to manipulate the data in a variety of different ways
makes possible an analysis of neuroanatomical systems in a
way that cannot be done by artists. The accuracy of this
reconstruction will be most valuable in our ability to
visualize neuronal groups in the three-dimensional context.
The use of color coding has further permitted us to analyze
multiple types of data simultaneously in the same section. We
can add nuclear boundaries or remove them as the analysis
requires. By rotation of the orientation of the sections one
can separate out and view the location of functionally
distinct and immunocytochemically distinct neuronal
populations. Figures 4 and 5 show that the computerized
rotation and tilting of the serial reconstruction of cells is

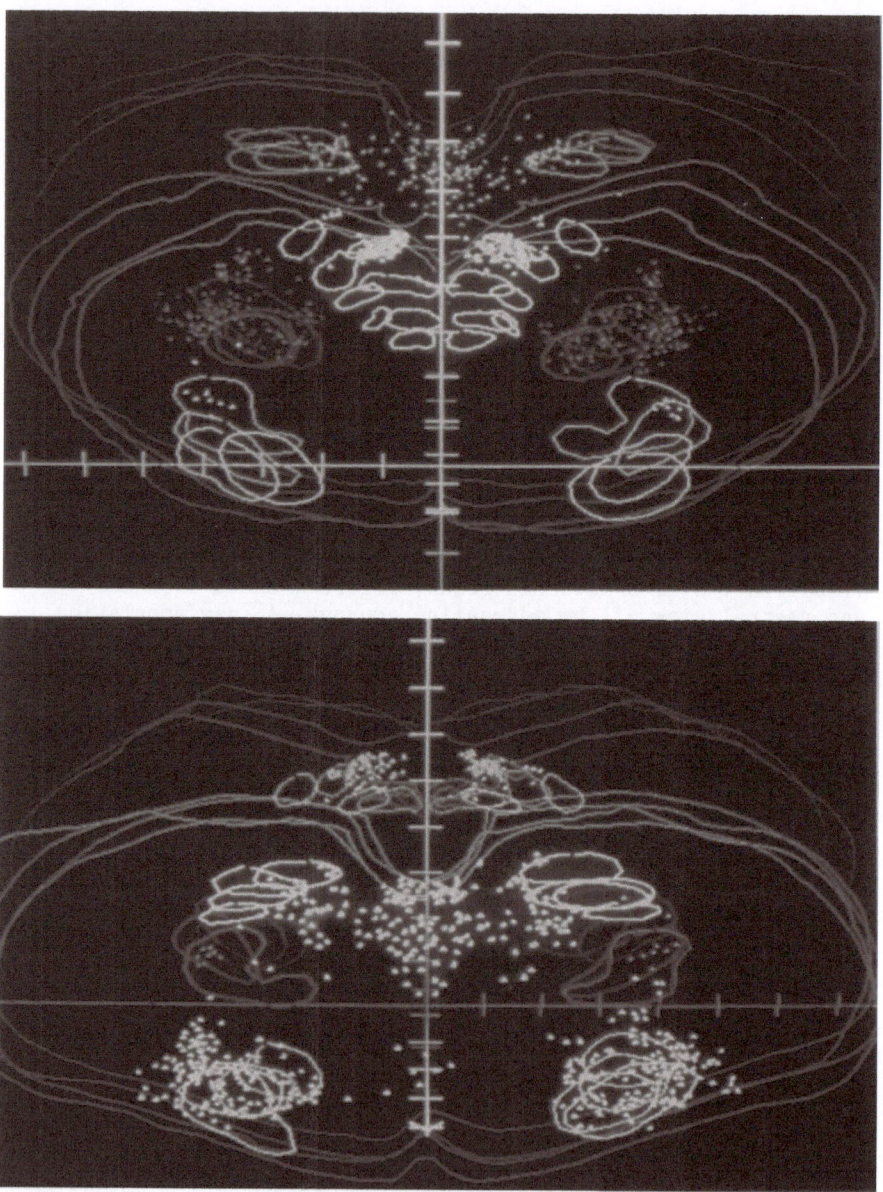

Figure 4: Three-dimensional reconstruction of adrenaline (A) neurons in the Cl and C2 cell groups. A. 30 degree tilt caudal view showing the Adrenergic neurons in the dorsal strip (ds) and dlnTS in the foreground and the Cl (ventral) and C2 (dorsal) cell groups in the background. B. Rostral, 30 degree tilt view of the same cell group with Cl and C2 cell groups in the foreground.

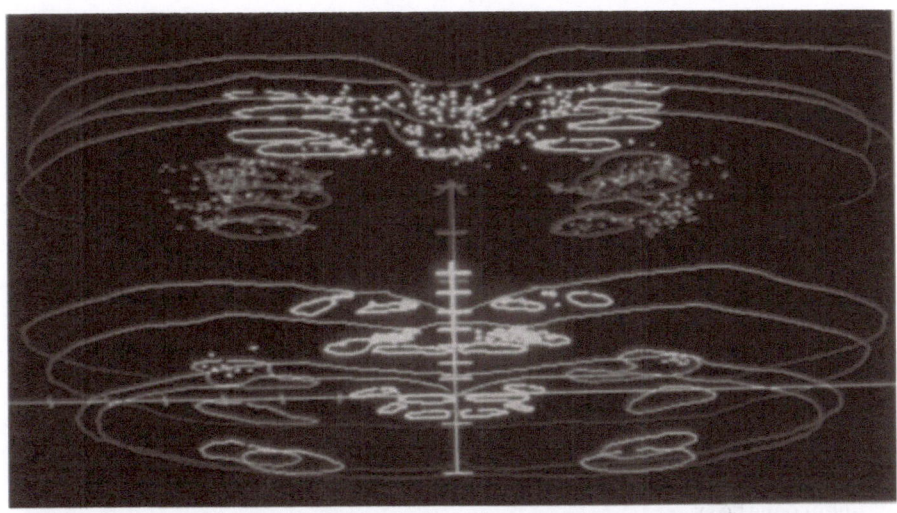

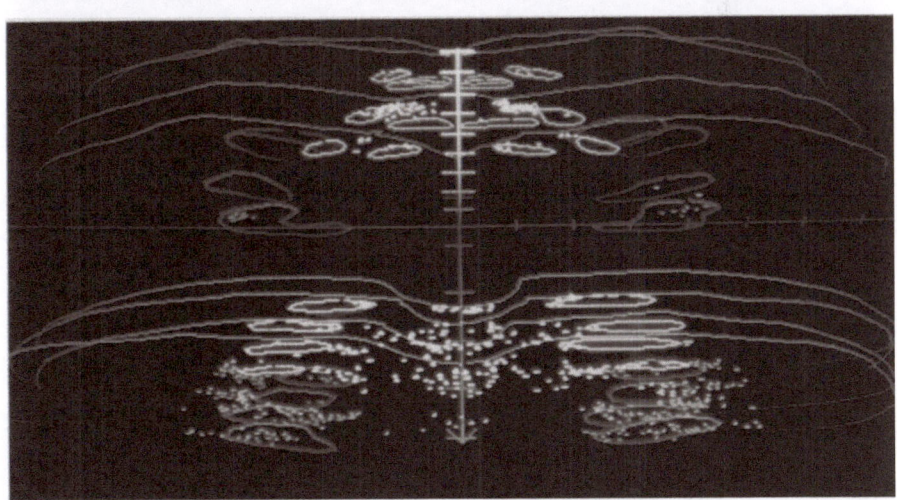

Figure 5: Three-dimensional computer reconstruction of
adrenaline (A) neurons in the medulla oblongata viewed from a
dorsal aspect (sections were tilted through 70 degrees in the
rostro-caudal plane). A: A caudal view showing very clearly the
separation between the caudal A cell group in the ds and dnTS
and the rostral dorsally located C2 adrenaline cell group. Note
the difference in organization of these two populations of
neurons. The C2 cell group appears blue in the background.
B. A rostral view of the same A cell groups with the sections
tilted up 70 degrees. The C2 group is in the foreground.

an extremely valuable tool in the separation and
identification of different populations of neurons. In order
to appreciate the view or the concept of the separation, the
overlap, and the rostral caudal extent of the cell group from
a series of classical neuroantomical drawings one has to
examine each of these drawings and do a mental computation of
where the cell groups are. A second series of drawings from a
separate series of sections cut on a different plane is the
only other method of showing the location of the different
cell groups in a different plane. The use of computer
reconstruction eliminates the need to do multiple experiments
since it is now possible to produce a three-dimensional
computer reconstructed image demonstrating all these features
in one single 3-D image. This requires only a few seconds of
viewing for the entire picture to be integrated by the human
brain and offers the single greatest advantage of
three-dimensional computer reconstruction for classical
neuroanatomy.

The nucleus of the tractus solitarius and the other
regions of the medulla oblongata provide an ideal anatomical
location wherein a large number of monoaminergic nerve cells
are distributed for us to examine the use of computer
reconstruction as compared to classical neuroanatomical
drawings. In this paper we have prepared a series of very
accurate drawings and then used these very same drawings to
produce anatomical maps stored in the memory of the computer
that can be rotated, tilted, viewed from different angles,
overlapped or viewed individually. Thus different neuronal
populations of cells have been plotted. In this region of
the brain there exist three populations of neurons, dopamine
containing, noradrenaline containing, and adrenaline
containing neurons. This data was fed into a computer, which
takes into account the rostro-caudal level, the medio-lateral
level and the orientation, thus enabling us to analyze this
data in a variety of ways. One can take this series of
sections and tilt them, view them from the top, below,
laterally, or medially,. So that different neuronal
populations can be viewed individually, and therefore, it is
possible to demonstrate individual cell groups in a very
precise and a meaningful manner.

The use of the computer in three dimensional neuroanatomy
in the future will be in our ability to take a neuron cell
population such as as exists in the nTS which is a
heterogenous population consisting of all the three different
monoamines and to be able to selectively view it from
different view points so that one can immediately visualize
at a glance in one section where the different neuronal

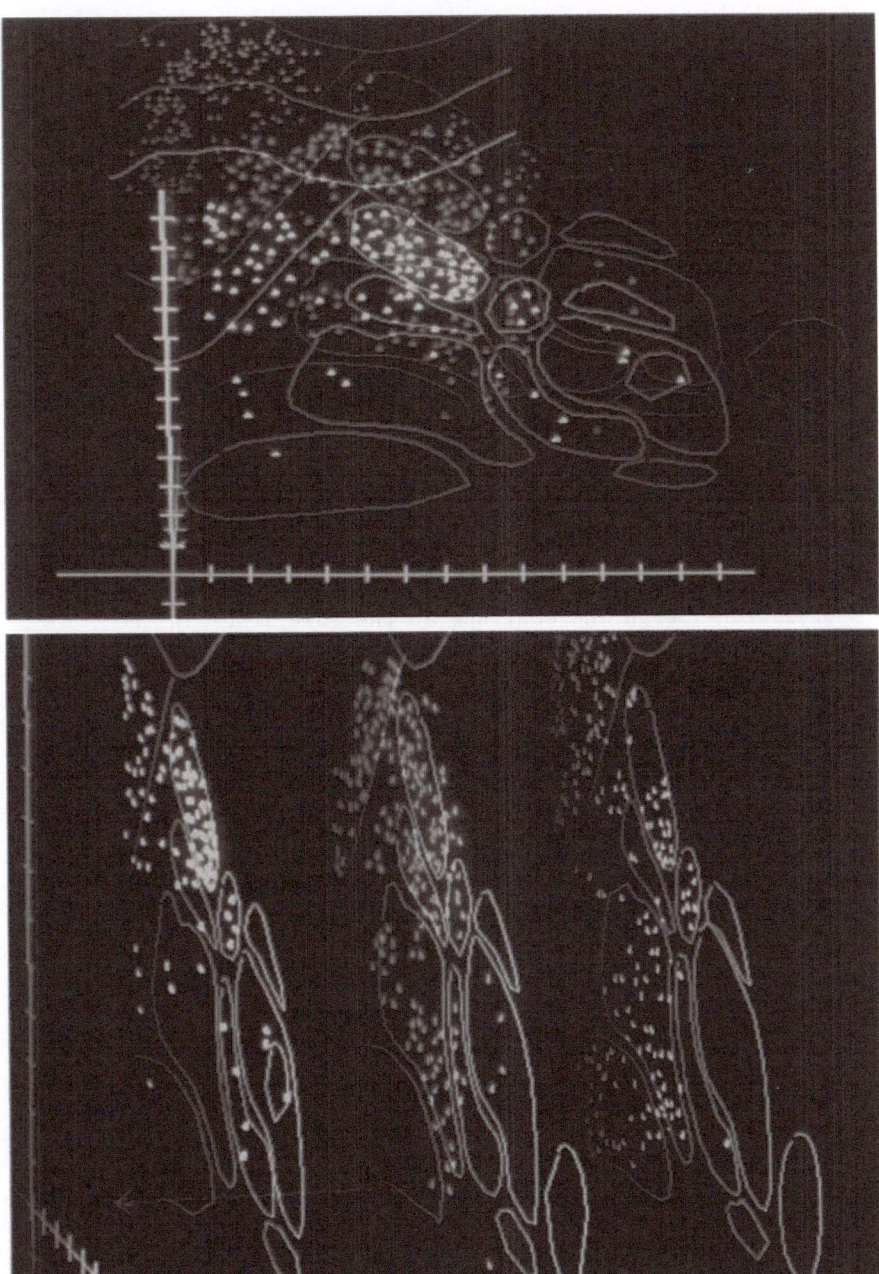

Figure 6: A. Frontal views of sections showing TH, DBH and PNMT immunoreactivity in the nTS (-0.25). Notice the overlap of cells in many subnuclei of the nTS. B. Side view tilted at 30 degrees. Cell color codes and nuclear boundaries are identical with Figs. 1 and 2.

populations lie. The precision by which this is done will
provide very valuable data on which electrophysiology and
other neuroanatomical studies can be planned and executed.
It is possible to obtain in very precise quantitative data
the dimensions of the nucleus subgroups, the topography maps,
histograms of the cell groups numbers (German et al, 1983).
This type of reconstruction will enable us to determine the
dendritic morphology, the cell size and other featues of
these neuronal populations when drawn at high power
magnifications.

Quantitation of these cell bodies as well as three
dimensional reconstruction with biological coordinates as
shown in this study have helped answer a large number of
questions. Where in the 3 dimensional space are adrenaline
neurons in the nTS located? Where in the 3 dimensional space
are dopamine and adrenergic neuronslocated? Do the C 1 and
the A 1 cell groups overlap and if they do , what is the
rostrocaudal level at which they overlap? Do the A 1 and A 2
cell groups overlap and if so what is the level of the
overlap? What is the extent of the overlap and do these cell
populations merge and continue at the same level in the
medio-lateral plane or do they diverge? What is their
relationship to other nuclear groups? These questions can be
answered by the analysis presented in this paper.

Other questions such as those relating to the comparison
between different brains to the comparison between
genetically different and species differences can be added to
this type of analysis so that functionally distinct neurons
can be identified, abnormalities or differences can be
detected in a very precise manner. The accurate visual
imaging and quantitation of the spatial organization of CNS
neurons within the classical neuroanatomical framework of
cytoarchitecture and the ability to manipulate the data in a
variety of different ways makes this method a very powerful
tool in neuroanatomical and neuropathological studies of the
future.

ACKNOWLEDGMENTS

This research was supported by USPHS Grants HL 30991,
MH 25504, NIAAA 390 and DA 2338, a Grant from the Biol.
Humanics Foundation and the Swedish Medical Research Award
14X4246-10B.

REFERENCES

1. Armstrong, D.M., Pickel, V.M., Joh, T.H., Reis, D.J., Miller, R.J. (1981): Immunocytochemical localization of catecholamine synthesizing enzymes and neuropeptides in area postrema and medial nucleus tractus solitarius of rat brain. J. Comp. Neurol. 196 , 505–517.

2. Blessing, W. W., Chalmers, J. P., Howe, P. R. C. (1978): Distribution of catecholamine-containing cell bodies in the rabbit central neurons system. J. Comp. Neurol. 179 , 407–424.

3. Dahlstrom, A., Fuxe, K. (1964): Evidence for the existence of monoamine containing neurons in the central nervous system. I. Demonstration of monoamines in the cell bodies of brainstem neurons. Acta Physiol. Scand. 62 , (Suppl. 232), 1–55.

4. Fuxe, K. (1965). Evidence for the existence of monoamine neurons in the central nervous system. III. The monoamine nerve terminal. Z. Zellforsch., 65 , 573–596.

5. Fuxe, K. (1965). Evidence for the existence of monoamines in the central nervous system. IV., The distribution of monoamine nerve terminals in the central nervous system. Acta physiol scand., 64 , Suppl 247, 39–85.

6. Fuxe, K., Goldstein, M., Hokfelt, T., Joh, T. H. (1970). Immunohistochemical localization of dopamine-beta-hydroxylase in the peripheral and central nervous system. Res. Commun. chem. Path. Pharmacol., 1 , 627–636.

7. Fuxe, K., Goldstein, M., Hokfelt, T., Joh, T. H. (1971). Cellular localization of dopamine-beta-hydroxylase and phenylethanolamine-N-methyl transferase as revealed by immunohistochemistry. In O. Franko (ed.), Histochemistry of Nervous Transmission, Progress in Brain Research, Vol. 34 , Elsevier, Amsterdam, 1971, pp. 127–138.

8. Fuxe, K., Hokfelt, T., Ungerstedt, U. (1970). Morphological and functional aspects of central monoamine neurons. Int. Rev. Neurobiol., 13 , 93–126.

9. Geffen, L.B. (1982). Histochemical and cytochemical localization of monoamine (arylethylamine) neurotransmitters. In: Chan-Palay, V. and Palay, S. (eds.) Cytochemical Methods in Neuroanatomy , Alan R. Liss Inc. New York, 119–127.

10. German, D. C., Schlusselberg, D. S., McMillen, B. A., McDermott, K, Smith, W. K., Woodward, D. J. (1982). Asymmetries in Human Brain Dopamine Receptor Binding: Relationship to 3-Dimensionnal Reconstruction of Midbrain Dopamine Neuorns. Society for Neuroscience Abstracts, 30.3.

11. German, D. C., McDermott, K. L., Sanghera, M K., Schlusselberg, D.S., Smith, W. K. Woodward, D. J., Speciale S. G., Saper, C. B. (1983): Three-dimensional reconstruction of dopamine neurons in the mouse: strain differences in regional cell densities and pharmacology. Neurosci. Abstr. 9.

12. Goldstein, M. (1972). Enzymes involved in the catalysis of catecholamine biosynthesis. In R. N. Ubell (ed) <u>Methods in Neuorchemistry</u> , Vol. 1, Plenum Press, New York, 1972, pp. 317-340.
13. Goldstein, M., Agnoste, B., Freedman, L. S., Roffman, M. Ebstein, R. P., Park, D. H., Fuxe, K. and Hokfelt, T. (1973). Proc. II. Catecholamine Symposium, Strasburg.
14. Goldstein, M., Fuxe, K., and Hokfelt, T., (1972). Characterization and tissue localization of catecholamine synthesizing enzymes. Pharmacol. Reve. <u>24</u> , 293-309.
15. Goldstein, M., Fuxe, K., Hokfelt, T., and Joh, T. H. (1971). Immunohistochemical studies on phenylethanolamine-N-methyltransferase, dopadecarboxylase and dopamine-beta-hydroxylase. Experientia (Basel), <u>27</u> , 951-952.
16. Hartman, B. K., Zide, D., Udenfriend, S. (1972). The use of dopamine-B-hydroxylase as a marker for the noradrenergic pathways of the central nervous system in the rat. Proc nat. Acad. Sci. (Wash.) <u>69</u> , 2722-2726.
17. Hokfelt, T., Fuxe, K, Goldstein, M., Joh, T. H. (1973). Immunohistochemical studies of three catecholamine synthesizing enzymes: Aspects on methodology, Histochemie, <u>33</u> , 231-254.
18. Hokfelt, T., Fuxe, K., Goldstein, M., Johannsson, O. (1973). Evidence for adrenaline neurons in the rat brain. Acta physiol. scan. <u>89</u> , 286-288.
19. Hokfelt, T., Fuxe, K., Goldstein, M., Johansson, O. (1974): Immunohistochemical evidence for the esistence of adrenaline neurons in the rat brain. Brain Res 66:235-251.
20. Kalia, M., Sullivan, J.M. (1982): Brainstem projections of sensory and motor components of the vagus nerve in the rat. J. Comp. Neurol. <u>211</u> , 248-264.
21. Kalia, M., Fuxe, K., Hokfelt, T., Johansson, O., Lang, R., Ganten, D., Cuello, C., Terenius, L. (1984): Distribution of neuropeptide immunoreactive nerve terminals within the subnuclei of the nucleus of the tractus solitarius of the rat. J. Comp. Neurol. <u>222</u> , 409-444.
22. Lewis, M. H., Schlusselberg, D. S., Smith W. K., Hagler, H. K., Woodward, D. J., Buja, L. M. (1982). Three-Dimensional Cardiac Morphometry with Computer Graphics. <u>Computers in Cardiology</u> , Conference Proc.
23. Macagno, E. R., Levinthal, C., Sobel, I. (1979). Three-dimensionaal Computer Reconstruction of Neuronal Assemblies. Ann. Rev., Biophys. Bioeng   <u>8</u> , 323-351.
24. Olson, L., Fuxe, K. (1972). Further mapping out of central noradrenaline neuorn systems: Projections of the "subcoeruleus" area. Brain Res., <u>43</u> , 289-295.
25. Reis, D.J., Benno, R.H., Tucker, L.W., Joh, T.H. (1982) Quantitative immunocytochemistry of tyrosine hydroxylase in brain. In: Chan-Palay, V. and Palay, S. (eds.) <u>Cytochemical methods in neuroanatomy</u> Alan R. Liss Inc, New York, 205-228.

26. Schlusselberg, D. S., Smith, W. K., Culter, B. G., Woodward, D. J. (1982): A computer system for semi-automatic cell recognition in neuroanatomic studies. Neurosci. Abstr. 8, 644.

27. Schlusselberg, D. S., Smith, W. K., Lewis, M. H., Culter, B. G., Woodward, D. J. (1982): A general system for computer based acquisition, analysis and display of medical image data. Proc. ACM 18–25.

28. Smith, W. K., Schlusselberg, D. S., Woodward, D. J. (1981). A Computer System For Neuoranatomical Data Acquisition, Analysis, and Display. Society for Neuroscience Abstract 135.18

29. Sternberger, L. A. (1979): Immunocytochemistry. New York: J. Wiley.

30. Swanson, L.W., Hartman, B.K. (1975) The central adrenaline system. An immunocytochemical study of the location of cell bodies and their efferent connections in the rat utilizing dopamine – beta – hydroxylase as a marker, J. Comp. Neurol. 163 , 467–506.

31. Ungerstedt, U. (1971): Stereotaxic mapping of the monoamine pathways in the rat brain. Acta physiol. scand. 82 , Suppl 367, 1–48.

EFFECTS OF CHRONIC GM1 GANGLIOSIDE TREATMENT ON NIGRAL DOPAMINE CELL BODIES AND DENDRITES IN EXPERIMENTAL RATS USING IMAGE ANALYSIS — RELATIONSHIP TO THE PHARMACOKINETIC PROPERTIES

LUIGI F. AGNATI[1], KJELL FUXE[2], GINO TOFFANO[3],
LAURA CALZA[1], ISABELLA ZINI[1], LUCIANA GIARDINO[1],
FRANCO MASCAGNI[1] and MENEK GOLDSTEIN[4]

[1]Department of Human Physiology, University of Modena, Modena, Italy
[2]Department of Histology, Karolinska Institutet, Stockholm, Sweden
[3]Department of Biochemistry, Fidia Research Laboratories, Abano Terme, Italy
[4]New York University Medical Center, New York, USA

INTRODUCTION

Previous studies have shown that chronic treatment with the GM1 ganglioside can increase the survival of dopamine nerve cell bodies in the substantia nigra, mainly in the caudal part on the lesioned side, following a partial unilateral hemitransection (Agnati et al. 1983, 1984; Toffano et al. 1983). It was also demonstrated that chronic treatment with the ganglioside GM1 increase tyrosine hydroxylase immunoreactivity within the rostral-ly located dopamine nerve cells present close to the site of the lesion (Agnati et al. 1984). Also it was found that chronic GM1 ganglioside treatment increases the density of tyrosine hydroxy-lase immunoreactive dendrites in the zona reticulata of the lesioned side as shown at several rostrocaudal levels. These results were interpreted to indicate that chronic GM1 treatment can exert an excitatory metabolic action on dopamine nerve cells present close to the lesion which may lead to enhanced production of neuronal trophic factors diffusing out in the substantia nigra and causing an increased survival of the less severely lesioned dopamine nerve cells present within the caudal part of the sub-stantia nigra (see Agnati et al. 1984). In the present paper we have continued these studies by using the IBAS Image Analyzer (Zeiss-Kontron) to evaluate the action of chronic GM1 treatment on the dopamine cell bodies and dendrites in the substantia nigra of the lesioned and unlesioned side. In order to further understand the mechanism of action underlying the neurotrophic activity of GM1 treatment, the pharmacokinetic characteristics of GM1 has been studied and compared in control rats and in rats with a unilateral partial hemitransection.

MATERIAL AND METHODS

Male Charles River rats (180 g b.w. were used). Partial unilateral hemitransections were performed as previously described

145

(Agnati et al. 1983, 1984). Briefly a 4 mm wide knife was inserted
in the coronal plane close to the midline 1 mm caudal to the
bregma and lowered at an angle of 70° to the horizontal plane to
reach the ventral border at a König-Klippel level of A3200. This
level is located just in front of the substantia nigra. This large
mechanical lesion produces an axotomy mainly of the lateral
component of the meso-striatal dopamine system. A 4 week treatment
with the GM1 ganglioside was performed the first dose being
administered 2-3 hrs following the operation. The dose of the GM1
ganglioside was 10 mg/kg given i.p. once daily.

By means of the IBAS Image Analyzer effects of chronic GM1
ganglioside treatment were evaluated on the nigral dopamine cell
bodies and on the dopamine dendrites present in the zona compacta
and in the zona reticulata. The dopamine neuronal structures were
demonstrated by the use of tyrosine hydroxylase (TH) immunocytochemistry. The unlabelled immunoperoxidase method using the
peroxidase antiperoxidase complex was used and the procedure was
principally performed according to Sternberger (1979). For further
details, see Agnati et al. (1983). The TH antiserum was diluted
1:1000 and has previously been characterized by Markey et al.
(1980). After incubation with the rabbit PAP the antigen-antibody
complex was visualized by means of diaminobenzidine and H_2O_2. The
sections were defatted, coversliped and mounted in Entellan.

Studies on the ^{3}H-GM1 related radioactivity distribution in the
normal and experimental rat (partial unilateral hemitransection)
(see Table 1)

Table 1

^{3}H-GM1 DISTRIBUTION IN NORMAL AND HEMITRANSECTED RATS.

ANIMALS   :   CD-COBS rats (Charles River), 180 g body weight

LESION   :   mesencephalic hemitransection (20° inclination)

EXPERIMENTAL GROUPS :   1. CONTROL RATS   a) 2 hrs after ^{3}H-GM1 (n = 4)
                                                 b) 6 hrs after ^{3}H-GM1 (n = 4)

                        2. HEMITRANSECTED RATS   a) 2 hrs after ^{3}H-GM1 (n = 4)
                                                 b) 6 hrs after ^{3}H-GM1 (n = 4)

^{3}H-GM1 ADMINISTRATION :   SA = 1.0 Ci/mmol
                             Injected dose = 65 µCi/rat (about 0.9 mg/kg) i.p. in saline

KILLING :   heart perfusion with 100 ml Krebs-Ringer solution (+ 1000 I.U. % ml heparin)

SAMPLES :   a) BLOOD from the right atrium, centrifuged for plasma separation (Eppendorf Minifuge, 90 sec)
            b) URINE 6 hrs urinary escretion after ^{3}H-GM1 administration (collected in metabolic cages)
            c) SKELETAL MUSCLE ⎫
            d) STRIATUM   ⎬ dissected, rinsed in saline, weighted and dissolved
            e) MESENCEPHALON   ⎭ with 1 ml Protosol (NEN) at 37°C in glass vials

COUNTING :   biological samples were counted by liquid scintillation by adding 14 ml Beckman EP
             scintillation fluid and using a Beckman LS 1800 spectrometer

^{3}H-GM1 (1.0 Ci/mmol) was dissolved in saline and injected i.p. in a dose of 65 μCi/rat (about 0.9 mg/kg). Injections were made both in the intact control rats and in partially hemitransected rats (see above) and all rats were killed 2 or 6 hrs after the ^{3}H-GM1 injection. Number of animals in each group was 4. All rats were perfused with 100 ml Krebs-Ringer solution containing 1000 IU of heparin. Samples were taken from the skeletal muscle, the striatum and the midbrain. Furthermore, both the right and left midbrain and the right and left striatum were analyzed. The weight of the sample regions varied from 50-150 mg. These three regions were dissected out, rinsed in saline, weighed and dissolved with 1 ml protosol (NEN) at 37 ^{o}C in glass vials. Blood was collected from the right atrium, and centrifuged for plasma separation (Eppendorf minifuge, 60 sec.). Furthermore by means of metabolic cages urine was collected over a 6 hr period after the administration of ^{3}H-GM1. All biological samples were counted by liquid scintillation by adding 14 ml Beckman EP scintillation fluid to the protosol solution. A Beckman LS 1800 spectrometer was used. The experiment with ^{3}H-GM1 took place 3 days following hemitransection or the corresponding sham operation.

RESULTS

Pharmacokinetic analyses of ^{3}H-GM1 related radioactivity in intact and partially hemitransected rats

The distribution of ^{3}H-GM1 related radioactivity in plasma and urine as evaluated 2 and 6 hrs after the ^{3}H-GM1 injection are shown in fig. 1. The tritium radioactivity in plasma is significantly increased in the lesioned animals compared with the control animals as evaluated 2 hrs after the injection, while at the 6 h time interval there is no difference in ^{3}H-GM1 related radioactivity between the control and the lesioned group. At both time intervals the radioactivity in the plasma is significantly different from the zero value. In fig. 1 it is also shown that the urinary volume during the 6 hr time period is significantly reduced in the lesioned rats compared with the intact control animals. This action is probably responsible for the reduced excretion of ^{3}H-GM1 related radioactivity in urine found in the lesioned animals (fig. 1). It should also be noted that the radioactivity in plasma is substantially higher at the 6 h time interval than at the 2 h time interval. In line with these results it is also shown in fig. 2 that the radioactivity in muscle, striatum and midbrain is higher at the 6 h time interval than at the 2 h time interval. No difference is found between the lesioned and unlesioned side in the partially hemitransected rats in the striatum and in the midbrain. Of substantial interest is the observation that the radioactivity at the 6 h time interval is not higher in the striatum and midbrain of lesioned rats than that found in corresponding areas of the control rats in view of the breakdown of the blood brain barrier caused by the lesion. Also it

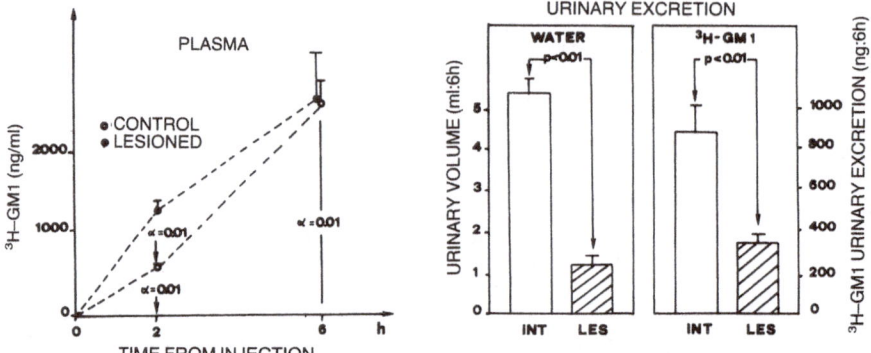

Fig. 1. The panel on the left shows the ^{3}H-GM1-related radioactivity in the plasma of control (intact) and lesioned (hemitransected) rats at various time intervals after the administration of 65 μCi/rat (about 0.9 mg/kg, i.p. dissolved in saline).
Means ± s.e.m. are given (n = 4). The statistical analysis has
been carried out by means of Dunn test for nonparametrical multiple
comparisons (α = experiment wise error rate; see Hollander and
Wolfe (1973)). The panel on the right shows the urinary volume as
well as the urinary excretion of ^{3}H-GM1-related radioactivity
during a 6 h period in control (intact) and lesioned (hemitransected) rats. Means ± s.e.m. are given (n = 4). Statistical
analysis was carried out by means of Student´s t-test.

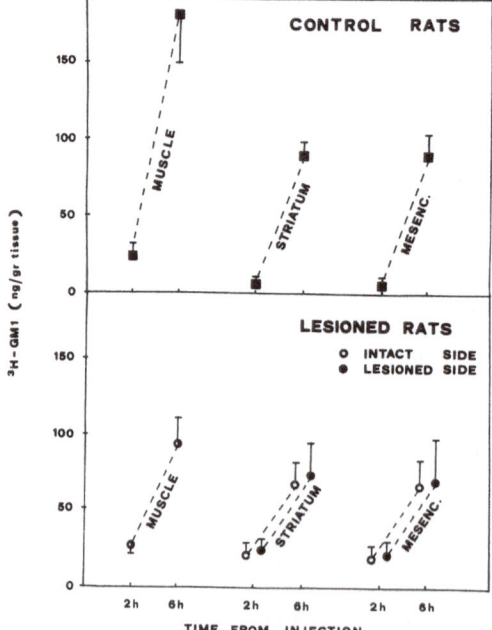

Fig. 2. Accumulation of
[3]H-GM1-related radio-
activity in various
tissues 2 and 6 h after
injections of 65 μCi/rat
(about 0.9 mg/kg, i.p.,
dissolved in saline) in
control (intact) and
lesioned (hemitransected)
rats. Means ± s.e.m. are
given (n = 4).

should be noted that in the lesioned rats the radioactivity in the
muscle tissue was not clearly higher than that found in the
striatum and in the midbrain, which is in contrast to the pattern
found in the intact rats.

Effects of chronic GM1 treatment on tyrosine hydroxylase immunoreactive dendrites and cell bodies in the substantia nigra of
experimental rats using image analysis

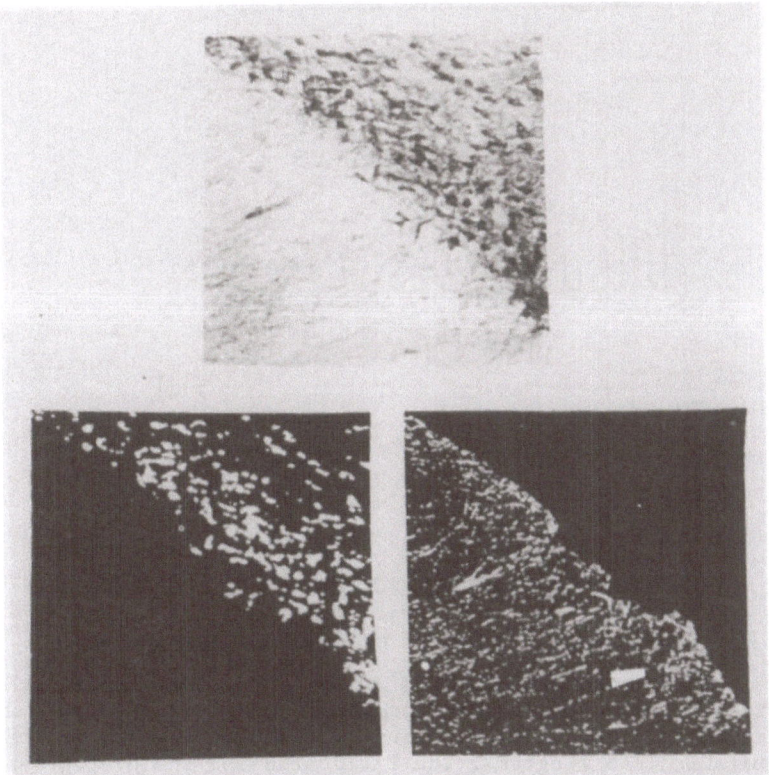

Fig. 3. The upper panel shows dopamine nerve cells in a field of
substantia nigra visualized by means of TH-immunoreactivity (PAP
technique). The lower left panel shows mainly the cell bodies
present in the pars compacta obtained from the image shown in the
upper panel by means of the IBAS system (see text). The lower
right panel shows the dendrites present in the pars reticulata
obtained from the image shown in the upper panel by means of the
IBAS system (see text).

L.F. Agnati, et al.

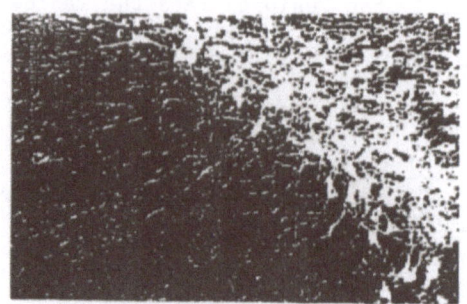

CELL BODIES

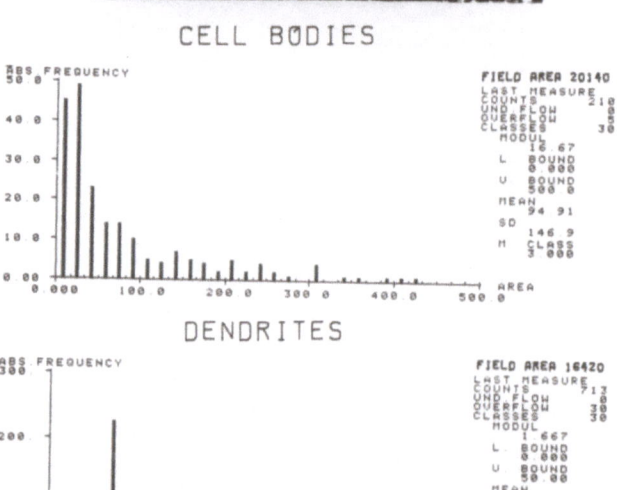

DENDRITES

Fig. 4. Morphometrical analysis of the DA nerve cells and dendrites present in the pars compacta and reticulata of sub-stantia nigra, respectively, using the procedure shown in fig. 3 and suitable IBAS programs. The field area and the histograms of the area of the single profiles have been evaluated (for further details, see text).

The present procedures to analyse dopamine cell bodies and dendrites in the substantia nigra in the IBAS Image Analyzer are illustrated in figs. 3 and 4. Firstly, the mean gray tone value of the unspecific background staining was determined by means of a gray tone histogram. The specific TH immunoreactivity was then obtained by erasing all the gray tone values equal or higher (less dark tones) than the mean gray tone value of the background

staining minus two standard deviations. Then, a binary elaboration
of the image was performed. Dopamine cell bodies and dendrites
could be selectively analyzed (see fig. 3) by removing by means of
interactive procedures either the zona reticulata or the zona
compacta. On these images the histograms of the profile areas
together with the field area for both the dopamine cell bodies in
the zona compacta and the dopamine dendrites in the zona reticulata could then be determined as shown in fig. 4. Fig. 4 exempli-
fies the elaboration carried out for the dopamine cell bodies and
dendrites in the border zone between the zona reticulata and zona

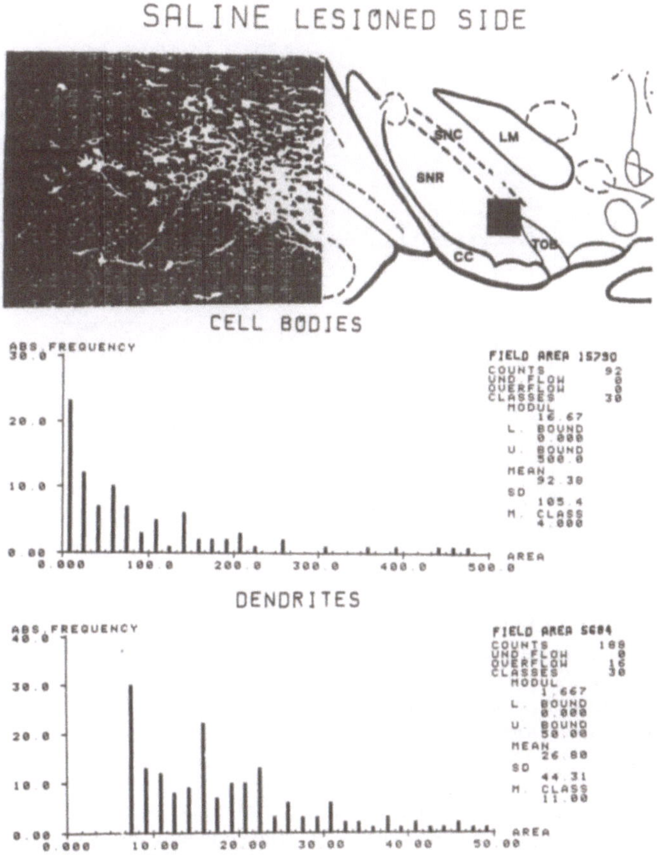

Figs. 5 and 6. Effects of saline and GM1 treatment on the TH-
immunoreactive profiles (cell bodies and dendrites) present on the
lesioned side of hemitransected rats. For further details see
legend to fig. 4 and text.

152   L.F. Agnati, et al.

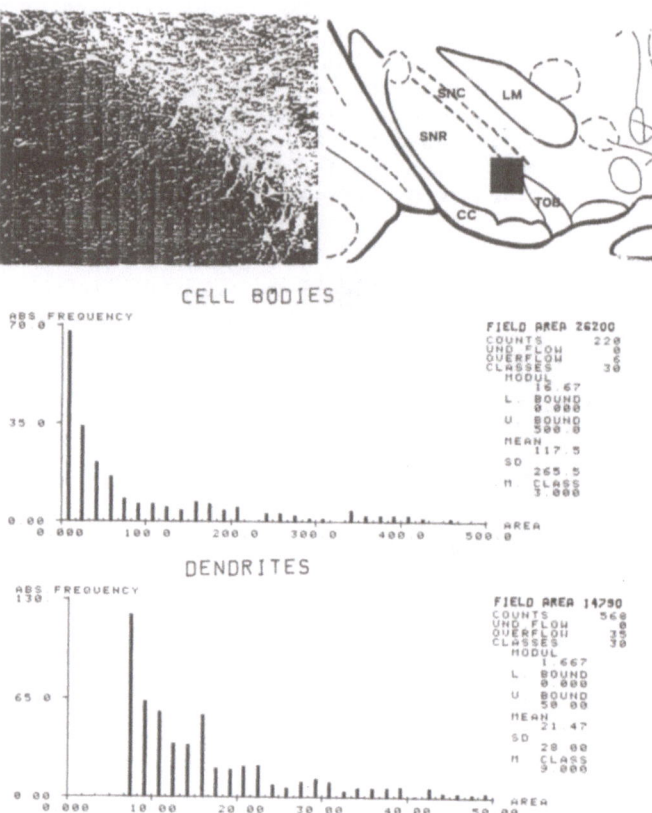

GM1 LESIONED SIDE

compacta in the medial part of the substantia nigra. The field
area gives an overall evaluation of the available TH immunoreactive surface. Thus, to study the density of TH positive neuronal
structures such as nerve cell bodies and dendrites, this parameter
can be very useful. Hence, it can be used in the analysis of the
degenerative and regenerative features of the nigral dopamine
nerve cells.

The above mentioned procedures have now been used in the
analysis of the dopamine nerve cell bodies and dendrites on the
lesioned side of partially hemitransected animals following saline
or GM1 chronic treatment. One of the most dramatic cases showing
the effects of GM1 on TH-immunoreactive nerve cells on the lesioned
side of the substantia nigra is shown in figs. 5 and 6. The field
area of the TH immunoreactive cell bodies in the zona compacta is
considerably increased (about 160 %) following GM1 treatment, when
compared with the value obtained following saline treatment. Also

the number of TH immunoreactive counts is substantially increased (about 230 %). An even greater increase (about 200 %) is observed in the field area of the TH immunoreactive dendrites in the zona reticulata following GM1 treatment when compared with the corresponding field area in saline treated animals. A larger increase (about 300 %) in the number of counts is also observed for the TH positive dendrites than for the TH immunoreactive cell bodies following chronic GM1 treatment. It can be noticed that the histogram profile of the immunoreactive areas for the cell bodies and dendrites are similar following saline and chronic GM1 treatment.

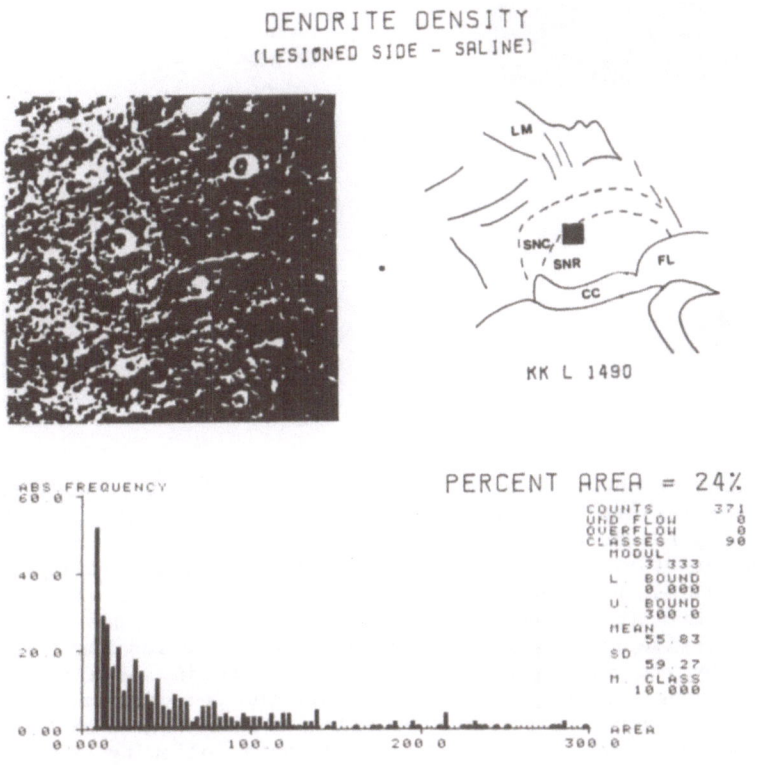

Figs. 7 and 8. Effects of saline and GM1 treatment on the TH-immunoreactive profiles (cell bodies and dendrites) present on the lesioned side of the pars compacta of the substantia nigra of hemitransected rats. Sagittal sections have been studied. In this case, since the size of the field under analysis could be different from one animal to the other the percent area has been used. Also the histogram of the area of single profiles is given. For further details, see legend to fig. 4 and text.

L.F. Agnati, et al.

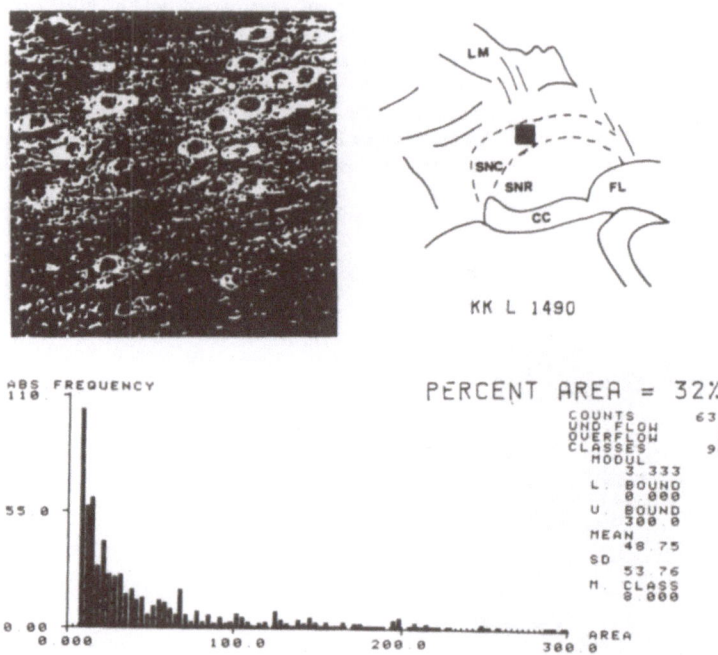

DENDRITE DENSITY

(LESIONED SIDE - GM1)

KK L 1490

ABS FREQUENCY
110

PERCENT AREA = 32%

COUNTS   631
UND FLOW   0
OVERFLOW   0
CLASSES   90
MODUL
        3.333
L. BOUND
        0.000
U. BOUND
        300.0
MEAN
        48.75
SD
        53.76
M. CLASS
        0.000

55.0

0.00
    0.000   100.0   200.0   300.0   AREA

By means of the IBAS Image Analyzer it has also been possible
to determine the dendrite density within the zona compacta itself
by encircling the dopamine cell bodies on the red, green and blue
(RGB) monitor by means of the light-pen and then initializing a
command on the menue table for their removal. Examples of the
results obtained in this analysis are shown in figs. 7 and 8. The
TH immunoreactive dendritic area in the sampled region has been
determined by the parameter "per cent area". It is shown that
chronic GM1 ganglioside treatment increases the per cent area from
24 % to 32 %. There is almost a doubling in the number of TH
immunoreactive counts within the zona compacta following chronic
GM1 treatment. The mean profile area is similar in the two experimental groups and also the histogram profile of the immunoreactive
areas is similar.

CONCLUSION

The present pharmacokinetic evidence indicates that i.p.
treatment with GM1 as performed acutely leads to an accumulation
of radioactivity in the striatum and midbrain in both intact and

in partially hemitransected rats with no differences between the
operated and unoperated side. Only for the intact animals is the
penetration of GM1 into the brain slower than into the muscle
tissue. The radioactivity measured at a 6 h time interval is
considerably higher than that found at the 2 h time interval
following the ^{3}H-GM1 injection. Thus, it seems likely that GM1 or
the GM1-metabolite can penetrate into the brain tissue and there
exert its trophic activity. The breakdown of the blood brain
barrier in the lesioned animals does not seem to be of crucial
importance, since there were no clear differences in the radioactivity found within the striatum and midbrain of the control and
lesioned animals at the 6 h time interval.

By means of the image analysis using the IBAS Image Analyzer
it has been further documented that chronic GM1 treatment can
substantially increase the density of dopamine dendrites both in
the zona compacta and in the zona reticulata of the substantia
nigra of the lesioned side. Also the number of nigral dopamine
cell bodies seems to be increased following chronic GM1 treatment
as previously described (Agnati et al. 1983 and 1984). Taken
together the present results further support our hypothesis that
GM1 via a direct action on the substantia nigra itself can increase the survival of the nigral dopamine nerve cell bodies with
a possible increase in the arborization of their dendrites following a partial unilateral hemitransection. It is also documented
that the IBAS Image Analyzer offers easily available programs to
evaluate the degenerative and regenerative processes in the nigral
dopamine cells.

ACKNOWLEDGEMENT

This work has been supported by a grant (04X-715) from the
Swedish Medical Research Council, by a grant (MH25504) from the
NIH, by a grant from Magnus Bergwall's Stiftelse, by a grant from
Knut & Alice Wallenberg´s Foundation as well as a grant from
Karolinska Institute Research Funds and by a CNR grant. For
excellent technical assistance we are grateful to Mrs. Beth
Andbjer, Mrs. Ulla-Britt Finnman, Miss Katarina Nilsson, Mrs. Siv
Nilsson, Mrs. Birgitta Nyberg, Miss Barbro Tinner, Mr. Mauro Ferri
and Mr. Carlo Brusiani. For excellent secreterial assistance we
are grateful to Mrs. Ulla-Britt Wedin and Mrs. Anne Edgren.

REFERENCES

Agnati, L.F., Fuxe, K., Calza, L., Goldstein, M., Toffano, G.,
Giardino, L. and Zoli, M. (1984). Computer assisted morphometry
and microdensitometry of transmitter identified neurons with
special reference to the mesostriatal dopamine pathway. II.
Further studies on the effects of the GM1 ganglioside on the
degenerative and regenerative features of mesostriatal dopamine
neurons. Acta Physiol Scand., Suppl. 532: 37-42.

Agnati, L.F., Fuxe, K., Calza, L., Benfenati, F., Cavicchioli, L., Toffano, G. and Goldstein, M. (1983). Gangliosides increase the survival of lesioned nigral dopamine neurons and favour the recovery of dopamnergic synaptic function in striatum of rats by collateral sprouting. Acta Physiol Scand., 199, 347-363.

Hollander, M. and Wolfe, D.A. (1973). Non-parametric statistical methods. Wiley, New York.

Markey, K. A., Kondo, S., Shenkman, L. and Goldstein, M. (1980). Purification and characterization of tyrosine hydroxylase from a clonal chromocytoma cell line. Mol Pharmacol 17, 79-85.

Sternberger, L.A. (1979). Immunocytochemistry. 2nd ed., Wiley, New York.

Toffano, G., Savoini, G., Moroni, F., Lombardi, G., Calzá, L. and Agnati, L.F. (1983). GM1 ganglioside stimulates the regeneration of dopaminergic neurons in the central nervous system. Brain Res., 261, 163-166.

DEVELOPMENT OF QUANTITATIVE METHODS FOR THE EVALUATION OF THE ENTITY OF COEXISTENCE OF NEUROACTIVE SUBSTANCES IN NERVE TERMINAL POPULATIONS IN DISCRETE AREAS OF THE CENTRAL NERVOUS SYSTEM: EVIDENCE FOR HORMONAL REGULATION OF COTRANSMISSION

KJELL FUXE[1], LUIGI F. AGNATI[2], MICHELE ZOLI[2], ANDERS HÄRFSTRAND[1], ROBERTA GRIMALDI[2], PASQUALE BERNARDI[2], MASSIMO CAMURRI[2] and MENEK GOLDSTEIN[3]

[1]Department of Histology, Karolinska Institutet, Stockholm, Sweden
[2]Department of Human Physiology, University of Modena, Modena, Italy
[3]Department of Psychiatry, New York University Medical Center, New York, USA

INTRODUCTION

We have recently developed several methods for the quantitative evaluation of coexistence of neuroactive substances in nerve cell bodies, namely the overlap method and the occlusion method (Agnati et al. 1982, Agnati et al. 1983, Fuxe et al. 1982, 1983, Agnati et al. 1984). In the overlap method the image analyzer can record the position and the perimeter of each nerve cell body visualized by means of the antiserum anti A in a Cartesian plan. Using an adjacent thin section or the same section following mild elution of anti A immunoreactivity the staining against antibody B is performed. The image analyzer then again records the position and the perimeter of the nerve cell bodies showing B immunoreactivity in the same Cartesion plan as used for the demonstration of anti A immunoreactivity. The positions in the Cartesian plan of the neurons are then compared by the computer and coexistence is considered to take place when an anti-A immunoreactive area and an anti-B immunoreactive area overlap by at least 30 %. Thus, a 30 % overlap is considered to be a threshold value below which coexistence is not considered to exist. When using thin adjacent sections a correction factor must be calculated since some A and B immunoreactive nerve cell bodies disappear as they move from one section to the other. Provided a mild elution technique can be used such as electrophoretic elution safe values of coexistence can be obtained with the overlap method. This method can also be performed using a binary coding of the images to reach the quantitative evaluation of three sets: the anti-A positive population, the anti-B positive population and the (anti-A) plus (anti-B) positive population, i.e. the unitary set. (see Jonsson this Symposium). The advantages of the overlap method is its ability to allow the assessment of coexistence of neuroactive substances in each individual nerve cell body of a nerve cell body population. (see Agnati et al. 1983a, 1984a). However, this method has the disadvantage of not being able to be used for studies on coexistence in nerve terminal populations.

We have developed another approach the so called occlusion
method, and in the present paper it will be shown that it can be
used for the quantitative determination of coexistence in nerve
terminal systems. (see figs. 2 and 7). We have previously introduced this method for determination of coexistence of neuroactive
substances in nerve cell bodies (Agnati et al. 1982, 1984). This
method gives an over all evaluation of coexistence in a nerve cell
body population. It is based on the analysis of 3 adjacent thin
sections stained in a random order with antiserum against A, B and
with antisera against both A and B. The method can be used on a
statistical basis which makes it possible also to use it for
determination of coexistence in nerve terminals. It has the
advantage that an X and Y axis can be avoided, and there is no
need for the use of elution techniques, which usually impair to
varying degree the immunoreactivity of the coexisting antigen.

MATERIAL AND METHODS

Male specific pathogen free Sprague-Dawley rats (150-200 g
b.w.) have been used. The rats were kept on a standard lighting
conditions (lights on at 6 a.m. and off at 8 p.m.) and given food
pellets and water ad libitum. To study the possible influence of
changes in pituitary adrenal activity on the coexistence of
neuropeptide Y (NPY) immunoreactive and phenyl-ethanolamine-N-
methyl transferase (PNMT) immunoreactive nerve terminal systems of
the hypothalamus and the locus coeruleus (see Results), some
animals have been adrenalectomized 14 days before sacrifice, while
others were sham operated. Half of the animals of the adrenalectomized group were treated with corticosterone immediately
following the operation (10 mg/kg; i.p.), and this group was
maintained on corticosterone substitution therapy for the entire
period of adrenalectomy (10 mg/kg; i.p. twice daily, last dose
given 2 hrs before decapitation). It has previously been shown
(Hökfelt et al. 1983) that NPY immunoreactivity is present in at
least some adrenaline cell bodies in the medulla oblongata.

To develop the present methods for the quantitative evaluation of coexistence of neuroactive substances in nerve terminal
populations immunocytochemical procedures were used. Both the
indirect immunofluorescence method and the unlabelled immunoperoxidase method, using the peroxidase antiperoxidase (PAP) complex,
were employed. In the studies on the catecholamine nerve terminal
systems well characterized tyrosine hydroxylase and PNMT antisera
were used (Markey et al. 1980, Goldstein et al. 1978). To demonstrate NPY and CCK-like immunoreactivity in the nerve terminals
also previously well characterized antibodies were used (see
Hökfelt et al. 1983, Vanderhaegen et al. 1980). These two antisera
have been shown not to cross-react with a large number of neuropeptides. CCK-like immunoreactivity exists in certain subpopulations of mesolimbic dopamine neurons (Hökfelt et al. 1980). We
have therefore in the present study also analyzed possible co-

existence of tyrosine hydroxylase and CCK-like immunoreactivity in dopamine nerve terminal populations of the caudal part of the nucleus accumbens. For details on the indirect immunofluorescence methods, see Hökfelt et al. 1975 and for details on the unlabelled immunoperoxidase method, see Sternberger (1979). The indirect immunofluorescence method was employed to analyze coexistence in synaptosomal preparations (see Results). The antisera were used in dilutions 1:100 to 1:400. To demonstrated the antigen-antibody complex the FITC conjugated sheep antirabbit immunoglobulins (diluted 1:4) was used at 37 C^o for 30 minutes. Synaptosomal preparations were analyzed in a Zeiss fluorescence microscope with an appropriate excitation (BG12) and stop filter (Schott 50) or in a large Zeiss universal microscope coupled via a TV camera with a high sensitivity to fluorescence light to an IBAS Image Analyzer. Using the unlabelled immunoperoxidase method the antisera were diluted 1:500 to 1:1000 in PBS containing 1 % goat serum and 0.3 % Triton X-100. The rats had been perfused via the ascending aorta with a calcium free Tyrode's solution, in which Ca^{++} had been replaced by Mg^{++} ions followed by perfusion with 500 ml of 4 % paraformaldehyde dissolved in 0.1M of phosphate buffer (pH 7.4)). Vibratom sections were made and 30 μm thick sections were collected in small cups containing PBS. After extensive washing the vibratom sections of the forebrain were incubated with the antisera against CCK or TH for 2 days at + 4 C^o. Frontal sections of the pons containing the locus coeruleus were incubated with antisera to NPY or PNMT or both (occlusion method) in the same way as described above. The antibody antigen complex was demonstrated with a PAP complex. In the presence of H_2O_2 the peroxidase reacted with diaminobenzidine to produce a brown insoluble compound which could easily be demonstrated in the microscope. After the staining procedures the sections were defatted and coverslipped in Entellan.

RESULTS

The present methods

    Two types of methods have been developed using the principles of the occlusion method to quantitatively evaluate coexistence of neuroactive substances in nerve terminal systems. One method is based on the use of synaptosomal preparations obtained from discrete brain areas and the other method is based on vibratom sections showing immunoreactivity. These preparations were analyzed by means of an image analyzer (IBAS Image Analyzer, Zeiss Kontron). This latter method has the advantage of allowing determination of coexistence of nerve terminals in discrete nuclei or subnuclei of the brain such as the locus coeruleus or parts of it, while the first method has the advantage of allowing an over all evaluation of coexistence in nerve terminals in large parts of e.g. the cerebral cortex.

K. Fuxe, et al.

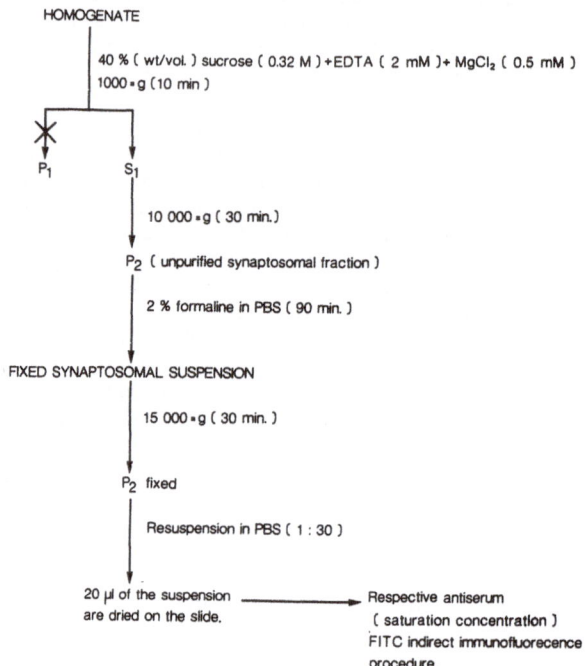

Fig. 1. Schematic illustration of the procedure to prepare
synaptosomes for the studies on coexistence using the occlusion
method. S1 is the first supernatant obtained. P_2 is the unpuri-
fied synaptosomal fraction, which was fixed in 2 per cent
formaline dissolved in phosphate buffered saline. For other
details, see text.

The synaptosomal method for the quantitative evaluation of
coexistence in nerve terminals

The biochemical procedure of preparing the synaptosomes for
immunocytochemical studies is shown in fig. 1 and summarized in
the following text.

1. Preparation of the brain homogenate.

Principally the homogenization was performed as described by
Whittaker and de Robertis and collegues (Whittaker 1959, de
Robertis et al. 1961 (homogenization in 0.32 M sucrose). It was
found that optimal results were obtained in the presence of EDTA 2
mM and $MgCl_2$ 0.5 mM. Without the addition of $MgCl_2$ various types
of neuropeptide immunoreactive synaptosomes could only be weakly

demonstrated in the immunocytochemical producedure. Depending upon the amount of brain substance to be used a normal potter (>150 mg of substance) or a small potter was used in order to obtain optimal results.

2. Fixation procedure

As seen in fig. 1. optimal results were obtained, when resuspending the partially purified synaptosomal fraction for 90 min in 2 % formaldehyde, dissolved in phosphate buffered saline . Small variations in the concentration of formaline was not found to be of importance for the demonstration of immunoreactivity in the nerve terminal systems. No satisfactory results were obtained when the fixation instead was performed at the last stage of the preparation of the synaptosomes, that is after drying the synaptosomal preparation on the slide. Also note that satistactory results were not obtained when formaline fixation was performed on the first supernatant (S1). Thus, immediate fixation of the supernatant (S1) led to marked increases in unspecific staining, and PNMT immunoreactivity could no longer be detectable with this technique. In contrast, tyrosine hydroxylase immunoreactivity appeared to be slightly stronger within the synaptosomes, when using an immediate fixation technique.

3. Choice of staining procedure

Both the indirect immunofluorescence procedure, using FITC or rhodamineisothiocyanate conjugated immunoglobulins, and the indirect immunoperoxidase procedure of Sternberger (1979) were used. The best results were obtained by using the FITC conjugated immunoglobulins, which allowed a clearcut distinction between specific and unspecific immunoreactivity. Thus, when using FITC labelled antibodies, the UV illumination will produce a rapid disappearance of the specific immunofluorescence. Furthermore, the specific and unspecific staining can be easily distinguished. It should also be pointed out that the specificity of the immunoreactivity was tested by incubating the synaptosomal suspension with preadsorbed antisera, using a neuropeptide concentration of 50 μg/ml. With this concentration of the neuropeptide it was found that the CCK immunoreactive and NPY immunoreactive dots could no longer be demonstrated.

Another problem to consider in the immunohistochemical procedure is to use saturating concentrations of the respective antiserum in order to obtain as true value as possible of the entity of coexistence.

EVALUATION OF THE COEXISTENCE IN SYNAPTOSOMAL PREPARATION

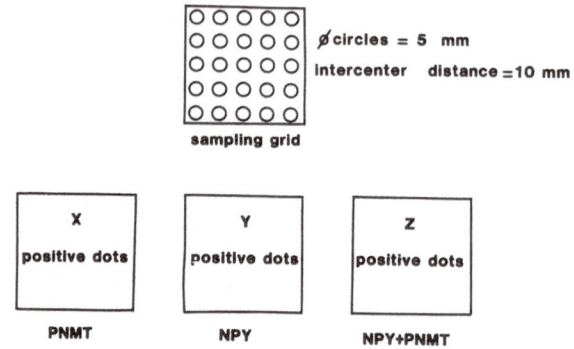

X + Y - Z = NPY-PNMT coexisting dots

unspecific dots/grid : 2-5 %

Fig. 2. Schematic illustration of the sampling procedure to study the coexistence in synaptosomal preparations using the occlusion method. By means of the sampling grid it is possible to obtain a systematic sampling of the immunoreactive dots present in the smear of the synaptosomal preparation. Ten samples were taken from each smear preparation present on the object glass. The principle of the occlusion method is illustrated. X equals the number of PNMT positive dots; Y equals the number of NPY positive dots; and Z represents the number of immunoreactive dots obtained, when the smear has been incubated with the antibodies against both NPY and PNMT. The number of coexisting dots is obtained by the formula in the figure. The number of unspecific dots has also been evaluated by this sampling procedure without incubation with a primary antiserum. The unspecific number of dots was found to be 2-5 per cent of the total number of positive dots. This sampling procedure has been performed on photomicrographs taken from the smear preparation. The magnification was 400 x.

4. Evaluation of coexistence in synaptosomal preparations

The evaluation procedure is illustrated in fig. 2. The fluorescence microphotographs are taken with a 25 times objective and the final magnification of the prints is 400 times. In each slide 10 fields are randomly selected in order to obtain a correct value for each slide on the number of immunoreactive profiles present. In each experimental group 4 slides were evaluated and the mean ± s.e.m. calculated for the number of immunoreactive

profiles. In the evaluation of the prints for immunoreactive
profiles a sampling grid was used (see fig. 2) with a circle
diameter of 5 mm and an intercenter distance of 10 mm. The number
of unspecific dots/grid was in the order of 2-5 % of the total
number of dots.

We have recently been able to measure the number of immunoreactive dots directly in the fluorescence microscope by means of
a TV camera highly sensitive to fluorescence light. The TV camera
was coupled to the IBAS Image Analyzer. It was found that following about 3 min of UV irradiation there was a marked disappearance
in the fluorescence intensity of the strong specific FITC immunofluorescence, while the weak unspecific fluorescence had completely disappeared. Furthermore, a few strongly autofluorescence
dots remained with an intact strong fluorescence intensity. Due to
the high sensitivity of the TV camera to the fluorescence light
the dots with the specific immunofluorescence could still be
clearly demonstrated within the image. Under these circumstances
it therefore became possible to obtain an index of the number of
specific immunoreactive nerve terminals by simply measuring the
number of dots within the sampled image using the measuring
programs in the IBAS Image Analyzer.

5. Experimental studies

The occlusion method was used to study coexistence of neuroactive substances in synaptosomal preparations as well as the
effects of adrenalectomy on coexistence of PNMT and NPY immunoreactivity in synaptosomal preparations of the hypothalamus.

The results are summarized in figs. 3, 4 and 5. In the intact
sham operated male rat the PNMT and NPY coexistence taken in per
cent of the total number of PNMT immunoreactive profiles is in the
order of 30 %. Following 14 days of adrenalectomy the coexistence
of PNMT and NPY is markedly increased to around 90 %. This change
is highly significant ($p < 0.01$, Mann-Whitney U-test). Furthermore,
when the restitution therapy is performed in the adrenalectomized
male rats (10 mg/kg i.p. twice daily, last injection 2 hrs before
killing) the coexistence of PNMT and NPY immunoreactivity is
markedly reduced and almost abolished. As also seen in the figs.
3, 4 and 5 analysis of the fluorescence also showed a marked
increase in the number of NPY immunoreactive profiles following
adrenalectomy. These results indicate that adrenalectomy can
increase the synthesis of NPY-like peptides in the nerve terminals
of the hypothalamus leading, as shown by the occlusion method, to
a significant increase in the degree PNMT and NPY coexistence. It
seems likely that the possible stimulation of NPY synthesis by
adrenalectomy takes place to a large extent within the PNMT
immunoreactive neurons in view of the marked increase in PNMT and
NPY coexistence. One mechanism underlying the possible stimulation
of NPY synthesis following adrenalectomy may be the marked increase of ACTH secretion and/or the disappearance of adreno-

K. Fuxe, et al.

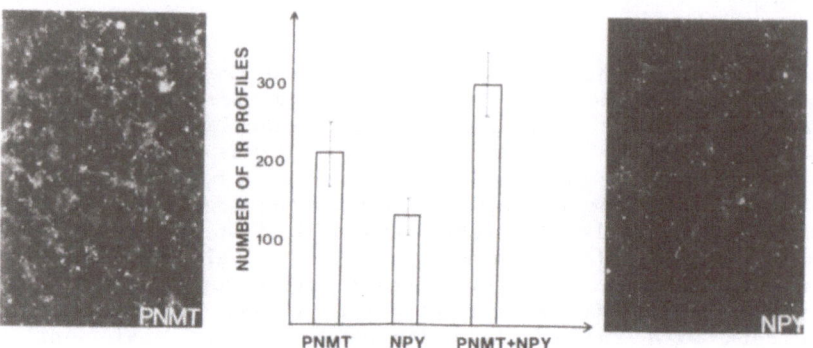

HYPOTHALAMUS

SHAM OPERATION

PNMT-NPY coexistence (% of PNMT)=29.3 ± 2.4

Fig. 3. Application of the occlusion method to study the coexistence in synaptosomal preparations. In this case the evaluation has been performed on PNMT and NPY coexistence in hypothalamic synaptosomal preparations. Using the occlusion method the PNMT and NPY coexistence is shown to be around 30 per cent expressed in per cent of the number of PNMT immunoreactive dots. Two representative microphotographs are shown in the figure.

cortical steroids. In line with this view it was found that high doses of corticosterone abolished the PNMT and NPY coexistence. This treatment will of course in addition to activation of corticosterone receptors also markedly reduce the secretion of ACTH. The importance of corticosterone in regulating NPY synthesis is also supported by our recent observations of the existence of glucocorticoid receptor immunoreactive nerve cell bodies within the areas of the adrenaline cell groups Cl and C2 of the rostral medulla oblongata (see Fuxe et al. 1984). Thus, the possibility should be considered that there exist glucocorticoid receptors within some adrenaline nerve cell bodies <u>i.a.</u> regulating the synthesis of NPY. In agreement with this view are the observations shown in fig. 6. By means of the image analyzer it can be demonstrated that the NPY immunoreactive area within the nerve cell population is significantly increased compared with sham operated controls. These analyses were performed within the adrenaline cell group Cl.

HYPOTHALAMUS

ADRENALECTOMY

PNMT-NPY coexistence (%of PNMT)= 89.5 : 10.5

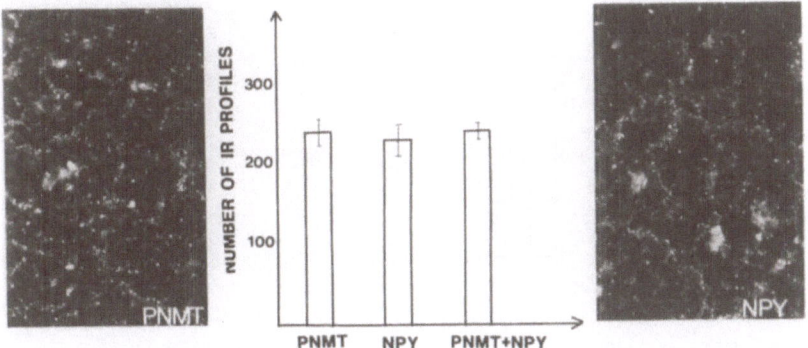

Fig. 4. PNMT and NPY coexistence has been studied in synaptosomal preparations of the hypothalamus of the adrenalectomized rat. The coexistence is shown to be around 90 per cent. The microphotographs illustrate the substantial increase in the number of NPY immunoreactive dots found in the hypothalamic synaptosomal preparations following adrenalectomy. This observation is quantitatively shown in the mid panel of the figure.

HYPOTHALAMUS

ADRENALECTOMY + CORTICOSTERONE TREATMENT

PNMT-NPY coexistence (%of PNMT)=0%

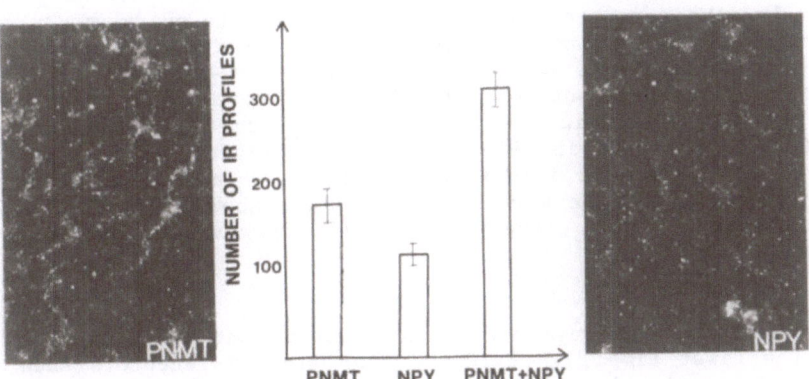

Fig. 5. PNMT and NPY coexistence is shown in synaptosomal preparations of the hypothalamus of the adrenalectomized rats treated chronically with corticosterone. For further details and treatment, see text. The coexistence are now shown to be around zero per cent.

166   K. Fuxe, et al.

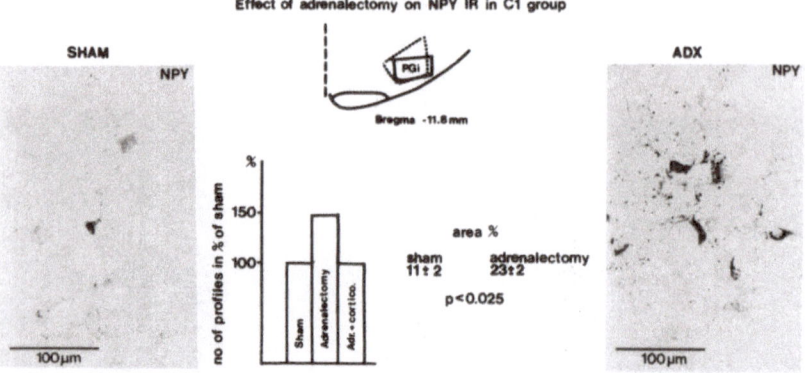

Fig. 6. Effects of adrenalectomy on the NPY-like immunoreactivity
in nerve cell bodies present within the area of the adrenaline
group C_1 of the rostral part of the medulla oblongata (see
schematic drawing in the middle panel). The schematic drawing
represents the ventral half of a frontal section of the medulla
oblongata at the bregma level indicated. TGI = nucleus para
giganto cellularis reticularis. The number of NPY immunoreactive
profiles are reported in per cent of the sham operated control
group. Following adrenalectomy there is approximately a 50 per
cent increase in the number of the immunoreactive cell bodies.
Also an overall evaluation of the NPY immunoreactivity present in
the nerve cell bodies has been performed using the IBAS Image
Analyzer. After a discrimination procedure making it possible to
remove in an objective way the unspecific background staining (see
fig. 7), the parameter "area per cent" was chosen to study the
amount of NPY-like immunoreactivity present in the shamoperated or
adrenalectomized rat. Using the Mann-Whitney U-test it was found
that the "area per cent" was significantly increased by about 100%
in the adrenalectomized group compared with the sham-operated
group. The increase in NPY-like immunoreactivity observed within
the C_1 group area is illustrated in the left and right panel of
the figure, where the microphotographs are shown of the NPY-like
immunoreactivity in the sham and adrenalectomized group
respectively.

     The above findings are of substantial interest, since they
for the first time give evidence that steroid receptors such as
the glucocorticoid receptors may modulate chemical neurotransmission by preferentially producing an inhibition of the comodulator transmission line, in this case NPY. In this way the
glucocorticoid receptors make possible a marked increase in
synaptic plasticity. By reducing activity in the comodulator
transmission line, different types of adrenaline postsynaptic

responses can be modulated such as the decoding mechanisms at the α-2 adrenergic receptors, since NPY has been shown to increase the density and to reduce the affinity in α-2 adrenergic receptors in membrane preparations of the medulla oblongata (Agnati et al. 1983b). Thus, the hormone modulation of brain function may in part take place by altering the entity of coexistence of neuroactive substances in certain groups of neurons: the pattern of release of multiple messangers within the individual synapses is thus changed, leading to a change in function and thus in synaptic plasticity.

II. Development of procedures for the use of the occlusion method to quantitatively determine the coexistence of neuroactive substances in nerve terminals in sections

Again it should be emphasized that the occlusion method can be used to study the entity of coexistence in nerve terminals, since it is a statistical procedure, in which a random sequence of 3 thin adjacent sections can be studied to reveal e.g. tyrosine hydroxylase, cholecystokinin and tyrosine hydroxylase plus cholecystokinin immunoreactivity. The various steps of the present procedures and assumptions are shown in fig. 7. The analysis is made possible by the use of the IBAS Image Analyzer. The analysis starts by defining the background by initiating the program for the gray tone histogram. The image is then discriminated by using the mean gray tone level of the background minus 2 standard deviation. In this way a specific area of cholecystokinin, tyrosine hydroxylase and cholecystokinin plus tyrosine hydroxylase immunoreactivity can be assessed in the 3 adjacent sections. The area of the single terminal can then be determined by analysing the histogram of the diameter of the various single objects making up the specific immunoreactive area in the respective sections. The minimum value can be assumed to represent the diameter of a single terminal. The minimal value in the diameter of the single terminal was 1.5 mm as seen in fig. 9. As seen in fig. 7 the number of terminals with coexistence can then be obtained by dividing the immunoreactive area for CCK plus tyrosine hydroxylase immunoreactivity with the minimum value of the area of the objects making up the specific area of immunoreactivity. The per cent of coexistence is obtained by expressing the number of terminals in which coexistence is present in per cent of the number of tyrosine hydroxylase immunoreactive terminals.

This new procedure has been used to study the entity of coexistence of CCK and tyrosine hydroxylase immunoreactivity within the caudal and medioventral part of the nucleus accumbens in the male rat and also to analyse the entity of coexistence of NPY and PNMT immunoreactivity in nerve terminals of locus coeruleus in sham operated and adrenalectomized rats. It is seen in fig. 8 that the entity of coexistence within the selected area of the caudal nucleus accumbens is in the order of 80 %. It will now

K. Fuxe, et al.

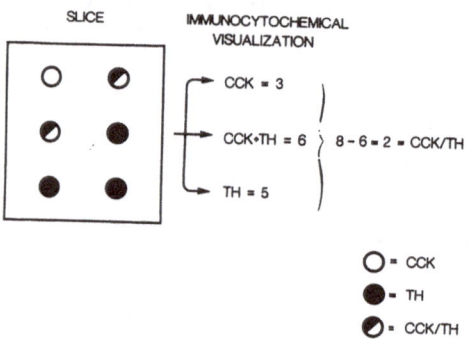

The occlusion for the quantitative determination of
coexistence in terminals

Steps of the procedure and assumptions

1. Background   : Graytone histogram   ($\bar{X}$: SD)
2. Discrimination by using the graytone level   ($\bar{X}$-2 SD)
3. Assessment of the specific area for CCK   (SA_{CCK})
4. Assessment of the specific area for TH   (SA_{TH})
5. Assessment of the specific area for CCK・TH   ($SA_{CCK・TH}$)
6. Evaluation of the specific   area of distribution
   of the single objects :
   The minimum value is the area of the single terminal   (A_T)

$$SA_{CCK}・SA_{TH}-SA_{CCK・TH}=SA_{CCK/TH}$$

$$SA_{CCK/TH} / A_T = N_{CCK/TH} = \text{Number of terminals with coexistence}$$

$$SA_{TH} / A_T = N_{TH} = \text{Number of TH positive terminals}$$

$$(N_{CCK/TH} / N_{TH})・100 = \text{Percent of coexistence}$$

SLICE   IMMUNOCYTOCHEMICAL
                    VISUALIZATION

CCK = 3

CCK・TH = 6   }   8 - 6 = 2 = CCK/TH

TH = 5

○ = CCK
● = TH
◑ = CCK/TH

Fig. 7. Schematic illustration of the procedures used to determine
the coexistence in nerve terminal networks present in sections by
means of the occlusion method. SA = the size of the specific area
evaluated by means of the IBAS Image Analyzer. The specific area
has been obtained for TH, CCK and CCK + TH-like immunoreactivity.
A_T = area of the single terminal obtained by means of the
histogram of the single objects present in the image. The
principles of the occlusion method is also indicated in the lower
part of the figure.

be of substantial interest to analyse, if this coexistence can be
modulated by antidepressant and neuroleptic treatment in order to
further understand the mechanism of action of these drugs. It has
recently been shown by Agnati and collegues (1984b) that the
entity of coexistence within the dorsal part of the caudal nucleus
accumbens is reduced in the aged brain from 45 to 25 %.

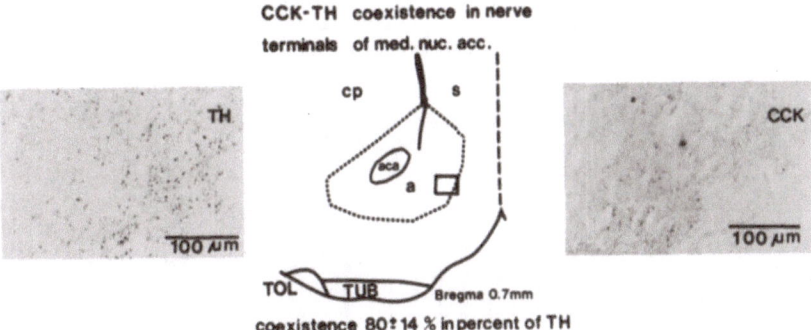

Fig. 8. Evaluation of the entity of CCK-TH coexistence in nerve
terminals of the medial part of the caudal nucleus accumbens, (see
schematical illustration in the middle part of the figure). Half
of a the frontal section is shown at the rostrocaudal level Bregma
= 0.7 mm. CP = nucleus caudatus putamen, SL = nucleus septalis
lateralis, a = nucleus accumbens, TUB = tuberculum olfactorium,
TOL = tractus olfactorius lateralis, ACA = anterior limb of the
commissura anterior. Rectangle shows the size of the sampled
region. The two microphotographs showing TH and CCK-like immunoreactivity from this region are present in the left and right part
of the figure. In this part of the nucleus accumbens the entity of
coexistence of TH and CCK-like immunoreactivity are shown to be
around 80 per cent using the IBAS Image Analyzer as shown in the
previous figure.

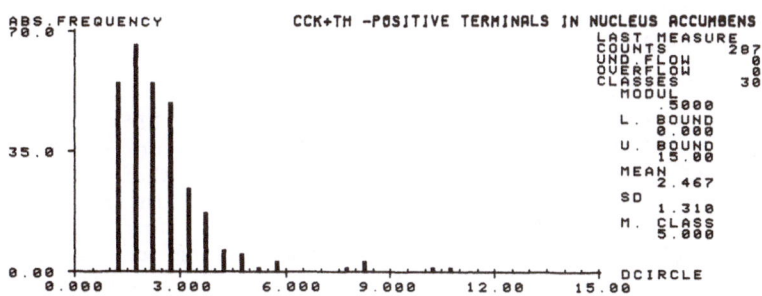

Fig. 9. Example of the histogram of the diameters of the immunoreactive nerve terminals present in the sampled regions of the
nucleus accumbens following incubation with antibodies against
both CCK and tyrosine hydroxylase. The first class of the
histogram gives the minimum value of the positive objects and
gives the diameter of the single terminal, which is 1.5 µm.

K. Fuxe, et al.

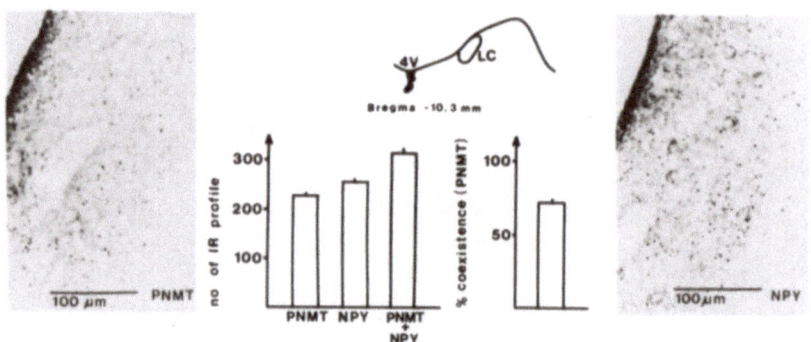

Fig. 10. Evaluation of NPY and PNMT coexistence in nerve terminals of the locus coeruleus using the occlusion method by means of the IBAS Image Analyzer. The locus coeruleus has been studied at the rostrocaudal level indicated in the midpanel part of the figure, showing the dorsal half of a frontal section of the pons at bregma level -10.3 mm. LC = locus coeruleus. 4V = fourth ventricle. In the left part of the middle panel the number of PNMT, NPY and PNMT + NPY immunoreactive profiles are shown. Means ± s.e.m. n= 3. In the right part of the middle panel the per cent coexistence is illustrated and shown to be in the order of 70 %. The coexistence is expressed in per cent of the number of PNMT immunoreactive profiles. Two microphotographs exemplifying the distribution of PNMT and NPY immunoreactive dots are shown in the left and right part of the figure.

The results obtained in the analyses of NPY and PNMT co-existence within nerve terminals of the locus coeruleus are summarized in fig. 10, 11 and 12. It is shown that the entity of coexistence of PNMT and NPY immunoreactivity is in the order of 70 and 80 % in sham operated and adrenalectomized rats, respectively. Thus, in contrast to the hypothalamus using the synaptosomal method, there exists in the sham operated animal a high degree of coexistence of NPY immunoreactivity in the locus coeruleus PNMT immunoreactive terminals. This may be part of the reason, why there is no significant increase in the entity of coexistence following adrenalectomy in the locus coeruleus in contrast to the case of the hypothalamus, where adrenalectomy markedly increases the entity of NPY and PNMT coexistence in synaptosomes. These results open up the possibility that the glucocorticoid receptors may exert a more powerful inhibitory control of NPY synthesis in the adrenaline cell groups projecting into the hypothalamus

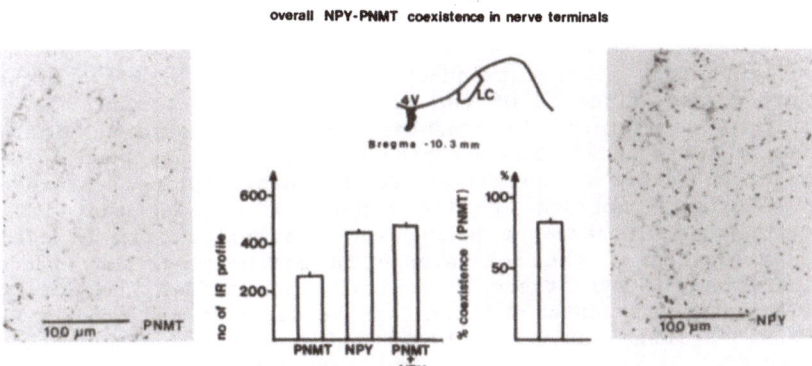

Fig. 11. NPY and PNMT coexistence is shown within nerve terminals of the locus coeruleus of the adrenalectomized male rat, using the occlusion method with the help of the IBAS Image Analyzer. For details, see text to figure 10. It is shown that adrenalectomy does not modify the entity of coexistence of PNMT and NPY within the nerve terminals of the locus coeruleus.

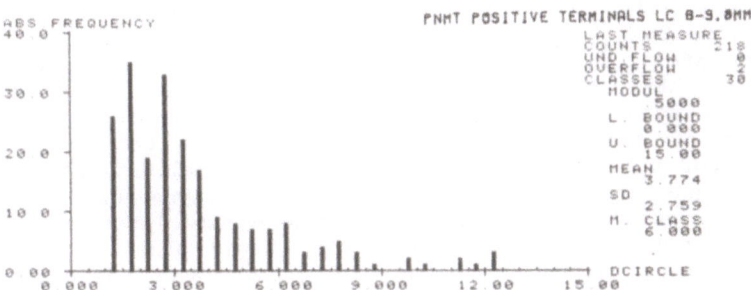

Fig. 12. Histogram of the diameters of the PNMT immunoreactive nerve terminals within the locus coeruleus of the sham operated rat. For further details, see text to fig. 9. Minimum value of the diameter of the single terminal as shown in the histogram is again 1.5 μm, giving the size of the individual nerve terminal.

(mainly the C1 group) than in those projecting into the locus
coeruleus (mainly the C2 group).

SUMMARY

The present paper introduces for the first time two methods,
based on the principle of the occlusion method to determine the
entity of coexistence in synaptosomes and in nerve terminals of
discrete brain areas. The synaptosomal approach has been used for
the demonstration of catecholamine and neuropeptide costorage in
nerve terminal networks in the hypothalamus using the FITC indirect immunofluorescence procedure. A sampling grid is used in
the evaluation of the coexistence in microphotographs, randomly
taken from the synaptosomal smear on each slide. The number of
immunoreactive synaptosomes can also be directly measured by means
of an IBAS Image Analyzer. The second method is based on the use
of the IBAS Image Analyzer to quantitatively determine coexistence
in nerve terminals of discrete nuclei present in sections using
again the principles of the occlusion method. By means of the IBAS
Image Analyzer it is possible to assess the specific area of
immunoreactivity of the respective antigen after a discrimination
procedure performed in a highly standardized way. The area of the
single terminal is obtained from the histogram of the single
objects building up the specific immunoreactive area. Thus, the
area of the single terminal is regarded to be the minimal value in
the histogram. Using these two new approaches it has been possible
to begin to analyze the cotransmission processes in neurons, in
this case the role of NPY in the adrenaline neurons. It could be
demonstrated by the use of the synaptosomal method that adrenalectomy markedly increases the entity of coexistence of PNMT and
NPY within PNMT immunoreactive nerve terminals, a response which
is completely abolished by restitution therapy with corticosterone. These results open up the possibility that steroid
receptors such as glucocorticoid receptors may make possible
synaptic plasticity by preferentially reducing activity in the
comodulator transmission line, leading to a qualitatively different response in the hypothalamic adrenaline synapses, since the
pattern of activity in the multiple transmission lines is changed.
Such a functional synaptic plasticity makes possible long-term
changes in the neuronal networks of hypothalamus without alterations in the synaptic homeostasis, controlling adrenaline
release.

ACKNOWLEDGEMENTS

This work has been supported by a Grant (MH 25504-10) from the
NIH, by a Grant (04X-715) from the Swedish Medical Research
Council and by a CNR International grant. For excellent technical
assistance we are grateful to Miss Barbro Tinner and for excellent
secreterial assistance we are grateful to Mrs. Anne Edgren and
Mrs. Ulla-Britt Wedin.

REFERENCES

Agnati, L.F., Fuxe, K., Locatelli, V., Benfenati, F., Zini, I., Panerai, A.E., El Etreby, M.F., Hökfelt, T. (1982). Neuroanatomical methods for the quantitative evaluation of coexistence of transmitters in nerve cells. Analysis of the ACTH- and beta-endorphin immunoreactive nerve cell bodies of the mediobasal hypothalamus of the rat. J Neurosci Methods 5, 203–214.

Agnati, L.F., Fuxe, K., Benfenati, F., Calza, L., Battistini, N., Ögren, S.O. (1983a). Receptor-receptor interactions: possible new mechanisms for the action of some antidepressant drugs. In: Frontiers in Neuropsychiatric Research. (eds. E. Usdin, M. Goldstein, A.J. Friedhoff & A. Georgotas). MacMillan Press.

Agnati, L.F., Fuxe, K., Benfenati, F., Battistini, N., Härfstrand, A., Tatemoto, K., Hökfelt, T., Mutt, V. (1983b). Neuropeptide Y in vitro selectively increases the number of $\alpha 2$-adrenergic binding sites in membranes of the medulla oblongata of the rat. Acta Physiol Scand., 118, 293–295.

Agnati, L.F., Fuxe, K., Benfenati, F., Zini, I., Zoli, M., Fabbri, L., Härfstrand, A. (1984a). Computer assisted morphometry and microdensitometry of transmitter-identified neurons with special reference to the mesostriatal dopamine pathway. I. Methodological aspects. Acta Physiol Scand., Suppl. 532, 5–32.

Agnati, L.F., Fuxe, K., Giardino, L., Calza, L., Zoli, M., Battistini, N., Benfenati, F., Vanderhaeghen, J.J., Guidolin, D., Ruggeri, M., Goldstein, M. (1984b). Evidence for cholecystokinin-dopamine receptor interactions in the central nervous system of the dult and old rat. Studies on their functional meaning. New York Academy paper, in press.

Fuxe, K., Agnati, L.F., Ganten, D., Lang, R., Calza, E., Poulsen, K., Infantellina, F. (1982). Morphometrical evaluation of the coexistence of renin-like and oxytocin-like immunoreactivity in nerve cells of the paraventricular hypothalamic nucleus of the rat. Neurosci Lett 33, 19–24.

Fuxe, K., Agnati, L.F., Andersson, K., Calza, L., Benfenati, F., Zini, I., Battistini, N., Köhler, C., Ögren, S.O., Hökfelt, T. (1983). Analysis of transmitter-identified neurons by morphometry and quantitative microfluorimetry. Evaluation of the actions of psychoactive drugs, especially sulpiride. In: Special aspects of psychopharmacology (eds. M. Ackenheil & N. Matussek), pp. 13–32, Espansion scientifique francaise, Paris.

K. Fuxe, et al.

Fuxe, K., Gustafsson, J-Å., Agnati, L.F., Härfstrand, A., Yu, Z.Y., Wikström, A.C., Wrange, Ö, Zoli, M. (1984). Mapping out of glucocorticoid receptor immunoreactive neurons in the rat tel-and diencephalon using a monoclonal antibody against rat liver gluco- corticoid receptor. Endocrinology, submitted.

Goldstein, M., Lew, J.Y., Matsumoto, Y., Hökfelt, T., Fuxe, K. (1978). Localization and function of PNMT in the central nervous system. In: Psychopharmacology: A Generation of Progress (eds. M.A. Lipton, A. DiMascio & K.F. Killam. Raven Press, New York.

Hökfelt, T., Lundberg, J.M., Tatemoto, K., Mutt, V., Terenius, L., Polak, J., Elde, R., Goldstein, M.. (1983). Neuropeptide Y (NPY)- and FMRF-amide neuropeptide-like immunoreactivities in cate- cholamine neurons of the rat medulla oblongata. Acta Physiol Scand 117, 315-318.

Hökfelt, T., Fuxe, K., Goldstein, M. (1975) Applications of immunohistochemistry to studies on monoamine cell systems with special reference to nervous tissues. Ann NY Acad Sci 254, 407-432.

Hökfelt, T., Skirboll, L., Rehfeld, J.F., Goldstein, M., Markey, K., Dann, O. (1980). A subpopulation of mesencephalin dopamine neurons projecting to limbic areas contains a cholecystokinin-like peptide: Evidence from immunohistochemistry combined with retrograde tracing. Neurosci 5, 2093-2124.

Markey, K., Kondo, S., Shenkmann, L., Goldstein, M. (1980). Purification and characterization of tyrosine hydroxylase from a clonal chromocytoma cell line. Mol Pharmacol 17, 79-85.

De Robertis E., Pellegrino de Iraldi, A., Rodrigues de Lores Arnaiz, G., Salganicoff L. (1969). Electron microscope obser- vations on nerve endings isolated from rat brain. Anat Rec 139, 220.

Sternberger, L.A. (1979). Immunocytochemistry. 2nd ed., Wiley, New York.

Vanderhaeghen, J.J., Lotstra, F., De Mey, J., Gilles, C. (1980). Immunohistochemical localization of cholecystokinin and gastrin- like peptide in the brain and hypophysis of the rat. Proc. Nat. Acad. Sci., USA 77, 1190-1194.

Whittaker, V.P. (1959). The isolation and characterization of acetyl containing particles from brain. Biochem J 12, 694-706.

PRINCIPLES FOR THE CONSTRUCTION OF THE SOFTWARE FOR IMAGE ANALYSIS OF TRANSMITTER–IDENTIFIED NEURONS

P.L. FABBRI[1], L.F. AGNATI[2], K. FUXE[3], N. BATTISTINI[2], I. ZINI[2] and M. ZOLI[2]

[1]Instruments Centre, University of Modena, Modena, Italy
[2]Department of Human Physiology, University of Modena, Modena, Italy
[3]Department of Histology, Karolinska Institutet, Stockholm, Sweden

Quantitative analysis of the morphofunctional characteristics of neuronal networks is now possible using semiautomatic and automatic image analyzers. In the present paper we describe this quantitative approach with special reference to the software package developed on an assembled image analyzer based on a stand alone general purpose microcomputer. The hardware configuration is shown together with the basic and specific software used for receptor autoradiographic and morphometric analysis of transmitter–identified neurons.

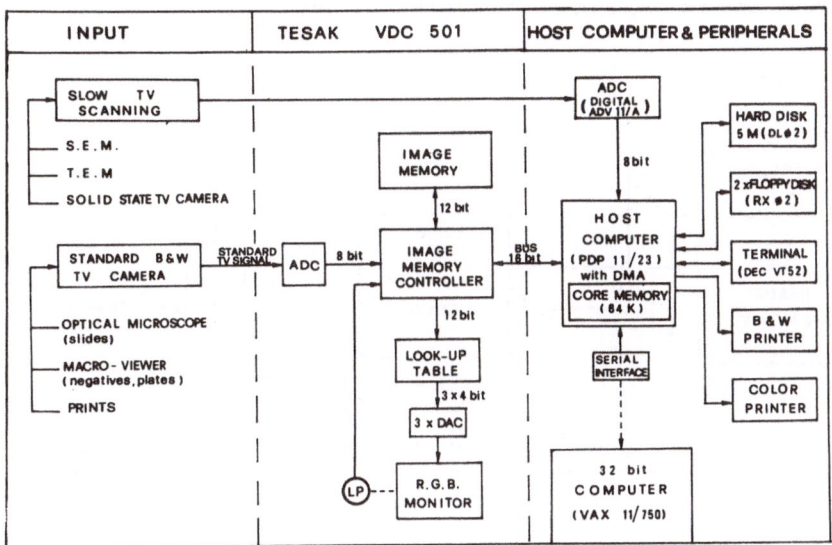

Fig. 1 Hardware configuration of the system.

The choice to use an assembled graphic computer instead of a commercially available image analyzer has been made initially by cost reasons. Later on we have realized that, after some initial troubles in writing basic programs and routines, it became easier and easier to modify and assemble them in different ways to find the best method to solve specific problems.

HARDWARE SPECIFICATIONS

The hardware configuration of the system shown in fig. 1 is divided into three main blocks: input, image memory and image memory controller, (host computer). The heart of the system is the image memory and the controller. This part, produced by an Italian factory, is called VDC 501 and, in the following, we will refer to it with the abbreviation VDC. It is simply a dedicated micro computer with a large RAM memory and some inherent microcoded programs, which enable it to execute fast speed simple elementary operations on the image memory like reading or writing a single pixel or vector, codifying colours, and displaying images on a TV monitor. The VDC is linked, through a parallel interface with the so called host computer, which supervises the whole system and drives the VDC to execute elementary or complex operations on the image memory. The organization of the image memory can be thought as 12 planes of 512 x 512 elementary units called "pixels". Each column of that three-dimensional structure represents a 12 bits value for the selected pixel information. As it will be seen later the first 8 bits are reserved for the intensity and the remaining four bits are available for overlaying. Also included in the VDC are the electronic parts needed for image

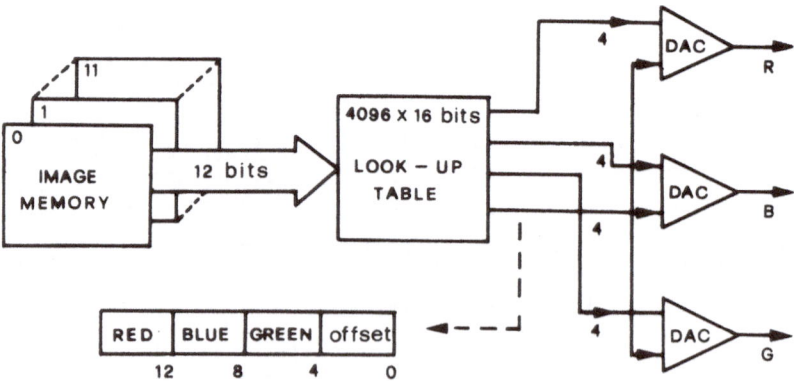

Fig. 2 Look-up table coding and colour display.

acquisition from a standard TV camera, image display and light pen control. During acquisition, the signal coming from the TV camera is converted in binary form by an internal 8 bits ADC (Analog to Digital Converter) and written in the image memory at standard TV frequency. At the same time and at the same frequency, the display electronics scans the image memory pixel by pixel and drive the monitor after the look-up table coding. In fig. 2 the look-up table (LKT) is schematically shown as a memory of 4096 locations of 16 bits. Each 16 bits location contains the three basic colour contri-

THE LOAD PROGRAM

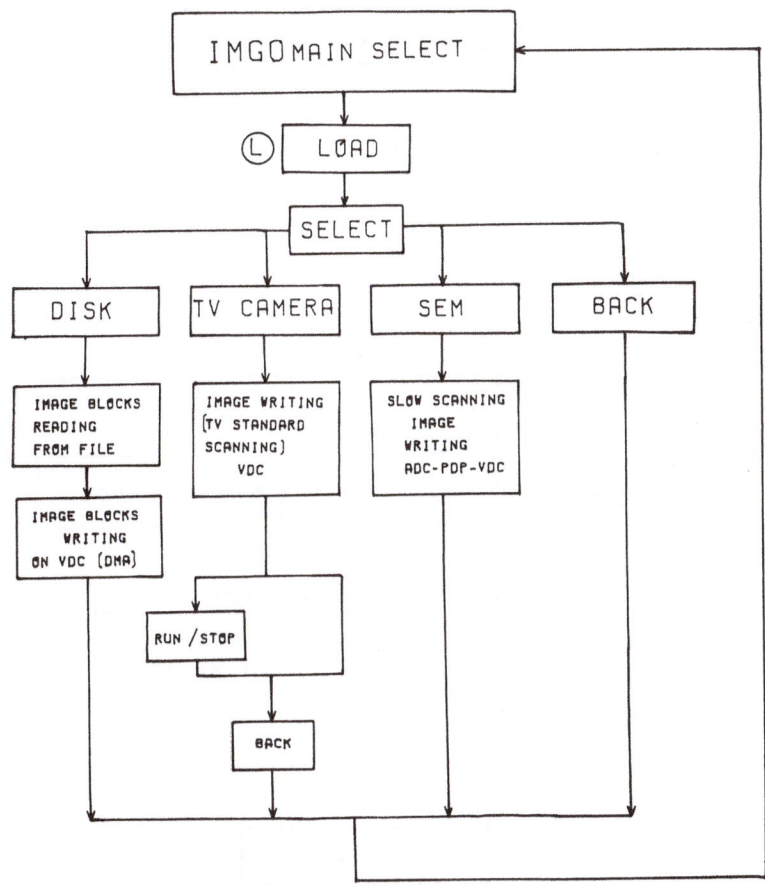

Fig. 3 LOAD-program flow-chart.

butions and a common offset.There is a point to point correspondence between the possible pixel intensities and the LKT addresses; so each pixel intensity addresses one and only one LKT location, which might have been previously programmed to obtain the desired colour or gray tone. In this way 4096 different colours and 256 gray tones can be obtained. Image acquisition can also be made through a slow scanning device such as an Electronic Microscope (both Scanning and Scanning-Transmission Microscope) or a solid state digital scanning camera. This type of input has been implemented on our system through another ADC converter directly connected to the computer bus. The acquisition is driven by a specific routine and the scan time may be selected by software. Slow scan acquisition is surely

$$I_{xy} = f(T_{xy}, X, Y, I_o, G, K)$$

X, Y = pixel coordinates

T_{xy} = transmissivity of pixel X, Y

I_o = incident light intensity

G = gain of imaging system

K = system offset

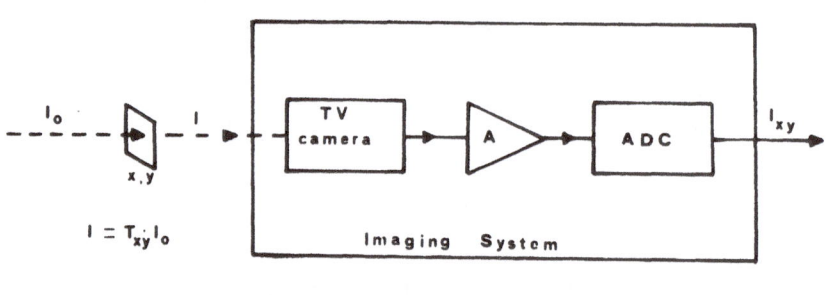

$$I_{xy} = G \cdot I + K$$

after shadow correction :

$$I_{xy} = T_{xy} \cdot I_o \cdot G + K$$

Fig. 4 Imaging system and transfer function.

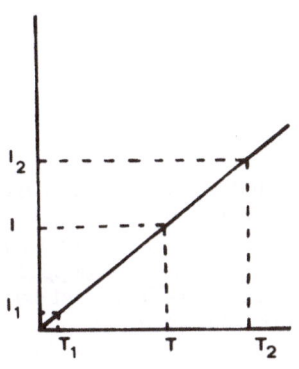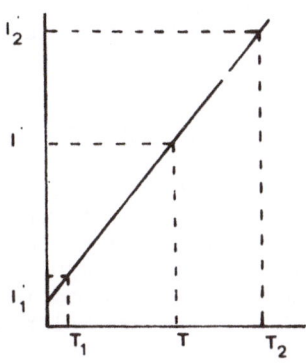

$$T_1, T_2 = \text{internal standards}$$

Standard conditions :

$$I = T \cdot I_o \, G \cdot K$$

Different conditions :

$$I' = T \cdot I_o' \, G \cdot K'$$

Linear Transform calculation :

$$I = A \cdot I' + B \qquad \text{1)}$$

$$I_1 = A \cdot I_1' + B \qquad \text{2)}$$

$$I_2 = A \cdot I_2' + B \qquad \text{3)}$$

from 2) and 3) :

$$A = \frac{I_2 - I_1}{I_2' - I_1'} \qquad B = I_1 - A \cdot I_1'$$

eq. 1) must be applied for each pixel to transform the whole image.

Fig. 5 Gray tone internal standards and linear transform.

the best choice because now high performance digital solid state
cameras are available with spacial and gray tone resolution comparable with a good microdensitometer.

SOFTWARE SPECIFICATIONS

The host computer is a general purpose DIGITAL PDP 11/23 with
RT-11 Operating System with Fortran and Assembler compilers. It is
a very flexible well diffused micro computer which can be used also
like a stand alone computer. A series of graphic routines supplied
with the VDC are grouped in a library which can be referenced by
both Fortran and Assembler programs. The basic structure of the
software we have written is of the easy to use "menu" type with several select levels linked together. The main program to which
other programs or routines are linked is called IMGO, and informs
the user about the available functions at any time and about which
character must be typed to execute one of them. Some of those functions may lead to another select sublevel and so on. Typing the L
character, for example, the LOAD program is started and the LOAD
select level is presented on the terminal where a new series of
options is available. Any option corresponds to an acquisition routine from the specified source as is shown in fig. 3.

In the case one had to analyze a series of autoradiograms, an
absolute scales is needed by which the measured intensity of a point
or a selected area depends only by its optical density. This is not
usually true as is shown in fig. 4; in fact the first step of the
autoradiographic image processing must be a SHADOW correction. Such
a program is available on the main select level and its working
principle is to create a correction matrix after a previous blank
acquisition and then to use it to correct each other image acquired
in the same background conditions. After such a correction the
measured intensity does not depend directly on the coordinates
but only on the transfer function of the imaging system as is shown
in fig. 4. Usually this transfer function is linear or can be easily
corrected to be linear. The next step is to compensate the changes
of that transfer function and incident light intensity which might
occur during acquisition of a large number of images. One can overcome that problem by using two internal standards to correct each
possible shift of the imaging system. The complete procedure is
shown in fig. 5. The intensities of the two internal standards are
measured once and for all under the so called reference conditions;
then two values are used, together with the new values obtained in
the working conditions, to calculate the two coefficients by which,
through a linear transform, the whole image can be corrected. The
result of the NORMAL correction is a new image which is the same as
if it had been acquired under the previously fixed reference conditions. In this way intensity variations between images are related

THE DENSIT PROGRAM

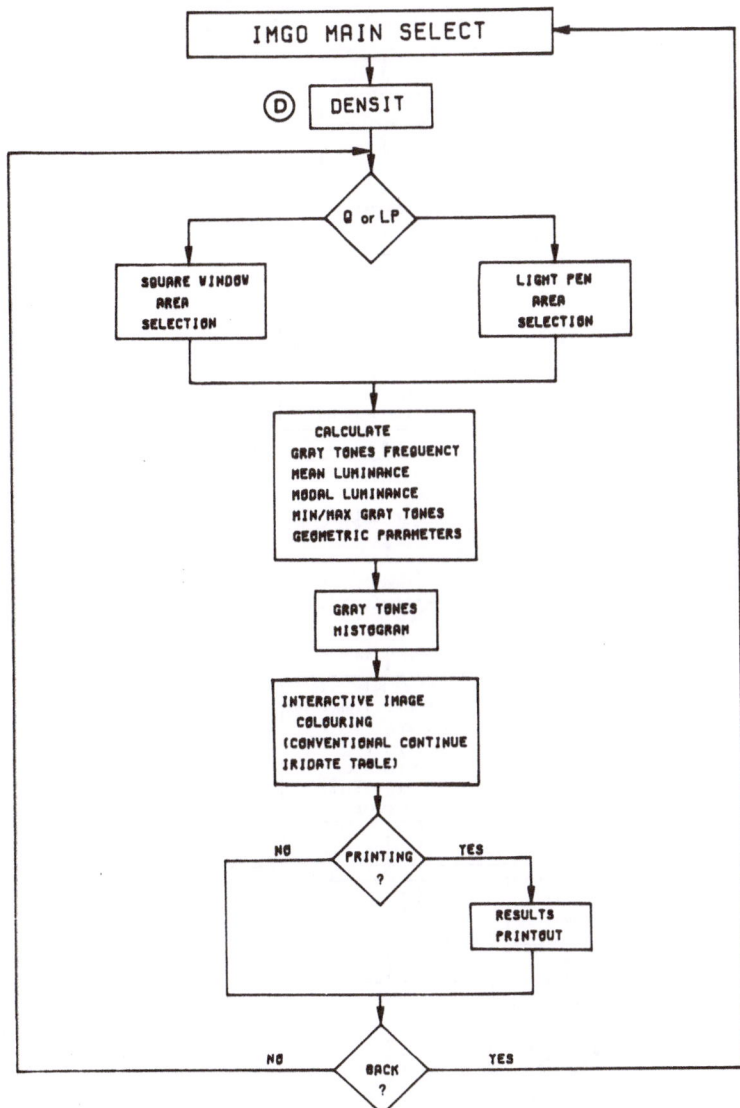

Fig. 6 DENSIT program flow-chart.

THE MØRPHØ PROGRAM

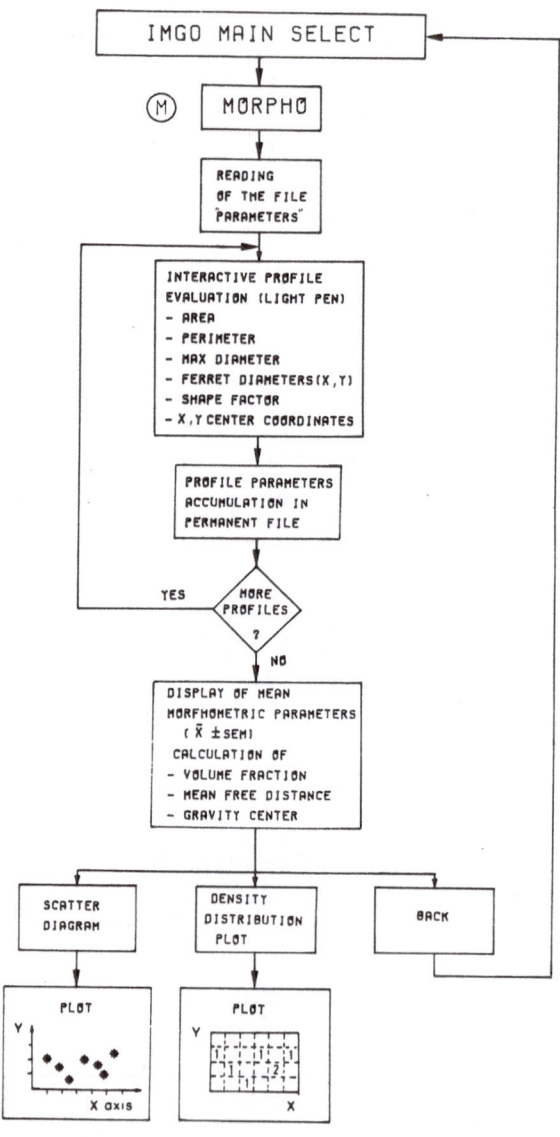

Fig. 7 MORPHO programs flow-chart.

only to optical density changes so that one is able to perform
qualitative and quantitative measurements on a large number of related autoradiograms without any disturbance from a possible shift
in the imaging system. For a quantitative evaluation a DENSIT program is available at the main select level giving the possibility
to measure, in a selected area, a number of densitometric parameters
as shown in fig. 6. The measured data can be stored and used to obtain through a calibration curve, the amount of specific ally bound
radio ligand per mm2. Once the intensity is directly related to the
amount bound false colour coding can be used to enhance contrast.
Then a colour scale can be generated and stored in which colours are
directly related to the concentration of locally bound radio ligand
allowing a simple and fast quantitative comparison between a series
of autoradiograms. This program is the COLOR program, which can be
selected from the main select level. The colour scale can be interactively created with few very simple commands and can be stored for
subsequent use. Otherwise a previously created colour scale can be
read from a disk and written into the look-up table. A reset function
tion writing in the look-up table, a 256 gray tone scale, is also
available.

A morphometric program called MORPHO has also been implemented
and can be used to study transmitter identified neuronal systems.
After the reference axis together with unit and scale factor have
been fixed and stored in a permanent file, this program can be
started. At the beginning a series of options are available, among
which there is a MANUAL function giving the possibility to select a
feature with the light pen and to achieve, through a calculation
routine, a series of morphometric parameters as it is shown in fig.
7. These parameters are stored and, when the whole cell group has
been measured, they can be used to calculate group parameters as
gravity center, mean free distance and so on. Scatter diagrams or
density distribution plots can also be obtained on the monitor and
printed on the graphic printer. When features are easily discriminated with respect to the background, it is possible to measure them in
automatic mode using the AUTO function. In this mode, after two gray
tone thresholds have been fixed, a specific routine automatically
contours features, count them, calculate individual parameters, and
store data in a permanent file. An automatic classification between
four previously defined classes is also possible.

Only the main parts of the software package have been described
in this paper. Also the main program IMGO is completely modular and
can be easily expanded with other special routines or programs to
fit new and specific requirements.

ACKNOWLEDGEMENTS

This work has been supplied by a grant from the NIH. For excellent secreterial assistance we are grateful to Mrs Gun Hultgren.

REFERENCES

Agnati L.F., Fuxe K., Locatelli V., Benfenati F., Zini I.,
Panerai A.E., El Etreby M.F. and Hökfelt T., (1982): J. Neurosci.
Meth., 5: 203-214.

Agnati L.F., Fuxe F., Zini I., Benfenati F., Hökfelt T. and
de Mey J., (1982): J. Neurosci. Meth., 6: 157-167.

Agnati L.F., Fuxe K., Hökfelt T., Benfenati F., Calzà L.,
Johansson O. and de Mey J., (1982): Brain Res. Bull., 9:45-51.

Agnati L.F., Fuxe K., Calzà L., Hökfelt T., Johansson O., Benfenati
F. and Goldstein M., (1982): Brain Res. Bull., 9: 53-60.

Agnati L.F.,Fuxe K., Calzà L., Benfenati F., Battistini N., Zini I.,
Fabbri P.L. and Goldstein M, (1983): In: Excitotoxins. (eds. K.Fuxe,
P. Roberts and R. Schwarcz). MacMillan Press, London.

IMAGE UNDERSTANDING AND THE CELL WORLD MODEL

L.W. TUCKER*, H.J. CORNEJO** and D.J. REIS

Laboratory of Neurobiology, Cornell University Medical Center, 411 East 69th St, New York, NY 10021, USA

INTRODUCTION

In recent years, image processing by computer has become increasingly important for a wide variety of fields ranging from analysis of aerial photography, industrial part inspection, and medical diagnosis, to robotics and the study of vision itself (IEEE, 1983). Substantial technological advances have also increased both the sensitivity, resolution and speed of detectors, and the computing power of the host systems. These accomplishments have been applied to the field of cytology (Preston, 1980) with the development of automated cell scanners (chiefly for blood cell analysis). For biomedical research, however, the real promise of computer vision systems—that is, the automated analysis of cells in stained tissue—has yet to be met.

In this paper we propose to examine briefly the application of computer vision to cytology and neuroanatomy. We describe a system currently under development which applies techniques from the field of Artificial Intelligence to the analysis of stained neurons in rat brain slices, and discuss directions for further research.

ROLE OF SEGMENTATION IN IMAGE UNDERSTANDING

The paradigm followed by a typical image analysis system is shown in Figure 1. A scanning device attached to a microscope first digitizes the image, making it suitable for processing. The system then segments the picture into regions which are subsequently recognized and analyzed according to a particular image domain. It is this latter phase, the division of the image into connected regions, that is perhaps the most challenging and crucial step in the analysis. An incorrect segmentation at this point might render any further analysis nonsensical and make image-understanding impossible.

---
*Supported by PHS Grant HL-18974
**Supported by PHS Grant HL-07378

L.W. Tucker, H.J. Cornejo and D.J. Reis

This scene-analysis paradigm is implemented for each application using those image enhancement, processing and segmentation techniques presumed most appropriate for the problem at hand. In automated blood cell scanning systems—such as Hematrak, Diff3, ADC-500, etc.—cells are differentiated from the surrounding background by techniques dependent upon the selection of an appropriate gray value threshold. Further analysis, identification, and classification are based upon the comparison of a number of parameters (absorbance at a specific wavelength, size, etc.) with standard values (Preston, 1980).

The relative success of commercially-available cell scanners is due in part to the histological characteristics of cell smears—relatively clear backgrounds with more or less sparse distribution of cells—which greatly simplify the initial segmentation into background and cell regions, thus allowing the system to concentrate on the problem of recognition and classification of the identified objects. In addition, as systems dedicated to one application, these scanners are often non-programmable and use specialized hardware to improve their operational speeds (Preston, 1980). All of this contributes to their reliable and accurate performance within their area of application, at speeds surpassing those of the human operator, trading flexibility and adaptability for efficiency.

In the case of tissue slices, however, the vision problem is complicated by uneven staining, cell texture, an often confusing background and an abundance of fragments. Segmentation using simple thresholding techniques, local homogeneity, or edges, may yield a number of regions of interest which

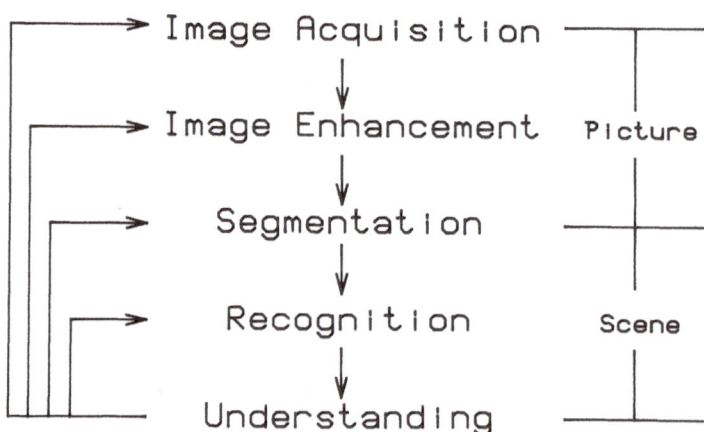

Figure 1.   Stages in image understanding.   Image information is acquired through a process of digitization and enhancement, and is then partitioned into a number of primitive image segments. These segments are combined to form objects that are matched against model object descriptions, thereby yielding an understanding of the scene. The system described here feeds back information from this last step to control each of these processes in a knowledge-driven manner.

cannot be directly matched to the target objects. Furthermore, segmentation depends greatly on choosing the correct threshold for the particular algorithm employed; an incorrect choice will often merge two or more individual objects in the image or split a textured object into an unintelligible array of small fragments.

Most image analysis systems in biomedical research laboratories therefore operate in a user-interactive mode, shifting the burden of the initial segmentation onto the user (Lindsay, 1977). While these systems are designed to make this interactive "image editing" process as "user-friendly" as possible, they leave the basic vision problem unsolved.

INTERPRETATION-GUIDED SEGMENTATION

An alternative to the sequential process shown in Figure 1 is that of interpretation-guided segmentation. Stored knowledge about the image domain being analyzed is used at the time of segmentation to produce regions that are meaningful to the task at hand. Tenenbaum and Barrow (1977) proposed such an integration of segmentation and interpretation in the analysis of outdoor scenes. In their scheme, a model of the object to be segmented (an air compressor) was used to guide the merging of picture elements (pixels) into regions. In much the same way, the characteristics of cellular parts and their topological relationships may be used to direct the segmentation of cell scenes. As the system learns more about the scene in question, the segmentation of the scene can be successively improved. This is the basic premise of the vision system described here.

The advantages offered by this approach are several. The total computational effort required to solve a particular scene-analysis problem is significantly reduced by restricting the application of computationally expensive image operators to only those areas of the picture that need it. In addition, the incorporation of scene-level information into the region-formation process should both improve the accuracy of the final segmentation, and increase the system's adaptability and flexibility. Finally, the use of knowledge at this level of the image understanding process allows the system to probe the environment for image detail instead of passively analyzing massive amounts of irrelevant data.

THE QUADTREE DATA STRUCTURE

When approaching the problem in this way, however, one is immediately struck by the immense difficulty of finding an object, whose size, shape and color may vary, somewhere within an image space represented by a two-dimensional array of 250,000 pixels (for a 512x512 picture). The size of the image space becomes an increasing problem as the decision-making process for each point becomes more involved. Some mechanism is needed to drastically reduce the size of the image space in such a way that it may be explored without the loss of picture detail. The quadtree data structure—a hierarchical organization of the image—provides such a mechanism.

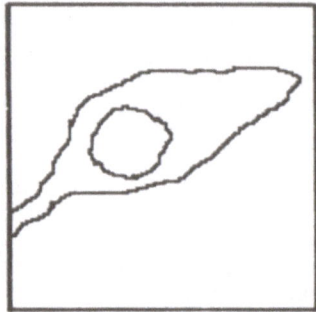 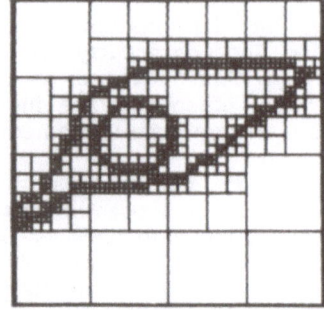

Figure 2. Quadtree segmentation of a neuron. Operator-segmented image showing background, cytoplasm and nucleus regions (left) and its corresponding quadtree tiling (right). Large areas have been represented using only a few large tiles, while a greater number of smaller tiles were required for border resolution.

For the purpose of this paper, a quadtree image may be considered to be a recursive subdivision of the image space into regularly shaped blocks of "tiles." Each such tile covers an area which is in some sense homogeneous (Klinger and Dyer, 1976) (Figure 2). Connected regions of similarly labeled tiles segment the image into discrete objects of interest. Using the quadtree representation, large areas of the picture may be represented using only a few such tiles. For the cell images studied here, many fewer tiles than pixels are required. This structure can therefore dramatically reduce the amount of computer storage and processing time needed for images of moderate complexity (Dyer, 1982), thus making it possible to consider segmentation schemes involving a more complex decision-making process.

SEGMENTATION BY SUCCESSIVE APPROXIMATION

The principal advantage of the quadtree data structure for the system described here is its ability to approximate the segmentation of an image. Thus, a system may approach the segmentation problem as one to be solved by successive approximation. Starting with an initial rough segmentation and interpretation of the scene, the quadtree is iteratively refined as the scene is better understood (Tucker, 1984).

This process is illustrated in Figure 3 for an image of a single neuron. The upper panels show the changing segmentation of the image as it is refined. Starting with the approximate region that most likely contains a cell, the system searches within the region's interior for the cell's nucleus, while at the same time refining the border. The lower panels show the corresponding quadtree image representation as it is altered by the refinement process. Notice that only along the cell's borders are tiles split into smaller components. Thus, for this image, many fewer tiles (850) were required than the pixels (16384) making up the original image. This greatly reduced the complexity of the image and the processing time required.

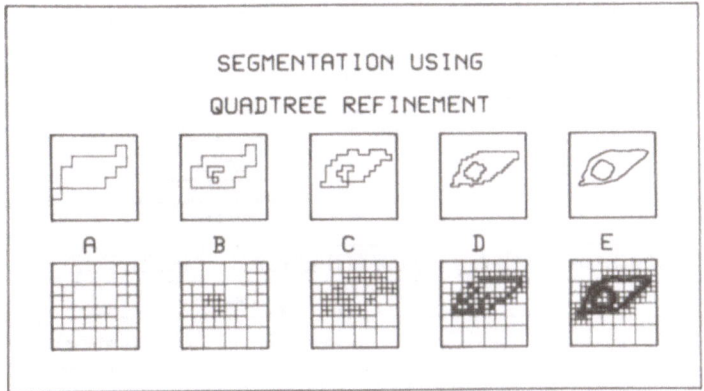

Figure 3. Successive refinement of a neuron. As the quadtree level is increased from 3 to 7, the corresponding segmentation was successively improved (top panels). At the same time, the number of tiles increases from 34 (A) to 850 (E).

REPRESENTATION OF CELL WORLD KNOWLEDGE

A central issue for artificial intelligence is how knowledge is represented and used to solve problems. In computer vision, knowledge is represented by a model of the "world" relevant to the scene being analyzed. This model is expressed as a set of formalized definitions, rules and relations. "Scene analysis" may be thus defined as the assignment of labels (nucleus, cytoplasm, etc.) according to this world model.

Neuroanatomical images, the immediate domain of interest for the system described here, are ideal subjects for this type of analysis. Basically two-dimensional, they have a number of relational characteristics which constrain the label assignment problem. These relationships between cellular components provide a natural decomposition of objects into parts. The scene is thus assumed to be divided into regions representing cells and background, while each cell is further decomposable into its component cytoplasm and nucleus (Figure 4). At the same time, this decomposition suggests rules for locating objects (Ballard 1982), such as "search for a nucleus within a region of cytoplasm."

Regions are matched with model elements according to their size, color or shape but must also be assigned in such a way that adjacent regions are not given incompatible labels (e.g., nucleus surrounded by background) (Barrow and Popplestone, 1971). The context therefore restricts the assignment of labels so that the system arrives at a coherent scene interpretation. As implemented by this system, a hierarchical graph-like structure (related to the representation of knowledge by frames (Minsky, 1975)) provides descriptive information about the objects in a uniform manner.

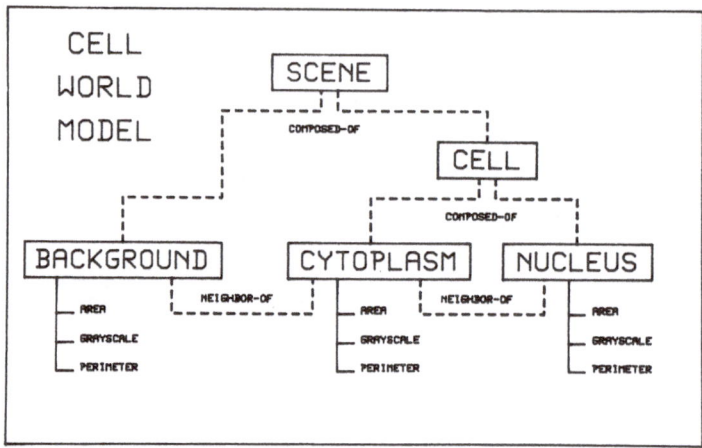

Figure 4. Hierarchical representation of knowledge into a model representing the cell world image domain. Both factual and relational characteristics are incorporated into the knowledge base.

CELL WORLD EXPERTS

As suggested earlier, a system may also use knowledge of the object domain to control the actual segmentation of the scene. This use of high-level information driving low-level processes has appeared in the VISIONS system of Hanson and Riseman (1978), and in a system for locating objects in response to user queries by Ballard and Brown (1982). Drawing upon image processing techniques appropriate for the segmentation of cell images, the system described here incorporates this procedural knowledge in the form of a number of semi-autonomous model-dependent agents or "experts" processes. These experts work together to simultaneously search for and recognize objects in the scene. Experts created for each model element, are implemented as a series of production rules (Shortliffe, 1976) in the following form:

IF (the situation is true) THEN (perform this action).

Examples of simplified rules to guide the segmentation of a cellular image might be:

IF (Cell N is missing a nucleus)
THEN (Search for a nucleus within the cytoplasm)

IF (Cell N has more than a single nucleus region)
THEN (Suppress all but the most likely nucleus region)

IF (Cytoplasm has a low resolution boundary)
THEN (Refine the boundary at a higher resolution)

The application of these rules by experts change not only the scene interpretation, but also the image segmentation itself in an iterative cycle until the picture is sufficiently understood. System control is exercised along the lines dictated by the cell world hierarchy: each expert is responsible for the execution of its subordinates. Sitting at the top of the hierarchy, the scene expert initiates and supervises the entire process, providing a top-down review mechanism for the system as a whole.

SYSTEM INTEGRATION

As shown in Figure 5, each of these components—the input image, quadtree segmentation, production-rule experts, cell world knowledge, and the emerging scene description—are required for the implementation of this computer vision system. Not shown is a user interface which permits the investigator to intervene so that new procedures may be tested, statistics gathered, and a record kept of the steps taken by the experts during execution. The system has been implemented on a PDP-11/45 minicomputer (Digital Equipment Corp.) interfaced to an Eyecom II image processing subsystem (LogE/Spatial Data, Inc.) under the UNIX (Bell Laboratories) operating system.

EXPERIMENTAL RESULTS

Experiments were performed using six images of single neurons immunostained for the catecholamine-synthesizing enzyme tyrosine hydroxylase (Benno et al., 1982). In each case, segmentation using a simple threshold or local region-homogeneity criterion resulted in a large number of regions that could not be immediately identified as corresponding to cellular parts. An operator-directed segmentation was performed for each cell image, and statistics gathered to set the parameters of the cell world model.

The system was presented with each original cell image and a record was made of the steps taken by the system during processing. At various points in the execution, copies were made of the image quadtree, representing the current state of the scene segmentation and interpretation. An example of the successive refinement of one of these cells is given in Figure 6. The input image is shown in Figure 6a and for comparison purposes the corresponding operator-segmented image in Figure 6b. It is worth noting that the user, in segmenting the image, did so using his knowledge of neurons and ignored much of the surrounding background.

The system started searching for regions that matched the expected objects at an intermediate level in the quadtree (16x16 tiles). Initially, the SCENE expert sampled the image to find likely tiles for BACKGROUND or CELL material (Figure 6c). The formation of connected components permitted the elimination of CYTOPLASM regions that were isolated or too small to be considered part of legitimate cells (Figure 6e). Once the region was accepted as containing a potential CELL, the evaluation by a CELL expert led to a search for the NUCLEUS within. Concurrently, the CYTOPLASM/BACKGROUND border was examined at a greater resolution

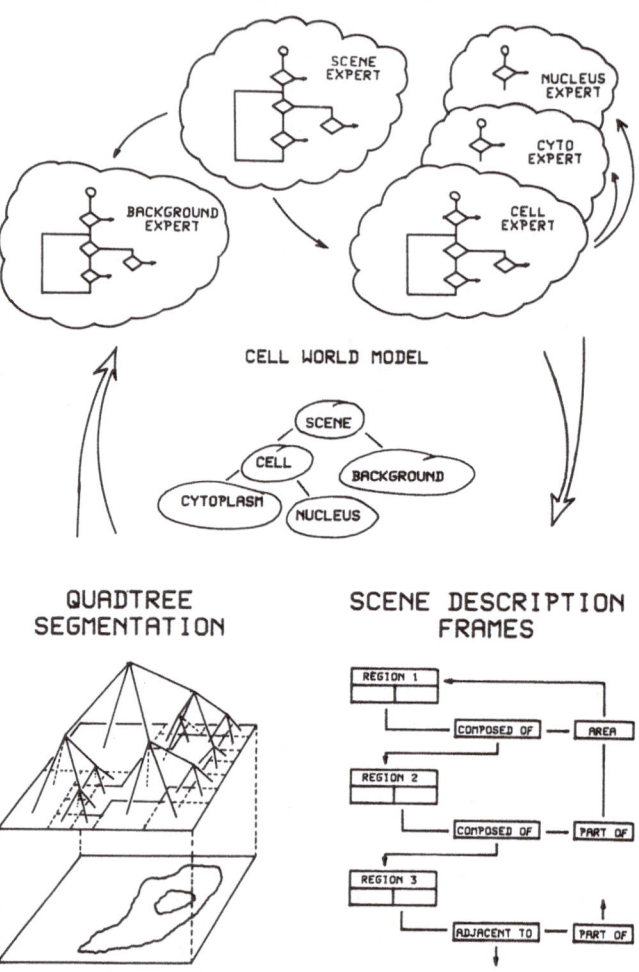

Figure 5. Representational schemes employed by the knowledge-driven vision system. Production-rule experts are derived from knowledge of the CELL world model. These experts guide the segmentation using quadtree tiling of the associated image, while scene-description relationships and information are maintained using frames (Minsky, 1975).

and shifted accordingly (Figure 6f). This border refinement process continued, resulting in the final picture (Figure 6g). Including an incorrectly indentified piece of CYTOPLASM, 96% of the picture matched that of the operator-segmented image. For illustrative purposes, the process was stopped before

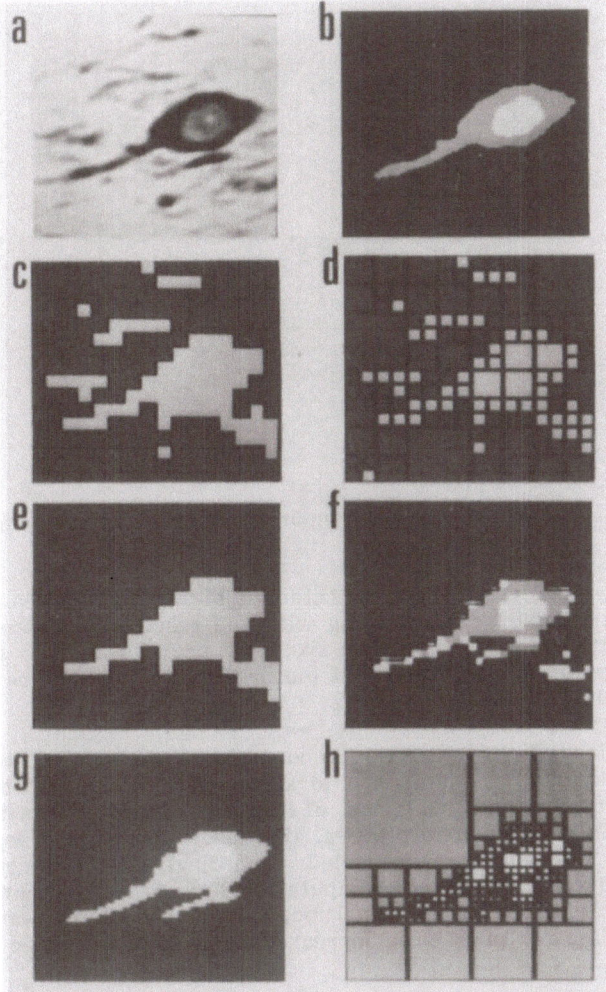

Figure 6.   Interpretation-guided segmentation of a neuron.   (a) Grayscale image, 128x128 resolution; (b) operator-segmentation; (c) system-identified background at 75% confidence level; (d) quadtree tiling used for background identification shown in (c); (e) current cytoplasmic region estimate; (f) nucleus identification and border refinement; (g) final system-segmented labeling; and (h) corresponding quadtree tiling.

the borders were completely refined showing that with as few as 346 tiles (compared with 16384 pixels of the original image) the system was able to accurately capture most of the cell's outline.

The process was repeated and carried to completion for all 6 cells in

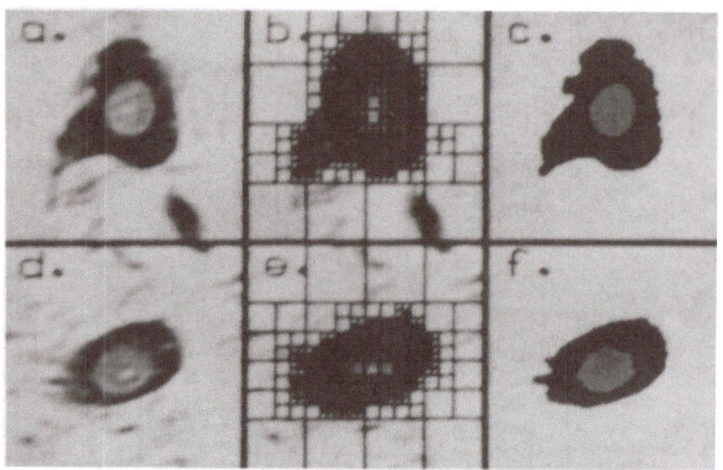

Figure 7. Segmentation of two neurons by quadtree refinement.

the test set; two of these are illustrated in Figure 7. For each cell, the original image is depicted (a,d), along with the resulting quadtree (b,e) and final segmentation (c,f). For the six such neurons tested, the system successfully labeled 97% of the total picture area, forming quadtrees that averaged 861 tiles.

Examination of the execution record shows that the system quickly eliminated background artifacts due to sectioning, focusing instead on the actual cellular region, and spent most of its time refining the borders of each cell. During this refinement process, the resolution of the segmentation increased and the quadtree expanded accordingly. The quadtree data structure greatly reduced the computational cost of refinement. As expected, for each increase in resolution, the number of tiles only doubled, whereas the effective pixel area increased each time by a factor of four (Tucker, 1984).

DISCUSSION

In all cases, the system correctly identified neurons; in two cases, however, the segmentation differed from that made by the user. In one of these cases, shown in Figure 6, this was due to the close proximity of a darkly stained cell fragment that caused a small bridge to form connecting what should have been two separate regions. In the other case, a darkly stained nucleus caused the system to merge half of its area with the surrounding cytoplasm.

We feel these results show that knowledge may be successfully used to direct the segmentation of simple neuron images. Incorporation of additional rules concerning object shape might have allowed the system to correct these segmentation defects. For this system to be of practical value, therefore, a larger number of rules would be needed to handle more complex situations, such as cells that are touching or overlap. This suggests the need for an additional component to facilitate the acquisition of additional knowledge.

In the ideal vision system, the system would be capable of learning about objects from examples provided by a user. Starting with a simple cell world model and a small set of expert processes, such a system would be able to modify its own rule base and thereby modify its behavior. This would be particularly important for experimental research laboratories where requirements of a vision system constantly change.

Winston (1975) showed how a system can learn structural descriptions from simple examples and counterexamples. For the most part, however, learning itself is poorly understood. Knowledge acquisition might be achieved in a more empirical way through interaction with a human expert. This requires that the system have knowledge about its own representational schemes and procedures. TEIRESIAS (Davis, 1978), a companion system for the medical consultation system MYCIN (Shortliffe, 1976), takes just such an approach. Through dialogue with a medical doctor, TEIRESIAS is able to accept new factual information regarding new drugs or recommended therapy and change MYCIN's knowledge accordingly. In much the same way, a vision system having knowledge about its own representation would be able to alter its image domain for new environments. Of greater concern for intepretation-guided segmentation, however, is the introduction of new procedural knowledge required for other image domains. How this might be accomplished, either automatically or through interaction with an operator/expert, is still an open question.

In conclusion, although model-matching provides a mechanism for the evaluation of a particular scene interpretation, the difficulty in establishing a useful set of visual pattern recognition and segmentation rules remains.

We have designed an image-understanding system that combines both low- and high-level processing in a unified manner. The quality of the final intepretation and of the resulting segmentation becomes then a function of the knowledge of the system. As more is learned about how low- and high-level information can be combined, the potential exists for generalizable agents capable of "learning" and acquiring procedural knowledge for segmentation directly from trial imagery.

REFERENCES

Ballard, D.H. and Brown, C.M. (1982). Computer Vision. Prentice-Hall, Englewood Cliffs.

Barrow, H.G. and Popplestone, R.J. (1971). Relational Descriptions in Picture Processing. Machine Intelligence, 6, 377-396.

Benno, R.H., Tucker, L.W., Joh, T.H. and Reis, D.J. (1982). Quantitative Immunocytochemistry of Tyrosine Hydroxylase in Rat Brain. I. Development of a Computer-Assisted Method Using the Peroxidase-Antiperoxidase Technique. Brain Research, 246, 225-236.

Davis, R. (1978). Knowledge Acquisition in Rule-Based Systems—Knowledge About Representations as a Basis for System Construction and Maintenance. In Pattern-Directed Inference Systems. (eds. D.A. Waterman and F. Hayes-Roth). Academic Press, New York.

Dyer, C. (1982). The Space Efficiency of Quadtrees, Computer Graphics and Image Processing, 19, 335-348.

Hanson, A.R. and Riseman, E.M. (1978). Segmentation of Natural Scenes. In Computer Vision Systems. (ed. A.R. Hanson and E.M. Riseman). pp. 129-163. Academic Press, New York.

IEEE (1983). Proceedings of the 1983 Computer Vision and Pattern Recognition Conference. IEEE Computer Society Press, Maryland.

Klinger, A. and Dyer, C.R. (1976). Experiments in Picture Representation Using Regular Decomposition. Computer Graphics and Image Processing, 5, 68-105.

Lindsay, R.D. (1977). Computer Analysis of Neuronal Structures. Plenum Press, New York.

Minsky, M. (1975). A Framework for Representing Knowledge. In The Psychology of Computer Vision. (ed. P.A. Winston). pp. 211-277. McGraw-Hill, New York.

Preston, K. (1980). Automation of the Analysis of Cell Images. Analytical and Quantitative Cytology, 2, 1-14.

Ranade, S. and Prewitt, J.M.S. (1980). A Comparison of Some Segmentation Algorithms for Cytology. Proceedings of the 5th International Conference on Pattern Recognition, Miami Beach, Florida, pp. 561-564.

Shortliffe, E.H. (1976). Computer Based Medical Consultations: MYCIN. Elsevier, New York.

Tenenbaum, J.M. and Barrow, H.G. (1977). Experiments in Interpretation-Guided Segmentation. Artificial Intelligence, 8, 241-274.

Tucker, L. (1984). Computer Vision Using Quadtree Refinement. Ph.D. Dissertation. Dept. of Electrical Engineering and Computer Science. Polytechnic Institute of New York, New York.

Winston, P. (1975). Learning Structural Descriptions from Examples. In The Psychology of Computer Vision. (ed. P. Winston). pp. 115-155. McGraw-Hill, New York.

QUANTITATIVE STUDIES OF RODENT AND PRIMATE NEOCORTEX: CENTRAL MONOAMINE AND PEPTIDE NEURONS

J.H. MORRISON and F.E. BLOOM

Division of Preclinical Neuroscience and Endocrinology, Research Institute of Scripps Clinic, 10666 North Torrey Pines Road, La Jolla, California 92037, USA

INTRODUCTION

The objectives of our cytochemical investigations of neurotransmitter systems have been to use cellular circuitry in the mammalian brain as a basis for functional insight. It has been our view that the most meaningful analysis of transmitter actions at the cellular level rely on an accurate determination as to the origin and targets of a specific transmitter containing system (Bloom, 1975). As a result, we have frequently resorted to detailed manual quantitative analysis of cytochemical features as possible indices to the identification of GABA (Iversen and Bloom, 1972) or noradrenergic terminals (Bloom and Aghajanian, 1968a; Koda and Bloom, 1977), or for the temporal development of synaptic systems (Aghajanian and Bloom, 1967; Bloom and Aghajanian, 1968b; Woodward et al, 1971). The utility of such enterprises, suggests that there is a highly useful purpose to be served by quantitative studies of neurotransmitter distribution and synaptic density. Acquiring similar data sets is time-consuming, labor intensive, and open to the problems of human error, thus limiting the application of such methods. A solution to this quandry can only be accomplished by computers.

In attempting to develop a computer-based system for the types of analyses that our work requires, we found that no existing instruments offered the combination of analytic strategies we wished to rely upon at a price we could afford. In designing our system (see below), we have been strongly influenced by our most recent experiences in achieving insight into the possible functions of monoamine and peptide transmitters in the primate neocortex, and the relevance of these transmitter systems to the pathology of senile dementias of the Alzheimers type (SDAT).

The present communication will, therefore, review some of these data to support our conclusion that an ideal computer based instrument for cytochemical quantitation should be able to

provide for: 1) provide a quantitative analysis of
cytochemically reactive axonal and dendritic varicosities across
a two dimensional area of brain;   2) provide accurate 2
dimensional mapping of cytochemically distinguished neuronal
perikarya or perikaryal-sized pathological entities;   and 3)
derive 3-dimensional constructions from the 2-dimensional maps,
and retain the real specimen landmarks to re-find any specific
cell-sized structure.   Ideally, we want the capacity to
distinguish the reactive elements on the basis of shape and size,
while retaining real specimen area coordinates so that adjacent
sections, reacted for different transmitter-related markers, can
be quantitatively compared in the same specimen.   Each of these
three uses has already been realized with manual analyses
(Iversen et al, 1983; Morrison et al, 1982a,c; 1983; Morrison
et al, 1984a,b) and earlier computerized analyses (Foote et al,
1980).   We will review the results of our more recent
applications of cytochemical analysis and then conclude with a
brief description of our own system now nearing completion.

Laminar, Regional, Developmental, and Functional Specificity of
Monoaminergic Innervation Patterns in Monkey Cortex

     Our understanding and appreciation of the noradrenaline
(NA)-containing central neurons and the noradrenergic innervation
of neocortex has progressed extensively since the initial
demonstration with the Falck-Hillarp method (Falck et al, 1962)
of a direct NA projection from the brainstem to the cortex (Anden
et al, 1966).   The suggestion of a direct ponto-cortical pathway
circumventing the obligatory synaptic pause in the thalamus
generated substantial controversy.   However, the existence of a
direct projection from brainstem to cortex is now an integral
part of neuroanatomical dogma rather than a challenge to it.   We
now recognize that the NA projection is merely one of at least
four major nonthalamic ascending projections to neocortex.

     The continued analysis of the NA coeruleo-cortical system in
the rat (see Morrison et al, 1981,1982b for review) has
demonstrated a rich network of NA innervation in all layers and
regions of the dorsal and lateral cortex in a relatively uniform
laminar pattern.   In both rodents and primates, a general
description of this system would view it as a tangential,
intragriseal afferent to neocortex, in which the organization of
termination patterns differs strikingly from the specific
thalamo-cortical and cortico-cortical systems.   In investigating
the degree of inter-regional variation of the monoamine
projections to neocortex, we moved from rodents to primates in
order to take advantage of the greater functional specialization
and landmarks of the gyrencephalic brain and compared the NA and
5-HT innervation of these systems in three primate species
(squirrel monkey, cynomologous, and rhesus (Morrison et al,
1984b).

We found that the specific density and pattern of NA or 5-HT innervation in a particular cortical locus varies systematically as a function of several factors: the cytoarchitectonic region, the cortical lamina, the species of animal, age of the animal, and the functional interaction of the region with other brain areas. The NA innervation of primate cortex exhibits a far greater degree of regional heterogeneity than the rat: most major cytoarchitectural regions exhibit much greater, but far less uniform fiber densities relative to the rodent. For this report we review only the laminar distribution of monoamine terminal densities within the visual cortex, perhaps the region of the primate brain with the most intensive specialization across cortical laminae. Other semiquantitative differences are reported in detail elsewhere (Morrison et al, 1982a,b; 1984b; Foote and Morrison, 1984).

Laminar specificity. The primary visual cortex of the primate brain has a relatively low density of NA innervation, while its 5-HT innervation, particularly in layer IV, is the densest of all primate neocortical regions. These two fiber systems exhibited a high degree of laminar specialization, and were, in fact, distributed across cortical layers in a complementary fashion: layers V and VI receive a moderately dense NA projection and a sparse 5-HT projection, whereas layers IVa and IVc receive a very dense 5-HT projection and are largely devoid of NA fibers (Morrison et al, 1982a;1984b). In addition, the NA fibers manifest a geometric order that is not so readily apparent in the distribution of 5-HT fibers. These patterns of innervation imply that the two transmitter systems affect different stages of cortical information processing: the raphe-cortical 5-HT projections may preferentially innervate the spiny stellate cells of layers IVa and IVc, whereas the coeruleo-cortical NA projection may be directed predominantly at pyramidal cells.

The old world monkeys receive an even denser 5-HT innervation of primary visual cortex than the new world monkeys, based on strictly visual comparisons of neuropil reactive elements. On the other hand, the overall density of the NA innervation is equivalent, or even slightly decreased in the old as compared to new world monkeys. Indeed, the variations of these 5HT patterns may correlate with the further laminar cortical specialization in the old world species. More specifically, a distinct band of tangential fibers is present at the IVc-ß - Va border in the old world monkeys which is not evident in the new world monkey.

Two important points emerge from these data: 1) with the possible exception of layer IVc- the various sublaminae of layer IV (the geniculo-recipient layer) receive the densest 5-HT innervation in both old and new world monkeys, and 2) the further laminar differentiation of visual cortex is coincident with the

more advanced phylogenetic level of the old world monkeys, as
reflected in further laminar differentiation of the monoamine
projections.

These observations suggest that coincident with the
extensive phylogenetic development and functional differentiation
of neocortex in the primate there is a parallel elaboration of
the ascending NA and 5-HT projections. Laminar complementarity
may result from the two systems terminating on different classes
of neocortical neurons, while regional specificity may result
from preferential participation in different aspects of behavior.
Such hypotheses will require detailed quantitative analysis under
stringently controlled conditions. With our computer based
system we intend to determine eventually whether patients dying
with psychotic depression may show any significant differences in
monamine innervation density or distribution.

Quantitative Analysis of Cytochemically Defined Neuronal
Perikarya

The detailed laminar or intraregional analysis and
quantitation of neuropil elements requires far higher analytic
resolution than would similar analyses of neuronal parikaryonal
number or distribution. Our colleague S.L.Foote spent
considerable effort in development of a method that could
assemble consecutive 2-dimensional maps of noradrenergic neurons
in the locus ceruleus (Foote et al, 1980). His data base can now
be used for detection of changes in cell number with experimental
treatments, or with retrograde mapping methods to determine, for
example, whether LC neurons with specific terminal fields show
any segregated distribution within the LC. Similarly, in
collaboration with Iversen (Iversen et al, 1983), we were able to
employ manual analysis of LC perikarya to demonstrate a specific
loss of LC only in aged subjects dying with neuropathologically
diagnosed SDAT.

More recently we have turned our attention to new prominent
classes of intracortical perikarya for which immunocytochemical
evidence of transmitter identification can be established. These
two intracortical peptide systems are the
somatostatin(SS)-containing neurons and the vasoactive intestinal
polypeptide (VIP)-containing neurons. Both systems will be
reviewed briefly for the purpose of documenting the advantages of
quantitative morphometric analysis in gaining possible insight
into normal function and the pathophysiological importance of
identified systems of transmitters.

Somatostatin

    Somatostatin (SS)-immunoreactivity is associated with
specific cortical neuronal populations and terminal fields (see
Morrison et al, 1983a,b). Some of these systems are far more
extensive in human and non-human primate than they are in rodent
neocortex (Morrison, unpublished results). The perikarya
labelled with antibodies against "SS28", a biologically active
large form of SS are distributed in two major laminar bands in
neocortex: layers II-III, and layers V-VI (Fig 1). In human and
non-human primate, a very extensive fiber pattern of SS
immunoreactivity is also found in cortical layers I-III (densest
in I), suggesting that SS neurons, especially the presumptive
small pyramids of layer III, could participate as the source of a
cortico-cortical system, as well in a separate functional role as
a heterogeneous group of intrinsic, intraregional cortical cells.
In addition, it is conceivable that an independent sub-cortical
SS-positive projection to layer I exists.

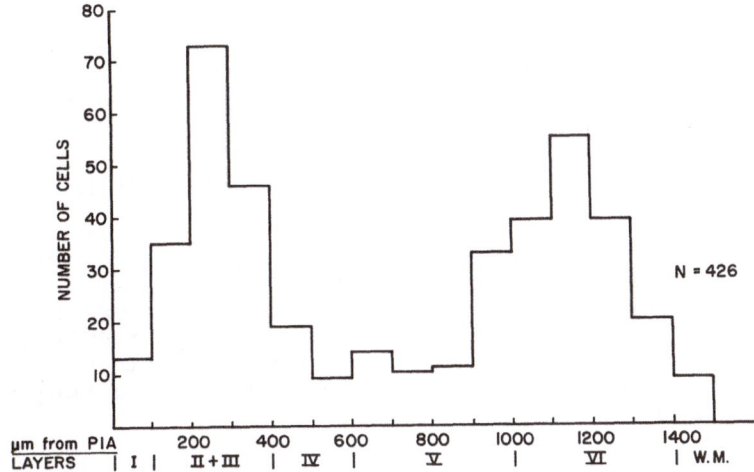

Figure 1. Histogram showing distribution of
somatostatin-positive cell bodies in primary
somatosensory cortex. Cell counts were derived from
photomontages of 550 μm wide strips of 50 μm thick
coronal sections. Total cells = 426 (Morrison et al,
1983b).

Using R.   Benoit's well characterized somatostatin (SS)
antisera, specific for either SS28, SS14, or SS28 (1-12) in rat
cortex, we demonstrated (Morrison et al, 1983a,b) that: 1) SS28
(1-12) is preferentially localized in neuronal processes in a
density which far exceeds that revealed by SS-14
immunoreactivity, whereas SS28 is largely confined to cell
bodies, and 2) somatostatin-positive cell bodies and terminals
exhibit a bilaminar distribution;   layer IV has a paucity of
labelled cell bodies and terminals, whereas both are present
superficial and deep to layer IV. The staining pattern suggests
that somatostatin may be present in locally projecting neurons,
as well as long projection systems such as the cortico-cortical
system.

In our more recent studies (Morrison, unpublished) we have
now determined that in primate cortex somatostatin shows similar
laminar specificity. Here, the terminal field in layer I is even
denser than in rat cortex, and there are also labelled cell
bodies in the sub-cortical white matter. Primate cortex shows
even more clearly the lack of arborization in the middle third of
cortex (deep III, IV and superficial V) and, compared to rat
cortex, possesses a greater degree of regional variation in
density of SS-staining by subjective estimation. The cingulate,
temporal and frontal cortices have the densest
somatostatin-positive cells of all neocortical regions, whereas
parietal cortex has an intermediate density and the primary
visual cortex has the least dense innervation.

Vasoactive Intestinal Polypeptide-Containing Cortical Perikarya

Using a sensitive immunohistochemical procedure, we
undertook a detailed morphological characterization of the
vasoactive intestinal peptide-positive (VIP-positive) neuron in
the cerebral cortex of the rat. VIP-positive neurons are present
in all regions of cortex, and are usually strongly bipolar,
possessing long, radially directed processes with very limited
branching in the tangential plane (Emson and Hunt, 1981;
Morrison et al, 1984a).

The laminar distribution of VIP-positive cells differs
throughout the cortex; labelled cells are must numerous in
layers II and III, but are present in all layers (see Figs.
2,3). The highest density of cells are in visual cortex,
(approximately 150% that of frontal cortex) and 80% of these
labelled cell bodies are contained within layers I-IV
(superficial 600 μm of cortex). In somatosensory cortex (Fig.
2) approximately 48% of cells are in layers II and III and 65% of
the total are found within layers I-IV. In the visual cortex
(Fig 3), approximately 50% of the labelled cells are in layers II
and III.

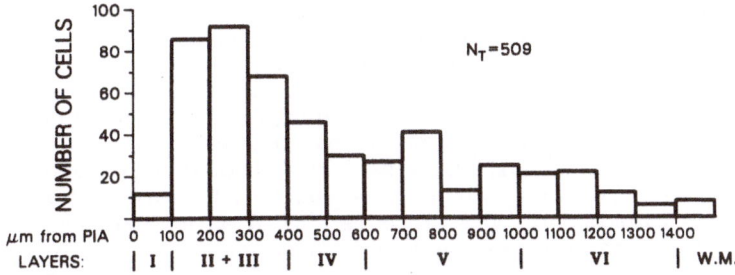

Figure 2. Histogram showing distribution of VIP positive cell bodies in primary somatosensory cortex. Cell counts were derived from photomontages of 550 μm wide strips of 50 μm thick coronal sections. Total cells = 509 (Morrison et al, 1984).

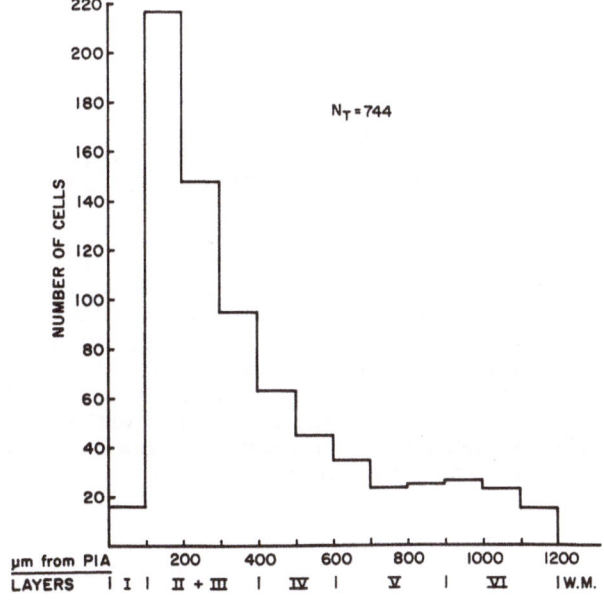

Figure 3. Histogram showing distribution of VIP positive cell bodies in primary visual cortex. Cell counts were derived from photomontages of 550 2 μm wide strips of 50 μm thick coronal sections. Total cells = 744 (Morrison et al, 1984).

In order to determine the density and 3-dimensional distribution pattern of these cells, we prepared serial tangential sections through the rat visual cortex. The distribution of all labelled cells in each section were mapped in the following manner. Prior to sectioning, a horizontal block of visual cortex was pierced by a fine gauge needle, orthogonal to the pial surface. After trimming and positioning on the cryostat chuck, serial 40 μm tangential sections through the superficial 600 μm of cortex were cut and stained for VIP immunohistochemistry. The pin hole served as a landmark for the re-imposition of adjacent sections. Sections after staining were immediately mounted on gelatin-coated slides. Photomontages were made of the corresponding region of each tangential section. Then, a transparency was made from each photomontage showing the position of major blood vessels, the pin hole, and each labelled cell body. The transparencies were then superimposed, aligned (using the pinhole and blood vessels as landmarks) and recorded on transparent acetate sheets, and the vertical dimension then compressed. These superimposed maps were then used to determine the statistical distribution of the VIP neurons within the compressed 2-dimensional area of visual cortical volume.

This analysis demonstrated that: 1) approximately 1% of the cortical neurons are VIP-positive; 2) their distribution is fairly uniform and statistically random, 3) there are no large areas (i.e. with a diameter greater than 100 μm) that lack a VIP-positive cell, 4) on the average, there is one VIP-positive cell per column of 30 μm diameter, and 5) the average nearest neighbor distance on the compressed display is 15 μm.

Given the morphological characteristics of VIP-positive cells, these data indicate that each VIP-containing cell occupies a similarly sized radial unit, contiguous with that of other VIP-positive cells, and which are partially overlapping throughout the neocortex. Within its own circumscribed radial domain, the VIP-positive bipolar neuron would be ideally situated to regulate the availability of energy substrates within specific transcortical cellular ensembles. The quantitative observations derived from these data are shown in Table 1.

TABLE 1

Mean Density	1 VIP Cell/737 μm^2
(N=567)	1 VIP Cell/30.6 μm diameter
Largest Surface Area	
Without a VIP Cell	60-90 μm diameter
Mean nearest neighbor	
(N=350)	14.8 $\pm$ 0.61 μm
Dendritic spread	
in Layers I and IV-V	60-100 μm

Legend for Table 1
    Distribution of VIP-positive cells in rat visual
cortex. Dimensions taken from Morrison et al, 1984a.
If mean nearest neighbor (m) is algebraically related
to the mean density (D) such that r = 2 D (m) , then r
is a reflection of the randomness of the distribution.
If the distribution is random, r = 1; r > 1 suggests a
non-random distribution, r < 1 suggests a clustered.
The r value in this case is 0.997 which suggests a
random, and non-clustered distribution. (Mean density
for the sample used in nearest neighbor analysis is 1
cell/880 μm^2.) A similar nearest neighbor analysis of
the distribution of a specific cell type was performed
by Winfield et calculations, because the free-floating
immunocytochemical staining procedure leads to minimal
shrinkage as compared to sections mounted and stained
conventionally. The dendritic spread refers to both
the coronal and sagittal planes.

    Four important quantitative details emerge from this
analysis: (1) the distribution of VIP neurons in visual cortex
is fairly uniform, and statistically termed random and
nonclustered, (2) there are no large areas (diameter greater than
100 μ) that lack a VIP-cell and 3) there is one VIP-containing
cell per 27 μm square or circle of 30.6 μ diameter, with the
average nearest neighbor distance being 15 μm. Thus, each
VIP-containing cell is identified with a unique radial volume,
which is generally between 15 and 60 μ in diameter, and overlaps
with the contiguous domains of neighboring VIP-positive cells.
Even though only 80% of the total of VIP containing cells are
represented in the cortical depth we analyzed, the minimal
dendritic arbor of cortical VIP bipolar neurons would include
virtually the entire cortical region. The emergence of an
approximate density of one VIP containing cell per 30 μ diameter
column is of interest for two reasons. First, Powell and his
colleagues (1981; Rockel et al, 1981) have recently shown that
across several different species and cortical regions, including
rat visual cortex, a cylinder of cortex 25 μ by 30 μ consistently
contains approximately 70 pyramidal cells, 30 small stellate

cells and 5 large stellate cells. On the average, this same
region would contain one VIP cell, in agreement with our estimate
that approximately 1% of the neurons in rat visual cortex are
VIP-positive. Second, a column of this size, the so-called
"mini-column", is viewed by some investigators as the smallest
unit of resolution in the anatomic and functional organization of
neocortex (Hubel and Wiesel, 1974; Mountcastle, 1981). The
concept of a hardwired mini-column has been controversial and
somewhat resistant to analysis. Given that the distribution of
the VIP-containing cells is random, our data cannot be taken as
evidence for the existence of anatomic or functional mini-columns
of a specific diameter. However, given their rough hypothetical
dimensions, the mini-column is likely to contain at least one
VIP-positive neuron. The morphological characteristics and
distribution of the VIP-containing cell are ideally suited for a
role in local, radial regulation of cortical activity. The
precise nature of this role is dependent upon the cellular effect
of VIP and the anatomic constraints of the circuits involving VIP
containing cells (Morrison and Magistretti, 1983).

Quantitative Distribution of Alzhemiers Neuropathology: Correlations with Transmitters

Recently, we initiated a collaborative study with Dr.
Joseph Rogers, of the University of Massachusetts, designed to
quantify the laminar and regional distribution of neuritic
plaques in SDAT cortical samples and relate their distribution to
that of various neurotransmitter systems. This project is still
in the very early stages: however, there are some preliminary
qualitative observations that are of great interest and relevance
to the characteristics needed in the design of a general purpose
cytological analysis system. Thus far, our observations are
based mostly on Bielschowsky-stained material; however, other
techniques, such as Von Braunmuhl's, have been used as well. The
areas that we have studied in greatest detail are the superior
frontal gyrus, the anterior cingulate cortex, the superior,
middle, and inferior temporal gyri, and the primary visual cortex
(area 17). The gradient in plaque density of these regions is:
cingulate > superior frontal > temporal neocortex > visual
cortex. The plaque density of cingulate cortex is several fold
higher than that of primary visual cortex, although in all five
patients studied thus far, there is extensive pathology in all of
the above regions.

Two important points emerge from these data on regional
distribution of plaques: 1) cortical regions that can be viewed
as association areas (frontal, and cingulate) have a higher
plaque density than those dominated by thalamo-cortical
projections (e.g. visual cortex), and 2) the above regional
gradient does not fit well with the primary gradients of
immunocytochemical or biochemical markers for the serotonergic,
noradrenergic, or cholinergic systems, but it does fit fairly

well with the regional distribution of somatostatin in primate cortex that was described earlier.

In visual cortex, as well as the other regions that we have studied, the laminae containing the most plaques are III and superficial V, and the laminae containing the least number of plaques are I and VI, with an intermediate density sublaminae of IV. The noradrenergic and somatostatin systems correlate best with this distribution whereas the cholinergic, laminar, pattern bears the least resemblance to the laminar pattern of plaque distribution. We do not mean to imply that plaque formation is a direct result of pathology involving only the noradrenergic or somatostatinergic systems. To the contrary, this illustration demonstrates that it is very unlikely that plaque formation is directly linked to the degeneration of any single transmitter system.

Plaque size also appears to be related to laminar location. The largest plaques are present in layers III and V, and the smallest plaques are often in layer IV. This pattern is particularly striking in temporal cortex. The large pyramidal cells with a broad dendritic area tend to be in layers III and V, whereas layer IV contains primarily densely packed, smaller, somatostatin-containing neurons. This finding demonstrates that it is very unlikely that plaque formation may result solely from the degeneration of cholinergic cortical projections and far more likely that such pathologic structures represent a complex of post-synaptic dendritic neurites and pre-synaptic axonal neurites.

A Computer-based System for Quantitative Transmitter Markers

The correlations described above are all very qualitative in nature, and as such should be interpreted with caution. Clearly, a rigorous quantification of transmitter-related markers along with pathological elements such as SDAT plaque distribution is needed for a complete and accurate description of the role of various transmitter systems in SDAT and other diseases. For example, based simply on dual illumination procedures, it has been possible to demonstrate in our preliminary SDAT cortical samples that somatostatin-positive profiles which appear to be pathologically degenerating, are often associated with an amyloid core. We have demonstrated this through the combined use of somatostatin immunohistochemisty and congo red counterstaining with cross-polarization optics. As with the other findings described above, methods are required that can provide quantitative documentation of the qualitative observations. We conclude by a brief description of our evolving system.

The System

Our system is being designed by Dr. Warren Young, and is centered around a graphics display unit (Vectrix VX384) capable of displaying 1672 by 480 pixels of resolution. Each pixel addresses 8-bit planes and can display in 512 separately assignable colors or gray monochrome levels. The VX384 is a self-contained graphics system which functions as the driver to an analog red, green, blue (RGB) monitor and as a frame memory to hold all digitized images. Functions include onboard graphic drawing commands for lines, dots, polygons, and primitive graphic manipulations such as rotation and translation.

The image from the microscope is captured first in unaltered analog form by an electronic image tube-type camera (Cohu 5000) with a modular, expandable, plug-in system, permitting a wide range of lighting conditions. Output is fed either to an image digitizer for computer analysis or to a video mixer for combination with other video signals or to a monitor. To trace a neural structure, the investigator watches a monitor display of the digital or analog representation of the microscopic live image (indistinguishable from each other) and traces over selected items with an electronic mouse on a bitpad (Summa Graphics MM1201). All data in terms of vectors drawn, the coordinates, and section identification are maintained by the computer. For automatic image analysis, a Colorado Video digitizer, Model 270A-1, digitizes the microscope image to 2048 by 480 pixels, with 672 by 480 usable resolution, each at 8 bits of resolution. Conversion time per field will vary from 2 to 20 seconds.

Position sensing and mechanical translation of the microscope axes are handled by Burleigh Instruments devices. Stage mounted optical sensors (EN-Series Encoders) yield relative positional measurements with 1 micron resolution. Mechanical translators (IW-Series inchworm) provide stage movement with absolute bidirectional accuracy of 1 micron. All encoders and translators are controlled by an intelligent command processor (CE-2000) which accepts abbreviated commands from the host processor and codes them into the necessary electrical signals for the translators, or codes the information from the encoders into computer data for the host processor.

The central processor is an LSI 11/23 of a Plessey 6740 system. Online to the processor is an 80 megabyte Winchester, two RX02 double-density floppy drives, a streaming 20 megabyte tape backup system and 512 kilobytes of addressable memory. Pascal-2 from Oregon Software is the main controlling language. Large data files can be maintained (up to the limit of 80 megabytes, presently). Standard serial (EIA RS-232 protocol) and parallel (16 bit words) interfaces will be used to connect the Q-bus structure of the LSI 11/23 to the outlying processors

(VX384, 270A-1 digitizer, Burleigh CE-2000, and Summagraphics MM1201).

A GraphOn GO-140 terminal is the main input terminal for this computer system. Information is entered for program use by the operator using the keyboard. Such information includes filenames, database attributes, selection of operating modes, etc., from menu driven programs. The terminal also provides vital information back to the operator such as database statistics (the assigned attributes, model involved, cell locations, nuclei under study, numbers of cells, classifications in the database, etc.).

Two levels of data entry are actually being coded for. The symbolized level entails assigning symbols to the various cell types and specifying from the microscopic field the coordinates via the bitpad mouse. In essence the computer acts like a very efficient notebook, keeping track of all the assignments, the various attributes of the cells, their locations, and the information needed to recall and perform detailed analysis. The system will be capable of quantifying the distribution of labeled cells within a region of cortex, the detailed pattern of arborization of a single cell, and generating a measure of the density of terminal boutons in a given microscopic field. The second level of data entry is a vector based system whereby the operator enters, via a bitpad and mouse, the endpoints of a line that he defines as representative of a portion of the structure undergoing image capture. Again, the computer takes care of the housekeeping. Any magnification can be used, allowing the operator to define the confines of the computer session. A database may be created that contains macro information on nuclei organization. Another database may be created that contains microscopic information on the actual three-dimensional structure of the cells that make up the nuclei. All of the pertinent geometric information can be carried over between databases to allow very intricate anatomical analyses from a macro to a micro level.

REFERENCES

Aghajanian, G.K. and Bloom, F.E. (1967). The formation of synaptic junctions in developing rat brain: A quantitative electron microscopic study. Brain Res., 6, 716-727.

Anden, N.E., Dahlstrom, A., Fuxe, K., Larsson, K., Olson, L. and Ungerstedt, U. (1966). Ascending monoamine neurons to the telencephalon and diencephalon. Acta physiol. scand., 67, 313-326.

Bloom, F.E. (1975). The gains in brain are mainly in the stain. In The Neurosciences: Paths of Discovery. (eds. F. Worden, G. Adelman, and J. Swazey). MIT Press, Cambridge, pp. 211-227.

Bloom, F.E. and Aghajanian, G.K. (1968a). An electron microscopic analysis of large granular synaptic vesicles of the brain in relation to monoamine content. J. Pharmacol. 159, 261-273.

Bloom, F.E. and Aghajanian, G.K. (1968b). Fine structural and cytochemical analysis of the staining of synaptic junctions with phosphotungstic acid. J. Ultr. Res. 22, 361-376.

Emson, P.C. and Hunt, S.P. (1981). Anatomical chemistry of the cerebral cortex in the organization of the cerebral cortex. In The Organization of the Cerebral Cortex. (eds. F.O. Schmitt, F.G. Worden, G. Adelman, S.G. Dennis). MIT Press, pp. 325-345.

Foote, S.L., Loughlin, S.E., Cohen, P.S., Bloom, F.E. and Livingston, R.B. (1980). Accurate three-dimensional reconstruction of neuronal distributions in brain: Reconstruction of the rat nucleus locus coeruleus. J. Neurosci. Meth., 3, 159-173.

Foote, S.L. and Morrison, J.H. (1984). Postnatal development of laminar innervation patterns by monoaminergic fibers in Macaca Fascicularis primary visual cortex. J. Neurosci., in press.

Iversen, L.L. and Bloom, F.E. (1972). Studies of the uptake of ^{3}H-GABA and ^{3}H-glycine in slices and homogenates of rat brain and spinal cord by electron microscopic autoradiography. Brain Res., 41, 131-143.

Koda, L.Y. and Bloom, F.E. (1977). A light and electron microscopic study of noradrenergic terminals in the rat dentate gyrus. Brain Res., 120, 327-335.

Morrison, J.H. and Magistretti, P.J. (1983). Monoamines and peptides in cerebral cortex: Contrasting principles of cortical organization. Trends in Neurosciences, 6, 146-151.

Morrison, J.H., Molliver, M.E., Grzanna, R., and Coyle, J.T. (1981). The intra-cortical trajectory of the coeruleo-cortical projection in the rat: a tangentially organized cortical afferent. Neurosci., 6, 139-158.

Morrison, J.H., Foote, S.L., Molliver, M.E., Bloom, F.E. and Lidov, H.G.W.. (1982a). Noradrenergic and serotonergic fibers innervate complementary layers in monkey primary visual cortex: An immunohistochemical study. Proc. Natl. Acad. Sci. 79, 2401-2405.

Morrison, J.H., Foote, S.L., O'Connor, D., Bloom, F.E. (1982b). Laminar, tangential, and regional organization of the noradrenergic innervation of monkey cortex:

dopamine-ß-hydroxylase immunohistochemistry. Brain Res. Bull., 9, 309-319.

Morrison, J.H., Benoit, R., Magistretti, P.J., Ling, N., and Bloom, F.E. (1982c). Immunohistochemical distribution of pro-somatostatin related peptides in hippocampus. Neurosci. Lett., 34, 137-142.

Morrison, J.H., Benoit, R., Magistretti, P.J., and Bloom, F.E. (1983). Immunohistochemical distribution of prosomatostatin related peptides in cerebral cortex. Brain Res., 262, 344-351.

Morrison, J.H., Magistretti, P.J., Benoit, R. and Bloom, F.E. (1984a). The distribution and morphological characteristics of the intracortical VIP-positive cell: An immunohistochemical analysis. Brain Res., 292, 269-282.

Morrison, J.H., Foote, S.L. and Bloom, F.E. (1984b) Laminar, regional, developmental, and functional specificity of monoaminergic innervation patterns in monkey cortex. In Press.

Mountcastle, V.B. (1981). An organizing principle for cerebral cortex. In Brain Mechanisms and Perceptual Awareness. (eds. O. Pompeiano and C. Ajmone Marsan). Raven Press, N.Y., pp. 1-19.

Rockel, A.J., Hiorns, R.W., and Powell, T.P.S. (1980). The basic uniformity in the structure of the neocortex, Brain, 103, 221-244.

Woodward, D.J., Hoffer, B.J., Siggins, G.R., and Bloom, F.E. (1971). The ontogenetic development of synaptic junctions, synaptic activation and responsiveness to neurotransmitter substances in rat cerebellar Purkinje cells. Brain Res., 34, 73-79.

IMAGE ANALYSIS OF CATECHOLAMINE FLUORESCENCE AND IMMUNOFLUORESCENCE IN STUDIES ON BLOOD VESSEL INNERVATION

TIM COWEN and GEOFFREY BURNSTOCK

Department of Anatomy and Embryology, and Centre for Neuroscience, University College London, Gower Street, London WC1E 6BT, UK

INTRODUCTION

Quantitative measurement of fluorescence has become increasingly valuable in neuroscience, particularly in the study of biogenic amines. Some semiquantitative methods have contributed valuable data on sympathetic nerves. These have generally involved either counting fluorescent nerve profiles in the light microscope, or making visual estimates of nerve density or fluorescence intensity (see for example, Malmfors, 1965; Dahlström et al., 1966; Bevan et al., 1972; Gerová et al., 1974; Thorbert et al., 1978). However, these approaches are limited in sensitivity and rely on subjective judgements.

A range of catecholamines and indoleamines can now be specifically identified, and efforts have been concentrated on the development of quantitative procedures which can measure the concentrations and distribution of these amines in nervous tissue (for review see Johnsson, 1971). Until recently, quantitative methods have been based on microscopic spectrofluorimetry, which was originally developed to identify the formaldehyde-induced fluorophors of different biogenic amines (Caspersson et al., 1966; Ritzen, 1966). This method has been successfully extended to measure the concentrations of a range of biogenic amines in different tissues, and good correlations have been shown between fluorescence intensity and amine concentrations in the autonomic nervous system (Schipper et al., 1978; Schipper et al., 1980a) and in the central nervous system (Tilders et al., 1974; Bacopoulos et al., 1975; Löfström et al., 1976). However, microspectrofluorimetry is not suitable for feature measurement, although attempts have been made to use it in this way (see Schipper et al., 1980).

Conventional image analysers can carry out sophisticated measurements of features detected by their particular image intensities (and shape), and therefore could provide both numerical information about the density and distribution of amine containing

213

tissue elements, as well as information about concentrations of amine fluorophors. This possibility has been explored in our own laboratory using a Quantimet 720 image analyser (Cambridge Scientific Instruments, UK) interfaced with a fluorescence microscope. We have developed methods which can measure the density and pattern of perivascular noradrenergic nerves and their varicosities (Cowen and Burnstock, 1980, 1982). Other groups are beginning to use similar methods (Amenta et al., 1983).

Our aim has been to provide an approach which is applicable to a wide range of tissues and problems. Recently, we have explored some further techniques which open up exciting new fields for the use of image analysis. The rapid expansion of immunofluorescent labelling techniques for the identification and localisation of neuropeptides, amines and other candidate messenger substances requires the development of appropriate quantitative methods. The results of these recent studies, where we have demonstrated the ability of image analysis to provide data from immunohistochemical preparations of blood vessels, will be described. One of the major problems of both immunohistochemical and catecholamine fluorescence images is the presence of variable amounts of non-specific background autofluorescence which often appears at a similar wavelength to that of the specific fluorescence label. In order to resolve this problem, we have recently made use of a new blue dye, Pontamine Sky Blue, and have tested its compatibility with both immunohistochemical and catecholamine fluorescence.

IMAGE ANALYSIS OF CATECHOLAMINE FLUORESCENCE

The fluorescence histochemical properties of adrenergic nerves are well suited for image analysis because of the high image contrast. The glyoxylic acid fluorescence histochemical method (Lindvall and Björklund, 1974; Furness and Costa, 1975) has been found to give good localisation of fluorophor and a more stable and intense reaction product than formaldehyde-induced fluorescence. Specimens can be stored deep-frozen for up to six months. However, measurements made on specimens stored for longer periods may be distorted by an increase in background autofluorescence. Glyoxylic acid is specific for the primary amines noradrenaline and dopamine. Assays on some of the blood vessels used in our studies have shown that dopamine is present in negligible quantities in these vessels (J. Lincoln, personal communication). Quantitative methods require the brightest possible images. Fluorescence microscopes used in our laboratories are equipped with exciter filters which select the 436 nm emission peak of the mercury light source. Although the wavelength of the excited light is rather higher than the accepted range for catecholamines (about 410 nm; Jonsson, 1971), the greater intensity of illumination at the higher wavelength more than compensates for this. In addition, the high numerical aperture of

INDIRECT METHOD

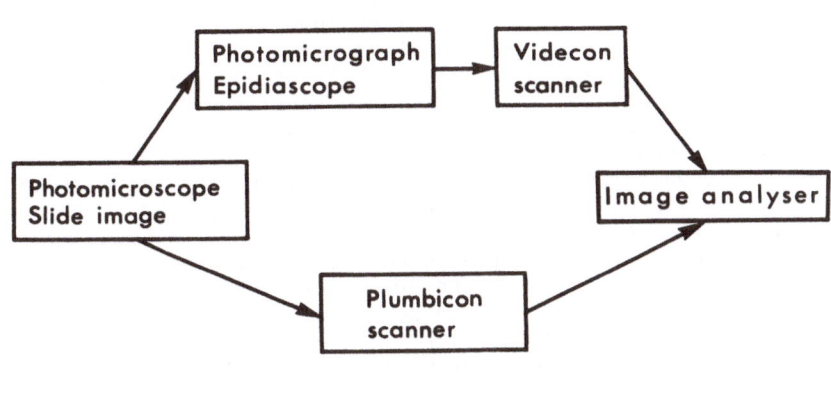

Fig. 1   Diagrammatic representation of image inputs to the
Quantimet 720 image analyser. From Cowen and Burnstock, 1980.

planapochromat objectives has been found to be essential for maximising image intensity. The Quantimet 720 image analyser came into
production about ten years ago. It remains a fast and flexible
machine, and although a new generation of analysers is currently
under development, the design principles are sufficiently similar
for data to be comparable. Image input is from a television
scanner interfaced to a fluorescence microscope or to an epi-
diascope. The scanner influences the intensity and resolution of
the image, and two basic types are used: plumbicon and videcon.
The plumbicon scanner is the most suitable for low intensity
fluorescence images because it gives high sensitivity, a linear
relationship (γ) of almost 1 between specimen and image brightness,
and also the ability to sum the intensities of successive images
(light integration). We have tested direct and indirect methods of
image input (Fig. 1) using different scanners, and have confirmed
that the direct method which uses the plumbicon scanner is able to
measure statistically significant differences between visually
similar nerve plexuses, where the indirect method can not (Cowen
and Burnstock, 1980). The link between scanner and microscope is
also important in maximising image brightness and resolution, both
of which are reduced by any magnifying lenses at this point. A
camera tube in which the image is focused directly onto the image
plane of the scanner gives five times the image intensity of a tube
containing the customary eyepiece lens. It should also be possible
to insert a monochromator and photomultiplier between the
microscope and scanner to allow intensity measurements of specific
wavelengths.

The Quantimet performs image analysis on the scanner signal and detection depends on image brightness. The detected image is monitored on a 720-line television screen composed of 5×10^5 picture points. The Quantimet has 64 detection levels and three range-selecting modes between black and white for detecting different images or parts of images. A simple calibration system has been developed to set the detection levels appropriate to each specimen. This is necessary because of the known variability of neuronal transmitter concentrations and of background autofluorescence in different tissues. An observer counts the intercepts made along a line by nerve fibres and by varicosities. The analyser is then set to count intercepts along the same line; the two detection levels that match the previously established numbers of nerve fibres and varicosities are noted (Cowen and Burnstock, 1980). These two detection levels are then used to scan the whole field. The sensitivity range is set manually for each image and is used in conjunction with the light integrating facility to optimise image brightness.

Total plexus density is measured at the first detection level by two parameters: (1) fluorescent area, which measures the total detected fluorescence, and (2) intercept density, which gives the number of times that the nerve plexus crosses the horizor:al lines of picture points which make up the screen of the image anlyser. The varicosities alone are measured at the second detection level, using two further parameters to give: (3) the number of varicosities, and (4) the total perimeter of all the varicosities, which is then divided by varicosity number to give the mean perimeter and thus the diameter of the varicosity population. The following additional information can be deduced from the measured parameters: (1) that fluorescent area is proportional to the total length or number of individual nerve fibres present, whilst intercept density is proportional to the total length or number of the nerve bundles which comprise the perivascular plexus; (2) that the fluroescent area divided by intercept density gives a figure that is proportional to the average size of the nerve bundles; and (3) that the proportion of varicosities to nerve fibres is given by the fraction - varicosity number divided by fluorescent area.

Two conflicting requirements influence the level of magnification selected: the sizes of nerve fibres (about 1 μm diameter) and varicosities (about 1-3 μm diameter) are near the limit of resolution of the light microscope and therefore require a high magnification, whereas whole-mount and sectioned fluorescence preparations generally require a certain depth of focus to include elements at all levels of the preparations, and therefore need a lower magnification. Our experiments have shown that a X25 planapo objective is the most suitable for catecholamine fluorescence. At this magnification a terminal nerve bundle containing 2 or 3 axons would be 4-5 times the size of the analyser pixel. Optimum resolution is achieved when image dimensions are about 10 x pixel size or more,

TABLE 1
IMAGE ANALYSIS OF NERVE DENSITIES IN THREE REGIONS OF THE RABBIT
EAR ARTERY

	Proximal region	p	Mid region	p	Distal region	Prox: distal:p
FA (%)	11.4+0.45	++	8.5+0.60		6.4+0.52	++
ID (mm^{-1})	66.8+2.28	++	50.6+3.79	+	37.6+3.29	++
DV (μm)	3.7+0.14		3.7+0.11		3.6+0.10	
NV (mm^{-2})	10,561+484		8,293+610		6,537+628	++

Image analysis of variations of noradrenergic nerve density in
three regions of the rabbit ear artery. Nerves were demonstrated
in whole-mount stretch preparations by glyoxylic acid fluorescence
histochemistry. FA = fluorescent area; ID = intercept density;
DV = diameter of varicosities; NV = number of varicosities.
Figures given are mean ± SEM, n=8, +p<0.05; ++p<0.01. From Griffith
et al., 1982.

consequently the compromise magnification used to give the
necessary depth of focus causes a certain amount of blurring; also
data obtained using different objective magnifications are not comparable.

APPLICATIONS

    Image analysis has been used to demonstrate regional variation
in perivascular noradrenergic nerve density and the changes that
occur during development and ageing of noradrenergic nerves in
several arteries of the rabbit.

    Image analysis was used to examine differences of nerve den-
sity between the proximal, mid and distal regions of the central
ear artery of the rabbit (Griffith et al., 1982). The results are
summarised in Table 1. Marked differences in the density of innervation were shown between the proximal and distal regions. The
innervation of the proximal region was about twice as dense as that
of the distal region, and there were approximately 10,500 and 6,500
varicosities per mm^2 surface area of vessel wall in the two
regions, respectively. These varicosities had approximately the
same mean diameter throughout the length of the vessel. Assay by
high performance liquid chromatography showed that the noradrena-

line contents per gram wet weight of tissue in the proximal and
distal regions were 1.03 and 0.94 μg respectively. It was esti-
mated that the nerve endings contained 2.8×10^{-14} and 1.2×10^{-14}g of
noradrenaline per varicosity in the proximal and distal ear artery,
respectively. Comparison of these results with those from the
adult stage of the study on development (see below) has shown that
the ear artery has a nerve density similar to that of other
arteries in the rabbit.

The rabbit ear artery is the only vessel for which numerical
values of nerve density produced by an alternative method are
available (Bevan et al., 1972): fluorescent nodes (varicosities)
were counted in tangential 5-6 μm-thick sections of the
adventitial-medial junction. Node density was $34,700+1,590$ mm^2,
and assay procedures gave noradrenaline concentrations of $3.04+0.17$
μg per g wet weight of tissue. Calculations from these data give
noradrenaline concentrations of 1.45×10^{-14} g per varicosity, which
is intermediate between the values for proximal and distal regions
of the same artery demonstrated by the study made in this laboratory (see above). No correction was made for the effects of
shrinkage of the vessel wall during preparation in the study of
Bevan et al. (1972). Our experience with this and similar arteries

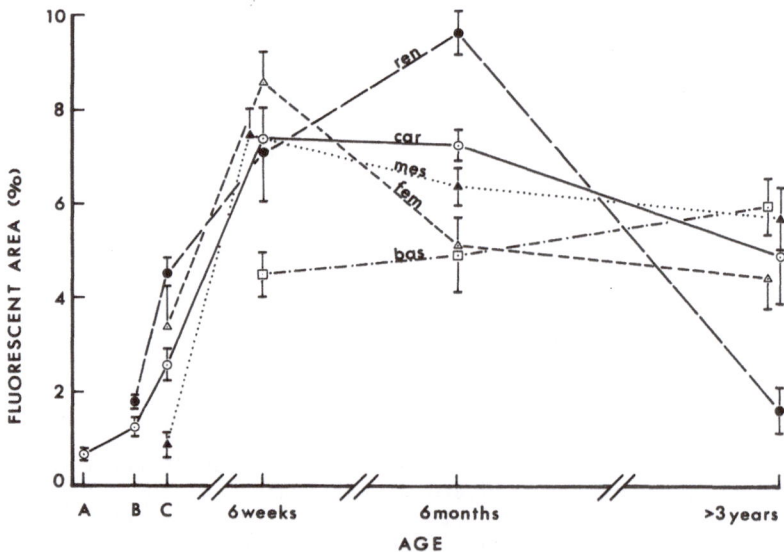

Fig. 2   Image analysis of fluorescent area of perivascular
noradrenergic nerves in five arteries of the rabbit at six stages
of development. Nerves were demonstrated in whole-mount stretch
preparations by glyoxylic acid fluorescence histochemistry.
Figures given are mean+SEM. (A) 25 days in utero, n=8; (B) 1 day
before birth, n=7; (C) 1 day after birth, n=6; 6 weeks, n=6; 6
months, n=7; 3 years, n=7. From Cowen et al., 1982.

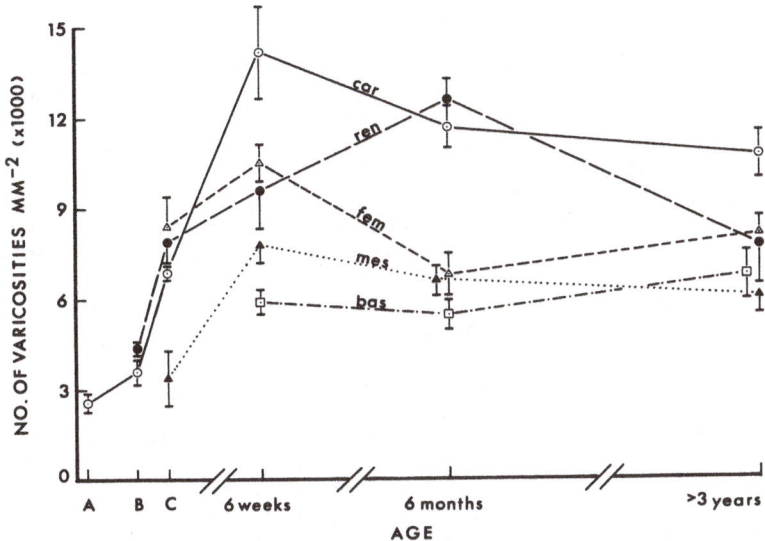

Fig. 3   Image analysis of the number of varicosities of perivascular noradrenergic nerves in five arteries of the rabbit at six
stages of development.   Nerves were demonstrated in whole-mount
stretch preparations by glyoxylic acid fluorescence histochemistry.
Figures given are mean+SEM.   (A) 25 days in utero, n=8; (B) 1 day
before birth, n=7; (C) 1 day after birth, n=6; 6 weeks, n=6; 6
months, n=7; 3 years, n=7.   From Cowen et al., 1982.

suggests that both length and circumference would shrink to about
50% of their in vivo values due to the effects of vessel elasti-
city and the preparative procedures.   This would introduce a four-
fold increase in node density measurements expressed in terms of
surface area of the vessel wall.   Correction of the measured node
density for shrinkage gives 8,675+398 varicosities per mm^2.   This
value is almost identical with the mean value found by Griffith et
al. (1982) of 8,464+574 varicosities per mm^2.   This calculation
illustrates the importance of shrinkage correction.

     The results of our study of perivascular noradrenergic nerves
during development and ageing are summarised in the graphs (Figs. 2
and 3).   Five arteries were taken from the rabbit at six age-stages
between twenty-five days in utero and three years old (old age),
and were prepared for image analysis by the methods previously
described.   Image analysis showed rapid and statistically significant increases in nerve growth during early life.   The pattern of
growth was similar in all the vessels studied: it consisted of a
period of outgrowth of axons; a period of rapid increase in density
and formation of varicosities; and a later period of more gradual

nerve growth. The timing of these stages varied greatly between
the different vessels. The larger vessels, i.e. the carotid, renal
and femoral arteries, had a well-developed innervation at birth,
whilst the innervation of the smaller mesenteric and basilar
arteries was sparse. The early development of perivascular innervation in the rabbit is probably related to the need for the regulation of blood pressure and also of body temperature in the
immediate postnatal period. In most of the vessels studied, the
final period of nerve growth took place between birth and six weeks
but extended to six months in the renal artery. The increase in
vessel wall surface area, together with the increase in the density
of nerve fibres per unit area, indicated a very large increase in
total fibre length over the whole surface area of the vessel wall.
The increase in total fibre length varied from vessel to vessel,
but in the mesenteric artery by six weeks it was as much as sixty
times the density seen at birth.

Two types of nerve loss were demonstrated in different blood
vessels: (1) during postnatal development and (2) in old age. In
the femoral artery there was a 40% reduction of nerve density
(fluorescent area) between puberty (six weeks) and adulthood (six
months) (Fig. 2); the proportion of varicosities to nerve fibres
remained constant so it may be inferred that whole nerves,
including endings, and perhaps also cell bodies, had degenerated.
Adrenergic nerve loss has also been reported in the postnatal development of the renal artery (Gallen et al., 1982) and in the
femoral artery of the dog (Doležel, 1973). The second type of
nerve loss was seen in the carotid and renal arteries between
adulthood and old age, and in the latter vessel involved an 80%
reduction in nerve density (Figs. 2 and 3). We found that the proportion of varicosities to fibres increased considerably over this
period in both vessels (e.g. by over three times in the renal
artery), in contrast to the first type of nerve loss in the femoral
artery, where this proportion remained constant. The small loss of
varicosities relative to fibres suggests that many of the axons
were still intact, but that the levels of axonal noradrenaline had
become reduced in old age making the axons invisible.

The proportion of varicosities to nerve fibres in all the
nerve plexuses at the six week and six month stages, varied little
when compared with the equivalent variations of nerve density and
varicosity numbers. This proportion was maintained in vessels of
contrasting type. Also, image analysis has shown that the same
ratio was stable throughout the postnatal growth and subsequent
loss of nerves, in the guinea-pig renal artery (Gallen et al.,
1982). The generation of characteristic proportions of axons and
varicosities may be a feature of the fully differentiated nerve, or
a product of the interaction between the nerve and the blood vessel
wall. The processes of development and ageing in perivascular nerves have not been measured numerically previously, therefore no
other data are available for comparison.

IMAGE ANALYSIS OF FITC IMMUNOFLUORESCENCE

Preliminary studies in our laboratory suggest that perivascular nerves containing immunofluorescent neuropeptides and serotonin demonstrate locally-specific patterns of innervation during development and ageing in a similar way to the sympathetic nerves. We have, therefore, investigated the use of image analysis to quantify the distribution of nerve plexuses containing these substances.

Despite the greater sensitivity of the peroxidase-antiperoxidase technique (Sternberger, 1974), maximum contrast and resolution of the final image is achieved in immunofluorescence histochemistry using the indirect labelling technique of Coons et al. (1955), with fluorescein isothiocyanate (FITC) as the second label. A major problem in quantifying FITC is that it fades on exposure to UV light and during storage. An advance in this respect is the recent development of a commercially available anti-fade mounting medium (City University, Department of Chemistry, London, UK). Neuropeptide and 5-HT-containing perivascular nerve plexuses are, in our experience, distributed throughout the adventitia of the blood vessel wall in contrast to sympathetic nerves, which are localised at the adventitial-medial border. The greater depth of focus required by this distribution when studied in whole-mount stretch preparations has required the use of a X16 oil immer-

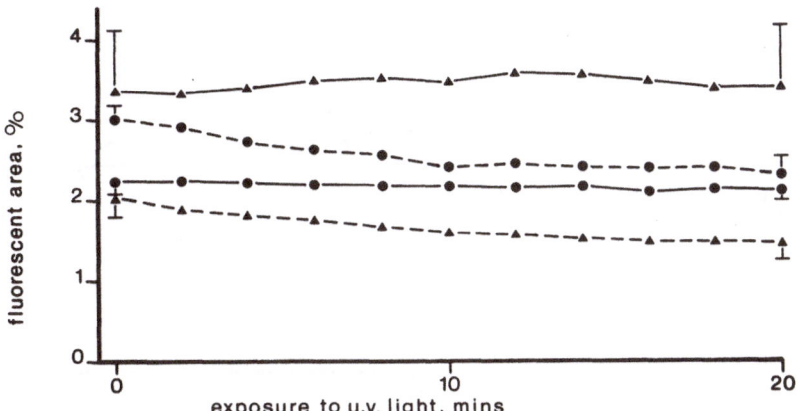

Fig. 4   Image analysis of the fluorescent area of perivascular substance P-like immunofluorescent nerves, labelled with FITC, in stretch preparations of the guinea-pig carotid artery. Uncorrected curves (---) show fading during 20 minutes of exposure to UV light in fresh specimens (●——●) and in specimens stored for 3 weeks in the freezer (▲——▲). Corrected curves (——) show the ability of the image analyser to compensate for fading on exposure to UV light in the fresh and stored specimens. From Cowen et al., 1984a.

sion objective lens which combines depth of focus with a high
numerical aperture and consequently excellent light-gathering
characteristics.

We have produced fading curves for substance P-containing
perivascular nerves in the guinea-pig carotid artery, labelled with
FITC. Two groups of arteries were studied, the first group was
examined immediately after labelling, the second group was
labelled, stored for three weeks in the freezer and then examined.
Both groups were exposed to UV light continuously for 20 minutes.
Measurements were made at the beginning of the exposure period and
thereafter every 2 minutes. The parameters of measurement and analyser set-up were the same as those used previously.

Considering first the fading on exposure to UV light, small
but insignificant reductions in fluorescent area were seen over the
20 minute period in the uncorrected curves of both groups of speci-

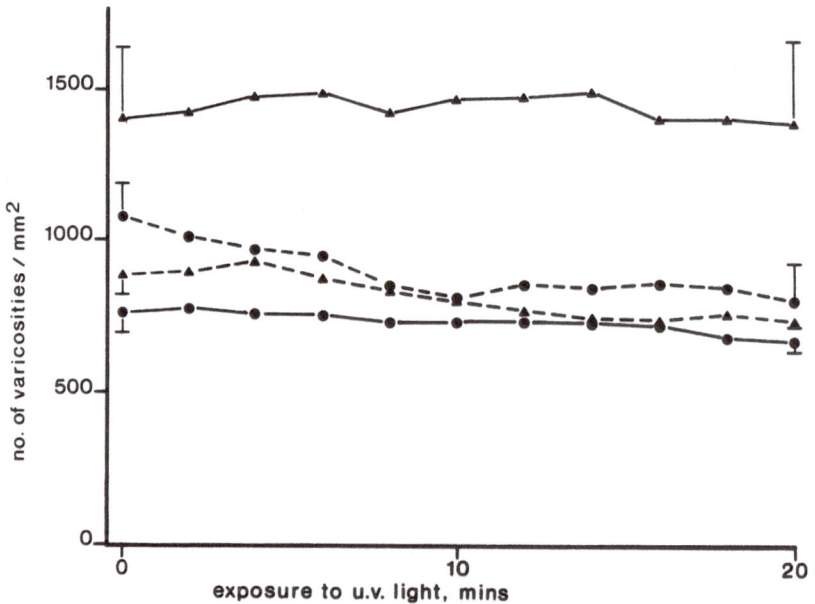

Fig. 5   Image analysis of the number of varicosities in perivascular substance P-like immunofluorescent nerves, labelled with
FITC, in stretch preparations of the guinea-pig carotid artery.
Uncorrected curves (---) show fading during 20 minutes of exposure
to UV light in fresh specimens (●—●) and in specimens stored for
3 weeks in the freezer (▲——▲). Corrected curves (———) show the
ability of the image analyser to compensate for fading on exposure
to UV light in the fresh and stored specimens. From Cowen et al.,
1984a.

mens (Fig. 4). We then tested the ability of the analyser to compensate for the fading of fluorescence over the same period of exposure to UV light by adjusting white level sensitivity prior to each measurement. The corrected curves (Fig. 4) showed that the analyser compensated in a remarkably consistent way for any loss of intensity caused by exposure to UV light. Measurement of varicosity numbers showed similar fading curves and consistent correction by adjusting the analyser's sensitivity (Fig. 5).

Comparing the specimens examined immediately after preparation with those stored for 3 weeks, it can be seen that despite some variability on the ordinate scale, there was no loss of fluorescence due to storage. The mean fluorescent area for the three fresh specimens was 2.6+0.15% whilst that for the three-week stored specimens was 2.7+0.49%. Varicosity numbers were 918+90/mm^2 and 1141+146/mm^2, respectively. The differences between these means were not statistically significant.

PONTAMINE SKY BLUE: A NEW STAIN FOR THE REDUCTION OF BACKGROUND AUTOFLUORESCENCE

We have become increasingly interested in the changes that occur in perivascular nerves in various autonomic diseases and old age in man. A major obstacle in this work has been the presence of high levels of non-specific background autofluorescence sufficient to obscure the specific neuronal labelling, particularly in human blood vessels (Fig. 6A). This problem exists for both catecholamine fluorescence and immunofluorescence.

Non-specific background autofluorescence is the product of the light-refractile qualities of some tissues, particularly elastic and collagenous connective tissue. Previous attempts have been made using Evans Blue to stain these connective tissue elements blue (de la Lande and Waterson, 1968). The dye-impregnated tissue then absorbs blue light and appears red under the UV fluorescence illumination. However, Evans Blue is a carcinogen (Lewis and Talken, 1982) and and alternative blue dye, Pontamine Sky Blue (PSB), has recently been shown to be compatible with catecholamine fluorescence histochemistry (McGinty et al., 1979, 1982).

We have tested the ability of PSB to reduce non-specific autofluorescence by a similar counter-staining process in preparations for catecholamine fluorescence and immunofluorescence histochemistry. Fig. 6B shows the noradrenergic nerve plexus of the same human blood vessel after staining with PSB and illustrates the successful reduction of background autofluorescence. Image analysis has been employed to compare nerve densities in fluorescence preparations with and without PSB staining. Dimethyl sulphoxide and pre-incubation in trypsin were used to assist penetration of the dye. The guinea-pig carotid artery, which has

Fig. 6 Whole-mount stretch preparations of human mesenteric vein. The perivascular noradrenergic nerves are demonstrated with glyoxylic acid fluorescence histochemistry. A. Specimen prepared with glyoxylic acid only; note the high non-specific background autofluorescence. B. Specimen prepared with glyoxylic acid and pontamine sky blue; note the low non-specific background autofluorescence. Scale bar 50 μm. From Cowen et al., 1984b.

relatively high autofluorescence, was compared with the guinea-pig mesenteric artery, with very low autofluorescence. Both arteries contain noradrenergic and substance P-containing nerve plexuses. Table 2 compares noradrenergic nerve densities in fluorescence preparations using glyoxylic acid alone and in combination with PSB. No significant differences were shown in the carotid or mesenteric arteries between the two groups of specimens.

Substance P-like immunofluorescent nerves were demonstrated in the two arteries using the indirect labelling technique (Coons et al. 1955), with FITC as the second label. Again, preparations were made with and without PSB staining and compared using image analysis. PSB staining was carried out after the second layer had been added and the same penetrating aids were used as before. Table 3 shows that there were no significant differences between the two groups of specimens in either the carotid or mesenteric arteries.

These techniques have been successfully applied to nerve plexuses of human blood vessels containing the neuropeptide VIP (Fig. 7). The methods are also compatible with immunofluorescence staining for 5-HT and for Neuropeptide-Y.

TABLE 2

Comparison of noradrenergic nerve densities with and without pontamine sky blue counterstaining.

MESENTERIC ARTERY (n = 4)	Glyoxylic acid	P	Glyoxylic acid + PSB
Fluorescent area (%)	4.4+0.25	NS	4.8+0.53
Intercept density (mm^{-1})	22+0.9	NS	25+2.0
Varicosity diameter (μm)	2.2+0.05	NS	2.1+0.12
Varicosity number (mm^{-2})	3188+295	NS	2858+321
CAROTID ARTERY (n = 4)			
Fluorescent area (%)	4.8+0.38	NS	5.8+0.49
Intercept density (mm^{-1})	23+1.8	NS	26+2.3
Varicosity diameter (μm)	3.1+0.15	NS	3.3+0.25
Varicosity number (mm^{-2})	4366+555	NS	3938+269

Comparisons of perivascular noradrenergic nerve density in stretch preparations of the mesenteric and carotid arteries of the guinea-pig stained with glyoxylic acid alone, and with glyoxylic acid and pontamine sky blue. Measurements are made using image analysis and expressed as means+S.E.M. NS = no significant difference between means using student 't' test. From Cowen et al., 1984b.

TABLE 3

Comparison of substance P-positive nerve densities with and without pontamine sky blue counterstaining.

MESENTERIC ARTERY (n = 4)	Substance P	P	Substance P + PSB
Fluorescent area (%)	1.9+0.08	NS	2.4+0.46
Intercept density (mm^{-1})	10+0.5	NS	11+1.8
Varicosity diameter (μm)	1.5+0.10	NS	1.7+0.24
Varicosity number (mm^{-2})	1726+263	NS	1879+171
CAROTID ARTERY (n = 4)			
Fluorescent area (%)	0.6+0.13	NS	0.8+0.15
Intercept density (mm^{-1})	2.3+0.36	NS	2.7+0.47
Varicosity diameter (μm)	2.4+0.08	NS	2.6+0.04
Varicosity number (mm^{-2})	471+87	NS	479+65

Table 3   Comparisons of perivascular substance P-like immuno-fluorescent nerves, labelled with FITC, in stretch preparations of the mesenteric and carotid arteries of the guinea-pig labelled with FITC alone, and with FITC followed by staining with pontamine sky blue. Measurements are made using image analysis and expressed as means+S.E.M.   NS = no significant difference between means using students 't' test.   From Cowen et al., 1984b.

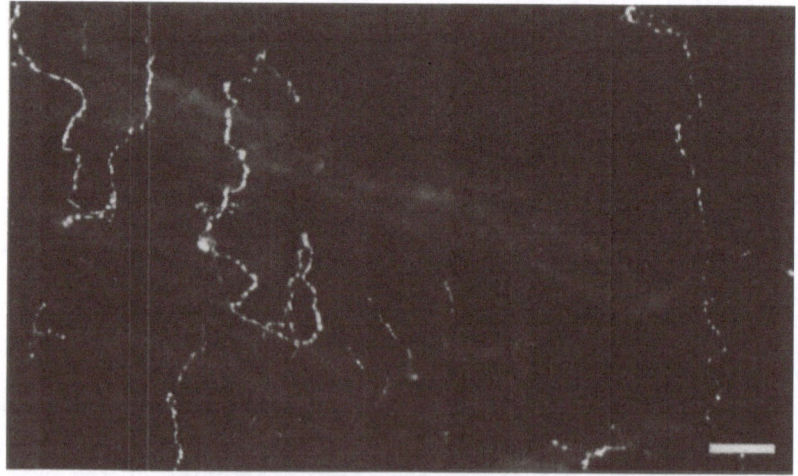

Fig. 7   Whole-mount stretch preparations of human mesenteric vein; the perivascular VIP-containing nerves are labelled with FITC and then stained with pontamine sky blue. Scale bar 50 μm.   From Cowen et al., 1984b.

FUTURE DIRECTIONS

The reduction of background autofluorescence using Pontamine Sky Blue may allow us to establish an objective method for setting analyser detection levels whereby non-varicose and varicose fluorescence intensities can be separated and distinguished from the background. The presence of highly variable autofluorescence has previously made this impossible. Characteristic brightness curves for the fluorescence associated with different neurotransmitters would form the basis for such a system, eliminating the subjective element involved in the present method. The new generation of image analysers with powerful computing capacity linked to more sensitive scanners may help in resolving these problems.

We are now using these techniques to investigate age-changes in peptide-containing perivascular nerves in animal and human tissues, autonomic neuropathies in man and in animal models and the possibility of reciprocal regulatory relationships between different groups of autonomic nerves.

* This work was funded by grants from the Medical Research Council (U.K.).

REFERENCES

Amenta, F., Mioni, M.C., and Napoleone, P. (1983). The autonomic innervation of the vasa nervorum. J. Neural Transmission 58, 291-297.
Bacopoulos, N.G., Bhatnagar, R.K., Schnute, W.J., and Van Orden, L.S. (1975). On the use of the fluorescence histochemical method to estimate catecholamine content in brain. Neuropharmacology, 14, 291-299.
Bevan, J.A., Bevan, R.D., Purdy, R.E., Robinson, C.P., Su, C., and Waterson, J.G. (1972). Comparison of adrenergic mechanisms in an elastic and a muscular artery of the rabbit. Circ.Res., 30, 541-548.
Caspersson, T., Hillarp, N-A., and Ritzen, M. (1966). Fluorescence microspectrophotometry of cellular catecholamines and 5-hydroxytryptamine. Exp.Cell.Res., 42, 415-428.
Coons, A.H., Leduc, E.H., and Conolly, J.M. (1955). Studies on antibody production. 1. A method for the 'histochemical demonstration of specific antibody and its application to a study of the hyperimmune rabbit. J.exp.Med., 102, 49-60.
Cowen, T., and Burnstock, G. (1980). Quantitative analysis of the density and pattern of adrenergic innervation of blood vessels. Histochemistry, 66, 19-34.
Cowen, T., and Burnstock, G. (1982). Image analysis of catecholamine fluorescence. Brain Res.Bull., 9, 81-87.
Cowen, T., and Burnstock, G. (1984a). Image analysis can compensate

for fading in measuring FITC-immunofluorescence. Histochemistry, submitted for publication.
Cowen, T., Haven, A.J., and Burnstock, G. (1984b). Pontamine sky blue: a counter-stain for background autofluorescence in catecholamine fluorescence and immunohistochemistry. Histochemistry, submitted for publication.
Dählström, A., Häggendal, J., and Hökfelt, T. (1966). The noradrenaline content of the varicosities of sympathetic adrenergic nerve terminals in the rat. Acta physiol.scand., 67, 289-294.
Doležel, S. (1973). Uber die Variabilität der adrenergen Innervation der groB en GefäBe. Acta anat., 85, 123-132.
Furness, J.B., and Costa, M. (1975). The use of glyoxylic acid for the fluorescence histochemical demonstration of peripheral stores of noradrenaline and 5-hydroxytryptamine in whole mounts. Histochemistry, 41, 335-352.
Gallen, D.D., Cowen, T., Griffith, S.G., Haven, A.J., and Burnstock, G. (1982). Functional and non-functional perivascular nerve-smooth muscle transmission in the renal arteries of the rabbit and guinea-pig: a developmental study. Blood Vessels, 19, 237-246.
Gerová, M., Gero, J., Doležel, S., and Konecny, M. (1974). Postnatal development of sympathetic control in canine femoral artery. Physiol.Bohemoslov., 23, 289-295.
Griffith, S.G., Crowe, R., Lincoln, J., Haven, A.J., and Burnstock, G. (1982). Regional differences in the density of perivascular nerves and varicosities, noradrenaline content and responses to nerve stimulation in the rabbit ear artery. Blood Vessels, 19, 41-52.
Jonsson, G., (1971). Quantitation of fluorescence of biogenic amines. Prog.Histochem.Cytochem., 2, 299-334.
de la Lande, and Waterson, J.G., (1968). Modification of autofluorescence in the formaldehyde-treated rabbit ear artery by Evans Blue. J.Histochem.Cytochem., 16, 281.
Lewis, R.J., and Tatken, R.L. (1982). Registry of toxic effects of chemical substances (1980 edition, vol.2) p.142. US Department of Health and Human Services.
Lindvall, O., and Björklund, A. (1974). The glyoxylic acid fluorescence histochemical method: a detailed account of the methodology for the visualisation of central catecholamine neurons. Histochemistry, 39, 97-127.
Löfström, A., Jonsson, G., Wiesel, F.A., and Fuxe, K. (1976). Microfluorimetric quantitation of catecholamine fluorescence in rat median eminence. II. Turnover changes in hormonal states. J.Histochem.Cytochem., 24, 430-442.
Malmfors, T. (1965). Studies on adrenergic nerves. Acta.Physiol.Scand., 64, Suppl.248, 1-93.
McGinty, J.F., Koda, L.Y., and BLoom, F.E. (1979). A novel fluorescent marker of CNS vasculature used in combination with monoamine histofluorescence. Neurosci.Abstr., 5, 344.
McGinty, J.F., Milner, T.A., and Loy, R. (1982). Association of sympathetic axons in denervated hippocampus to intracerebral vascu-

lature. 1. Fluorescence histochemistry combining glyoxylic acid and pontamine sky blue. Anat.Embryol.(Berl)., 164, 95-100.
Ritzen, M. (1966). Quantitative fluorescence microspectrophotometry of catecholamine-formaldehyde products. Exp.Cell.Res., 44, 505-520.
Schipper, J., Tilders, F.J.H., and Ploem, J.S. (1978).
Microfluorimetric scanning of sympathetic nerve fibres; an improved method to quantitate formaldehyde-induced fluorescence of biogenic amines. J.Histochem.Cytochem., 26, 1057-1066.
Schipper, J., Tilders, F.J.H., and Ploem, J.S. (1980).
Extraneuronal catecholamine as an index for sympathetic activity in the iris of the rat: a scanning microfluorimetric study. In:
Histochemistry and cell biology of autonomic neurones, SIF cells and paraneurons. Eds. Eranko, O., Soinila, S. and Paivarinta, H.
Raven Press, New York. pp. 745-751.
Schipper, J., Tilders, F.J.H., and Mulder, A.H. (1980).
Extraneuronal catecholamine in the iris of the rat: a consequence of nonsynaptic neurotransmission? Neuroscience, 5, 745-751.
Sternberger, L.A. (1974). The unlabeled antibody enzyme method. In:
Immunocytochemistry. Prentice Hall Inc. New Jersey. pp. 129-171.
Thorbert, G., Alm, P., Owman, Ch., Sjoberg, N-O., and Sporrong, B.
(1978). Regional changes in structural and functional integrity of myometrial adrenergic nerves in pregnant guinea-pig, and their relationship to the localisation of the conceptus. Acta physiol.scand., 103, 120-131.
Tilders, F.J.H., Ploem, J.S., and Smelik, P.G. (1974). Quantitative microfluorimetric studies on formaldehyde-induced fluorescence of 5-hydroxytryptamine in the pineal gland of the rat.
J.Histochem.Cytochem., 22, 967-975.

IMAGE ANALYSIS OF TRANSMITTER IDENTIFIED NEURONS USING THE IBAS SYSTEM

OLLE JOHANSSON and HÅKAN HALLMAN

Department of Histology, Karolinska Institutet, Stockholm, Sweden

INTRODUCTION

The introduction of advanced technology and a much more flexible hardware concept now means that image analysis systems can be used efficiently in diverse fields of scientific research and industrial inspection work. Thus, digital image processing has furnished mankind with some highly spectacular products: enhanced satellite photographs of the planets (including the earth), colourful representations of more distant celestial objects, aerial photography analysis, automatic material testing, industrial quality control and medical diagnosis such as images assembled from X-ray data to show a cross section of the human body (see e.g. IEEE, 1983). Technological advances have increased both the sensitivity, resolution and speed of the detecting systems as well as the computing power of the host systems. In the present chapter some studies using the Zeiss/Kontron IBAS interactive image analyser in conjunction with typical problems of neurobiology will be presented. The communication will review some preliminary data obtained regarding the distribution of transmitter identified nerve cells in a brain nucleus as well as morphometric analysis both at the light and electron microscopic level.

EXPERIMENTAL PROCEDURES

Hardware Specifications

The images were analyzed by a commercially available interactive image analysis system (IBAS, Zeiss/Kontron, Eching bei München, GFR). Via TV-monitors the investigator, with an electronic cursor, performs a 'dialogue' with the computer through which the measuring and evaluation program functions, interactive image editing as well as the control of various operations are activated by initializing commands at a 'menuefield'. The system used is highly user-friendly and enables the investigator either to interactively work with the image or allowing the computer to automatically process it. In this way, an interactive 'editing' of the image is possible e.g. for exclusion of artefactual areas and non-specific background, enhancing the structural content of the image, etc.

The hardware consists of two parts; IBAS 1 (host computer) and IBAS 2 (image processing unit). IBAS 1 operates as a stand-alone, semi-automatic data acquisition and data processing unit. In combination with IBAS 2, it functions as the 'host' computer and assists in the fully automatic image analysis evaluations.

The IBAS 1 is a microcomputer with a Z80A central processing unit, 64 kByte random access memory, 16 kByte graphics memory with a display-processor; 2 floppy disk drives (8"; capacity 1 MByte each); a B/W (black and white) monitor for user-dialogue with the IBAS 2 program as well as display of results of morphometric analysis and graphic information; and a digitizer tablet (280x280 mm, geometric resolution <0.1 mm) for manual structure tracing, to handle the menuefields and for interactive measurements and image editing.

The IBAS 2 is a fast, microprogrammable array-processor (cycle time 100 ns), connected to the IBAS 1 host computer by a 16 bit direct memory access interface. The processor is 'pipeline' structured with 16 bit arithmetic logic unit, hardware multiplier, 6 kByte microprogram memory, 4Kx17 bit cache memory and 1Kx16 bit register file. The digital grey image memory can be user configurated up to 744x512 picture elements (=pixels) in the present setup. The grey level resolution is 8 binary digits (=bits) per pixel, i.e. 256 grey levels. There is in addition 1 bit/pixel for graphic overlay. The video-digitizer and video-generator for real-time-digitizing of TV-signals with 8 bit resolution has a conversion rate of up to 20 MHz, the conversion time for an image is thus in the order of 0.1 sec. The RGB (red, green, blue)-output to the high resolution monitor (band width 14 MHz) is transformed in a user configurated 'look-up-table' where real-time grey level-transformations are made and pseudo-colour coding can be done.

The photographic images were obtained by using an ordinary 35 mm camera set-up at the B/W or RGB monitor. Kodak Tri-X or Ektachrome 400 film was used for photography.

Software Specifications

The standard software of the IBAS contains so many evaluation possibilities that practically any measurement problem can be solved. In combination with the free programmability of the IBAS 1 and 2, the possibility exists for the investigator to write his own programs in BASIC, FORTRAN or Microcode-Assembler. Due to space limitations, it is impossible to give all the details and possibilities regarding the IBAS 1 and 2 programs and thus, in the following a highly condensed protocol for the standard software is given:

IBAS 1

Basic program	Semi-automatic measurement of count, area, area differences, perimeter, length distance, orientation, projected lengths, maximum and minimum diameters, shape factors, coordinate locations, center of mass coordinates, etc. Display of results in individual data listing, sum values with basic statistics and graphic histogram.
Statistics program	Arithmetic, geometric and harmonic means, variance, skewness, excess, correlation, t-test, U-test, etc.
Distribution analysis	Auto-scaling and selective parameter histograms (lin/log), 2-parameter analysis (scatter diagram, double classification), etc.
Stereology program	Volume, density, surface density, specific surface, mean linear size, mean linear distance, mean curvature, mean diameter, mean volume. Area-, linear- and point analyses.
Data editor	Modification of stored data files, data channels and individual data including parameter arithmetic and manual input for further data processing.

IBAS 2

Image digitization	Addition, integration, realtime presentation with grey level transformation or pseudo-colour coding, storage of slow scan signals, storage and display of images with real colour.
Image enhancement	Linear and non-linear contrast enhancement functions, different filters for edge detection, texture enhancement, noise reduction, averaging, dilation and erosion in grey level images, shading correction.
Segmentation	Discrimination of grey ranges with one or two thresholds, multiphase-discrimination, adaptive thresholding, halo-effect-correction, grey level thinning algorithm (skeletonization), region growing algorithm and optimized segmentation functions for special applications.
Binary and multiphase processing	Erosion, dilation, opening (ouverture), closing (fermeture) with variable size and shape of the structuring element, superimposing of binary and grey level images, masking, logical binary image operations, skeletonization, automatic 'fill-in'.
Parameter definition	Feature-specific measurements: area, perimeter, convex perimeter, equivalent circle diameter, feret-diameter, maximum and minimum diameter, center of gravity, shape factors, orientation, etc. Field specific measurements: count, area percentage, area sum, perimeter sum, chord length distribution. Densitometric measurements: measurement of grey levels, transmission extinction, optical density, reflectivity. All measurements object or field specific.

Parameter arithmetic	Every arithmetic combination of basic parameters. Classification with regard to the values (e.g. for pattern recognition).
Display of results	Single value list or distribution analysis with histogram and classification list.
Interactive image editor	Cut, link, erase and automatic 'fill-in'. Selection of 'regions of interest'. Intensity measurements: histograms, field or object specific, local grey level distributions (point, matrix and profile measurement), interactive segmentation, interactive geometric measurements. Most functions can be executed not only in the binary picture, but also in the original grey level and the identified picture.
Grey level image operations	Picture arithmetic (add, subtract, multiply), coordinate-transformation, zooming, scrolling and copying of parts of the picture, copying of grey level into the overlay-plane, storage and loading of images to/from disk.

Comments on the Image Analysis Technique

There are several instances where morphometrical techniques are of great importance for functional studies. Quantitative morphological techniques have a high structure resolution, sensitivity and reproducibility and, in contrast to biochemical methods, leaves the tissue examined intact. However, biochemistry mostly allows for a more rapid handling with a great number of samples. Agnati, Fuxe and their collaborators have introduced new methods based on computer-assisted morphometry to objectively characterize transmitter identified neurons at the pre- and postsynaptic level (Agnati et al., 1982a, b, c, d, e, 1983; Fuxe et al., 1982). They have recently also summarized much of their elegant concepts regarding the criteria one has to set to assess different morphometric and stereologic parameters of transmitter identified neurons, such as criteria used for the description of a cell body group, description of the morphological features of nerve cell somata, dendrites and nerve terminals as well as the description of coexistence using the so-called "occlusion" and "overlap" methods (see Agnati et al., 1984; see also this Volume).

A study of the distribution of NADPH-diaphorase positive cell bodies in the caudate nucleus was made. Frontal sections were studied in transmitted light by help of a macro-zoom objective on the black and white TV-camera (RCA TC1005XC) allowing the entire caudate nucleus on one side to be studied. In the interactive measuring mode on IBAS 2 all cell bodies were marked and the gravity center coordinates calculated and stored, thereafter the coordinates of some points lying along the border of the nucleus was determined and stored. Morphometric analysis was also performed on the same sections at 250x magnification in the light microscope. A visual field containing cell bodies was digitized and stored, and a threshold was chosen to generate a binary picture containing the cell bodies. The dendrites and the axons was then cut off from the cell bodies in the interactive editing mode. Some morphological parameters were chosen (area, perimeter, projected size on the X and Y axes and form factor) and the cell bodies were selected out for measurement.

For the coexistence analysis study, cellular patterns in tissue sections obtained after immunohistochemistry for 5-hydroxytryptamine (5-HT) and substance P, respectively, were investigated (cf. Johansson and Hallman, 1984). The histochemical sections were investigated in the fluorescence microscope using the TV-camera. Adjustment for uneven background illumination and enhancement of contrast was made, thereafter a grey level threshold was chosen to select for the cells excluding most of the background in the population. After generation of the binary picture and median filtering to remove 'noise' elements, remaining non-specific background structures and artefactual areas in the image can be excluded from the field of view if necessary. The resulting binary images were assigned different levels of grey whereafter the images were added together. Overlapping structures then showed up in white while non-overlapping structures still had their original shade of grey (Fig. 8C; cf. Johansson and Hallman, 1984). It should be noted that in normal use the different grey levels are assigned sharply contrasting colours by help of the pseudo-colour feature of the IBAS.

Semiautomatic morphometric analysis of vesicles from enkephalin-containing nerve terminals in the adrenal medulla of the guinea pig was made on electron micrographs using the IBAS 1 unit. The outlines of the vesicles were traced on the digitizing board and the data (area, perimeter, maximum diameter, form factor, etc.) stored on a data disc for later statistical analysis.

Animals

Adult male albino rats or guinea-pigs were used. The animals were perfused via the ascending aorta with ice-cold 10% formalin for 20-30 min. After perfusion, the tissues were rapidly dissected out and immersed in the same fixative for about 90 min. After rinsing in ice-cold 5% sucrose in 0.1 M Sörensen's phosphate buffer for at least 24 h the tissues were quickly frozen and cut in 10-15 μm thick coronal sections on a Dittes cryostat (Dittes, Heidelberg, GFR). The sections were then processed for immunohistochemistry (see below).

Immunofluorescence Procedure

The indirect fluorescence histochemical technique of Coons and collaborators (see Coons, 1958) was used as adopted for transmitter histochemistry (see Hökfelt et al., 1973; Johansson, 1983). Cryostat sections of the various tissues were incubated with specific antisera or control sera (see below), diluted 1:100-1:400. The sections were then rinsed in 0.1 M phosphate buffered saline (PBS) for 15 min prior to incubation with sheep or swine anti-rabbit or anti-rat antibodies conjugated to fluorescein-isothiocyanate (FITC), diluted 1:4 or 1:10. All antisera contained 0.3% Triton X-100 (see Hartman et al., 1972). After another rinse in PBS for 15 min the sections were mounted in a mixture of glycerol and PBS (3:1) and examined in a Zeiss fluorescence microscope.

As controls, all antisera were routinely absorbed with the respective peptide in excess amounts (=blocked antisera). Only immunofluorescence which was abolished by such absorbtion was regarded as specific. In the coexistence situations the antisera were, in addition, tested for cross-reactivity in the immunohistochemical model. No evidence for cross-reactivity was observed.

Peroxidase-Antiperoxidase Procedure - Electron Microscopy

For ultrastructural immunocytochemical studies the animals were processed according to the technique of Johansson (1983). Briefly, they were perfused for 5 min via the ascending aorta with ice-cold 10% formalin containing 0.2 or 0.5% glutaraldehyde. After perfusion, the tissues were immersed in the same fixative for 90 min, followed by immersion for 1-3 days in 10% formalin solution, rinsed for at least 24 h in PBS with 5% sucrose added, sectioned on an Oxford Vibratome (San Mateo, California, USA; 50-100 μm thick sections) and processed for immunocytochemistry (see Sternberger, 1979).

The sections were incubated with 20% normal sheep serum for 15-30 min followed by antiserum or control serum, diluted 1:100-1:2,000, rinsed in PBS with 5% sucrose, incubated with sheep antiserum to rabbit immunoglobulin, diluted 1:50, rinsed as described above and incubated with horseradish peroxidase anti-peroxidase (PAP) complex, diluted 1:35, rinsed as described above,

followed by treatment for 1 or 2 min with a solution of 3,3'-di-
aminobenzidine tetrahydrochloride (DAB) and hydrogen peroxide
according to Graham and Karnovsky (1966). All sera were diluted in
PBS containing 1% normal sheep serum and 0.1% Triton X-100. After
treatment with DAB and hydrogen peroxide, the tissues were immersed
in 2% OsO$_4$, rinsed in Ringer's solution, dehydrated and embedded in
Epon 812. Semithin (2-3 μm) and ultrathin (silver-gold) sections
were cut on an LKB Pyramitome and a Reichert Om U3 ultramicrotome,
respectively. The semithin sections were examined in a Zeiss light
microscope using bright-field optics. The ultrathin sections,
uncontrasted or contrasted with uranyl acetate and lead citrate,
were examined in a Philips 300 electron microscope.

Controls used included blocked antisera, omitting the first
antiserum or only treating sections with DAB and hydrogen peroxide.
No evidence of unspecific reactions were observed.

Enzyme Histochemistry - NADPH-Diaphorase

Animals were processed according to Vincent and Johansson
(1983). In summary, they were perfused with 400 ml of ice-cold 10%
formalin. After perfusion the brains were removed and sections were
cut on a freezing-microtome (Leitz Wetzlar GmbH, Wetzlar, GFR) at
various thicknesses (20-200 μm). The free-floating sections were
stained for NADPH-diaphorase activity according to Scherer-Singler
et al. (1984). In this reaction, the reduced form of NADP (NADPH)
is generated endogenously by the action of malic enzyme which is
present in the tissue. The NADPH formed by this first enzyme can
then be used by the diaphorase system to reduce the dye nitro blue
tetrazolium to the visible, insoluble reaction product formazan.
After the reaction, the sections were rinsed in PBS and mounted
onto chrome-alum coated slides. The sections were coverslipped with
glycerol:PBS (3:1) and examined in a Zeiss light microscope using
bright-field optics.

As a control, the substrate was excluded in the enzymatic
reactions. No evidence for unspecific reactions were observed.

Comments on the Immunohistochemical Technique

Sections to be incubated for immunofluorescence and subsequently analyzed with image analysis may be obtained in several
ways: paraffin embedding and cutting on an ordinary microtome,
cryostat sections, freezing-microtome sections or Vibratome sections. We cut cryostat sections routinely at 10-15 μm, but under
optimal conditions a section thickness of 3-5 μm can be obtained.
Thin sections are important in experiments aimed at establishing
coexistence. A general rule seems to be that the quality of the
immunofluorescence staining obtained is partly dependent on section
thickness. Thus, it is easier to obtain good results with thick

sections, whereas the morphology of thin sections often is inferior. The problem with thicker sections (up to 40 μm) is the increase in background fluorescence. When using cryostat sectioning,
the cutting procedure is essential for good results. Generally, the
sections should be cut at a low speed. Humidity and temperature
inside the cryostat may influence the quality of the sections.

The dilution of the antisera used must be tested individually
for each antiserum and for each particular tissue and species. The
dilution of the second antiserum must also be tested, and the ratio
of specific against background fluorescence must be evaluated. Most
commercially available fluorophor-conjugated sera are nowadays of a
good quality, but their 'strength' varies considerably, which will
determine the dilution. It is our opinion that, at least in studies
on neuronal tissues and particularly the central nervous system, it
is an advantage to have a background high enough to allow recognition of land-marks and nuclei in the tissue both when examining
the tissue in the microscope as well as when examining micrographs.
This can be obtained by using the FITC conjugated antiserum, in
slightly higher concentrations.

RESULTS AND DISCUSSION

Studies of cell body distribution and individual cell morphometry

One population of forebrain neurons have been shown to contain
somatostatin and APP/NPY (Vincent et al., 1982b). Subsequently,
Vincent and collaborators (1982a, 1983) could also demonstrate that
within the striatum all of the somatostatin/APP/NPY-containing
neurons are selectively stained for NADPH-diaphorase activity.
Further fine structural analysis at the light and electron microscopic level has also been performed (Vincent and Johansson, 1983;
Johansson and Vincent, 1984) suggesting these neurons to be of the
medium-size type III nerve cells as defined by Dimova et al.
(1980). To elucidate further the distribution of these striatal
somatostatin/APP/NPY/NADPH-diaphorase neurons we have in the present article performed some preliminary studies using a newly developed procedural paradigm for the IBAS system allowing a rapid
analysis of whole sections (Johansson and Hallman, in preparation;
see also Johansson and Vincent, 1984).

A study of the distribution of NADPH-diaphorase positive cell
bodies in the caudate nucleus was made (Figs. 1-3). A statistical
analysis according to Agnati et al. (1984) did not indicate the
presence of of subgrouping within the nucleus.

In this context, it might be mentioned that in image analysis
it is very important to use histochemical techniques yielding a
high signal-to-noise ratio (Figs. 4 and 5; cf. Johansson and Backman, 1983) to allow a rapid automated analysis. Concomitant ana-

O. Johansson and H. Hallman

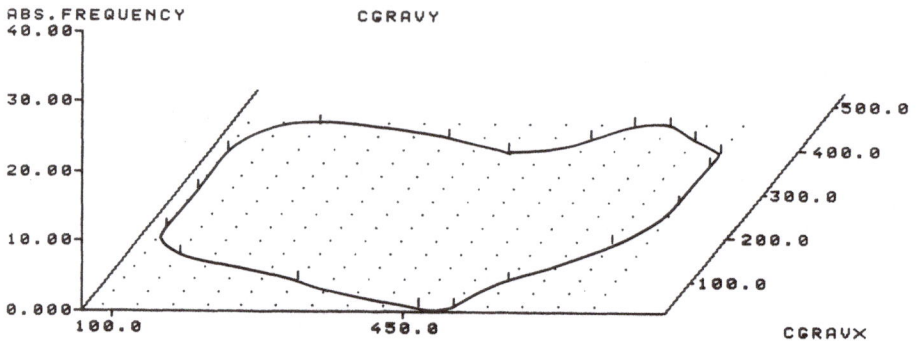

Fig. 1. Caudate nucleus border. Medial = up; ventral = left.

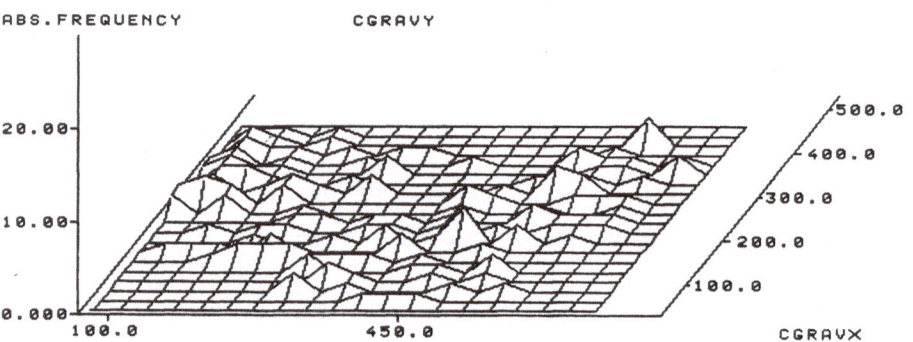

Fig. 2. 'Three-dimensional' histogram showing the number of NADPH-diaphorase positive cell bodies in 170x115 µm squares.

```
CGRAVY  UN 1 2 3 4 5 6 7 8 9100V
        . . . . . . . . . . . .OV
      . 3 1 2 . . . . . 2 1 .10
      . 5 3 1 1 2 1 . 1 2 1 . 9
      . 5 5 . 1 2 . . 2 2 4 . 8
      . 4 2 2 2 1 . 1 3 5 . . 7
      . 6 3 2 1 . . 4 2 1 3 . 6
      . 1 . 2 2 2 1 2 3 1 . 5
      . 1 . 3 1 1 3 6 3 1 1 . 4
      . . 3 3 4 3 1 2 . . . . 3
      - . . . 2 2 2 1 3 . . . 2
      . . . . 1 . 1 2 1 1 . . 1
      . . . . . . . . . . . .UN
CGRAVX
```

Fig. 3. Matrix showing the number of NADPH-diaphorase positive cell bodies in 340x230 µm squares 'superimposed' on the caudate nucleus.

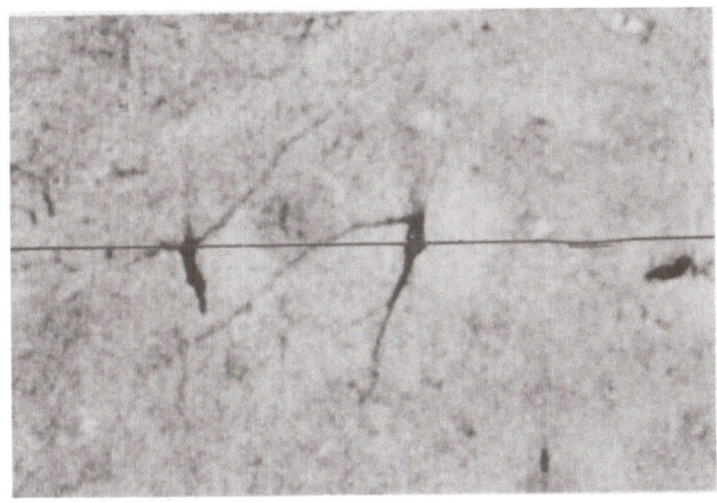

Fig. 4. NADPH-diaphorase positive cells in the caudate nucleus, part of a scanning line for a grey value profile runs across the picture.

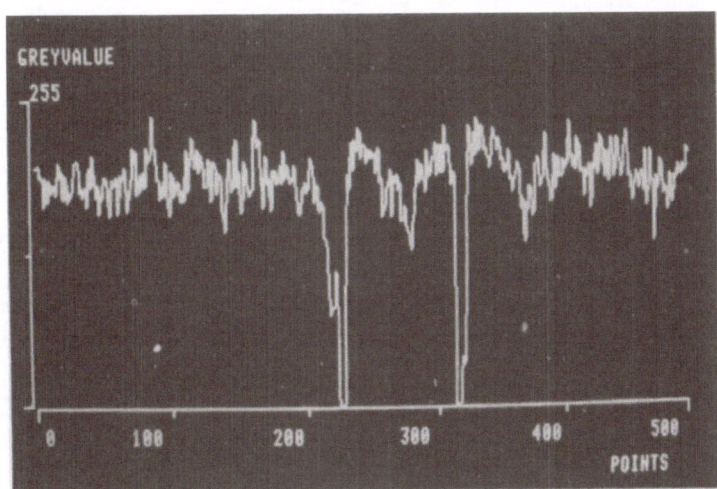

Fig. 5. Grey value profile from the NADPH-diaphorase stained section. The two cell bodies in Fig. 4 are clearly visible as two dips in the profile line.

lysis of individual cell bodies in a NADPH-diaphorase stained section (Table 1) showed no evidence of different size classes in the
cell body population (Fig. 6). There were, however, indications of
a preferential orientation of the cell bodies in a medio-lateral
direction (Fig. 7).

Table 1. Morphometric analysis of NADPH-diaphorase positive cell
bodies in nucleus caudatus[a].

Area[b]	Perimeter[c]	Proj. X[c]	Proj. Y[c]	Form factor[d]
72+3.8	41+0.8	11+0.5	13+0.6	0.55+0.021

[a] Mean $\pm$ s.e.m. (n=51); [b] μm^2; [c] μm; [d] Relationship between area
and perimeter where a circle has the maximum value of 1

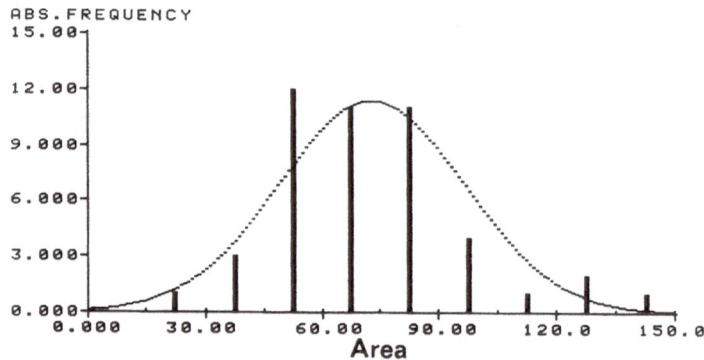

Fig. 6. Histogram showing the distribution of the area parameter of
Table 1. The Gaussian curve calculated from the area data is plotted into the histogram.

Studies of transmitter coexistence in single cells

A growing number of examples of coexistence have been observed
suggesting that coexistence may, in fact, be a common phenomenon in
the nervous system (see Hökfelt et al., 1980a, b, 1982; Lundberg
and Hökfelt, 1983). The substances may represent fairly well established transmitters, but their role is uncertain at the present
time. The results so far are largely of morphological nature and
functional considerations are mostly still not fully understood,

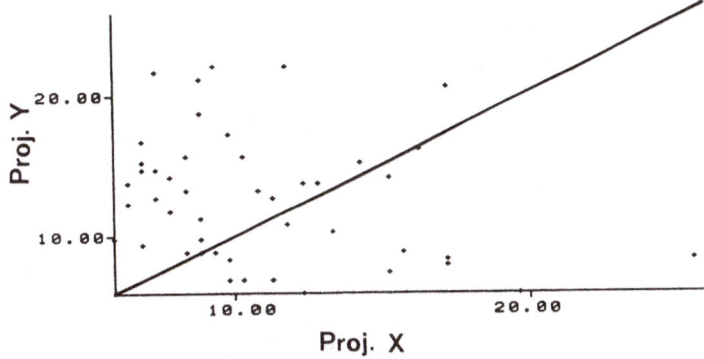

Fig. 7. Scatter diagram showing the sizes of the cell bodies projected on the X and Y axes (Proj. X and Proj. Y in Table 1).

although there are studies e.g. in the periheral nervous system
suggesting important interactions between two neuroactive compounds
(Lundberg, 1981).

In the article by Johansson and Hallman (1984), a new approach
to this problem is presented with the aim to use immunohistochemistry for investigating possible coexisting patterns of different neuroactive substances, such as peptides and monoamines, in
combination with recent advances in computerized image processing
technology. In the present investigation, we tried the image analysis program on tissue sections processed for the immunohistochemical localization of 5-HT and substance P according to Johansson and collaborators (1981) (Fig. 8A and B). The results were most
promising and it was easy to detect the respective subpopulation of
cells representing an unequivocal coexistence situation (Figs.
8A-C). In conclusion, computerized image analysis may represent an
excellent tool for evaluating possible cellular coexistence situations. The method allows for a rapid, non-biased and exact scanning
of tissue sections where a possible coexistence may be found.

Ultrastuctural analysis of neurons

Peptide neurons have been studied at the electron microscopic
level in order to elucidate the exact subcellular localization of
the peptides as well as the synaptology of these neurons. A large
number of recent electron microscopic immunocytochemical studies
demonstrate a number of peptides at the ultrastructural level,
mainly by using the PAP technique of Sternberger and colleagues
(see Sternberger, 1979). In order to further characterize the
peptide immunoreactive neurons at the ultrastructural level we have
used the morphometric capacity of the IBAS system to measure
different parameters such as area, perimeter and form factor (see

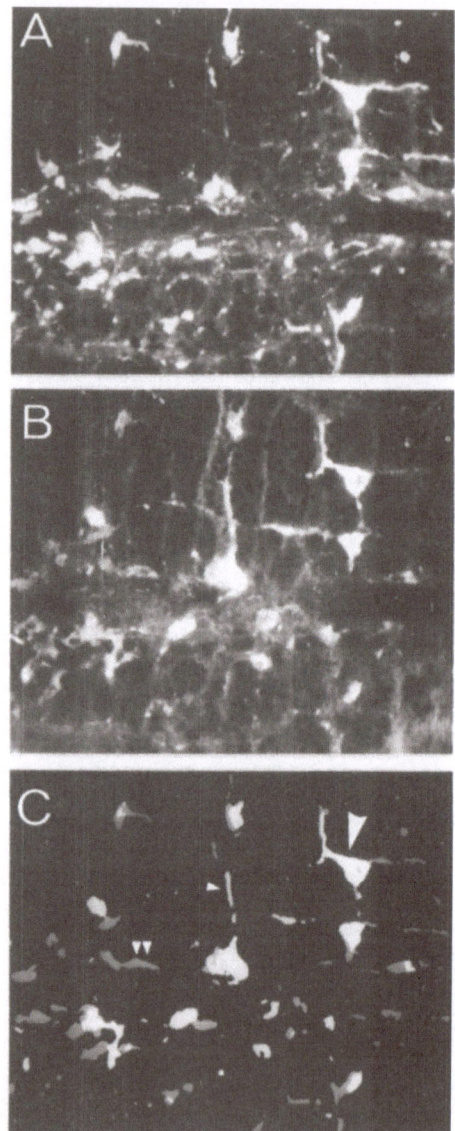

Fig. 8. (A) 5-HT immunofluorescent cell bodies in nucleus raphe obscurus. (B) The same section consecutively stained for substance P immunofluorescence. (C) Addition of the two binary images formed from (A) and (B). 5-HT fluorescence = grey (double small arrow heads); substance P fluorescence = light grey (single small arrow head); unequivocal coexistence = white (large arrow head).

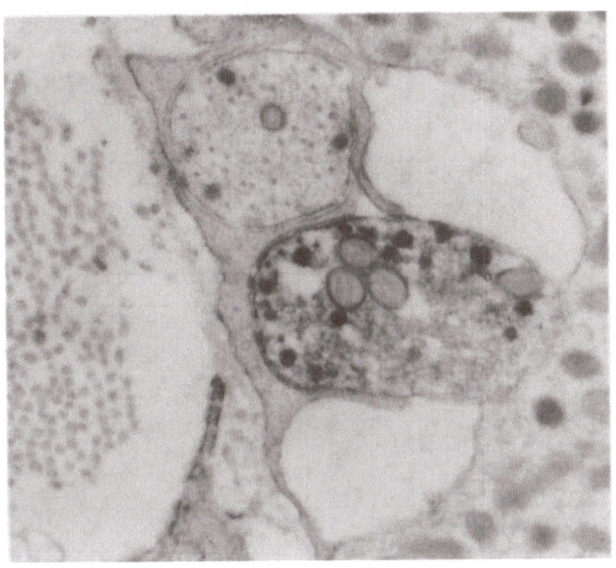

Fig. 9. Electron micrograph showing an enkephalin containing nerve terminal in guinea pig adrenal medulla.

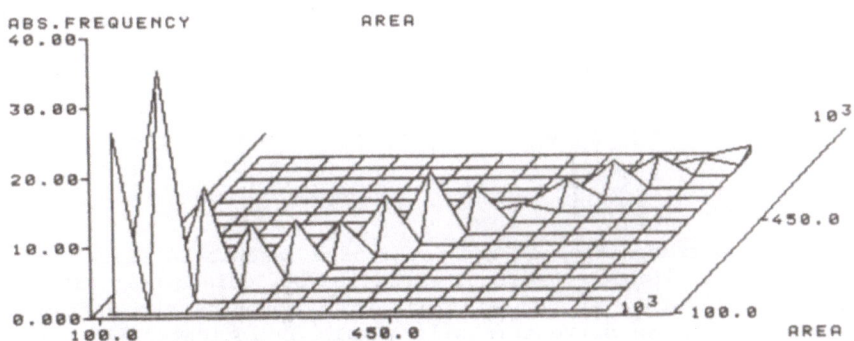

Fig. 10. 'Three-dimensional' histogram showing the size distribution of all vesicles measured in the electron micrographs of enkephalin containing nerve terminals in guinea pig adrenal medulla (note the use of Å^2).

O. Johansson and H. Hallman

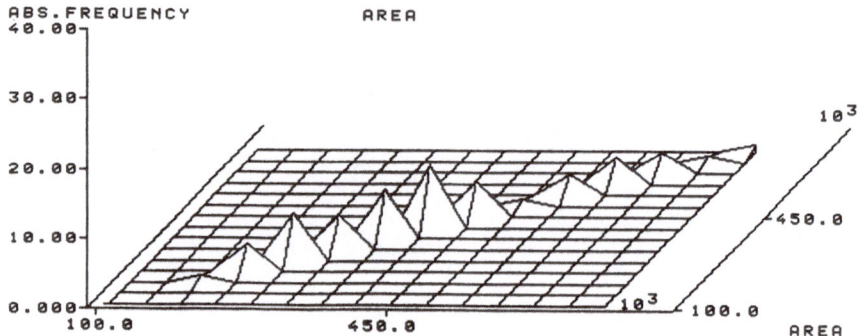

Fig. 11. 'Three-dimensional' histogram showing the size distribution of the dense core vesicles in the enkephalin containing nerve
terminals.

Table 2. Morphometric analysis of synaptic vesicles from enkephalin
containing nerve terminals in guinea pig adrenal medulla[a].

	Area[b]	Perimeter[c]	Diameter[c]	Form factor	
Large vesicles	4114+143	2322+44	730+14	0.94+0.006	(n=46)
Small vesicles	903+41	1102+15	453+7	0.95+0.005	(n=77)

[a] Mean+s.e.m.; [b] nm^2; [c] Å

Table 2) as examplified on the methionine-enkephalin immunoreactive
nerve fibers in the adrenal medulla of guinea pig (Fig. 9;
Johansson et al., in preparation). The morphometric analysis of
vesicles in these nerve terminals (Table 2) indicated the presence
of two size classes among the large dense-core vesicles (Figs. 10
and 11).

The same kind of procedure has also been used for the morphometric description of somatostatin immunoreactive nerve terminals
in the central nervous system using the immunocolloidal gold technique (Johansson et al., 1984; Foster and Johansson, 1985).

CONCLUSION

In the present study, the versatility of the IBAS instrument for analysis of transmitter identified neurons in histological sections is demonstrated by analysis of the distribution of a cell population and morphometric analysis of individual cells as well as subcellular structures.

ACKNOWLEDGEMENTS

This work was supported by grants from the Swedish Medical Research Council (14X-07162, 04X-2887, 12P-6965), the Swedish Council for Planning and Coordination of Research, Magnus Bergvalls Stiftelse and funds from Karolinska Institutet. We thank Prof. L.F. Agnati for help regarding the statistics and expert knowledge in the field of image analysis. Ms A. Peters and Ms W. Hiort are gratefully acknowledged for skilful technical assistance and Ms E. Björklund for typing the manuscript.

The antisera used in the present studies were prepared by Dr. P.C. Emson, MRC Neurochemical Pharmacology Unit, Department of Pharmacology, Cambridge, United Kingdom (substance P antiserum), and Drs. H.W.M. Steinbusch and A.A.J. Verhofstad, Department of Anatomy and Embryology, University of Nijmegen, Nijmegen, The Netherlands (5-HT antiserum).

REFERENCES

Agnati, L.F., Fuxe, K., Benfenati, F., Zini, I., Zoli, M., Fabbri, L. and Härfstrand, A. (1984). Computer assisted morphometry and microdensitometry of transmitter identified neurons with special reference to the mesostriatal dopamine pathway. Methodological aspects. Acta physiol. scand., Suppl. 532, 5-36.

Agnati, L.F., Fuxe, K., Calza, L., Benfenati, F., Battistini, N., Zini, I., Fabbri, L. and Goldstein, M. (1983). Characterization of striatal ibotenate lesions and of 6-hydroxydopamine induced nigral lesions by morphometric and densitometric approaches. In Excitotoxins (eds. K. Fuxe, P. Roberts and R. Schwarcz). Macmillan Press, London.

Agnati, L.F., Fuxe, K., Calza, L., Hökfelt, T., Johansson, O., Benfenati, F. and Goldstein, M. (1982a). A morphometric analysis of transmitter identified dendrites and nerve terminals. Brain Res. Bull., 9, 53-60.

Agnati, L.F., Fuxe, K., Hökfelt, T., Benfenati, F., Calza, L., Johansson, O. and De Mey, J. (1982b). Morphometric characterization of transmitter-identified nerve cell groups: Analysis of mesencephalic 5-HT nerve cell bodies. Brain Res. Bull., 9, 45-51.

Agnati, L.F., Fuxe, K., Locatelli, V., Benfenati, F., Zini, I., Panerai, A.E., El Etreby, M.F. and Hökfelt, T. (1982c) Neuroanatomical methods for the quantitative evaluation of coexistence of transmitters in nerve cells. Analysis of the ACTH- and beta-endorphin immunoreactive nerve cell bodies of the mediobasal hypothalamus of the rat. J. Neurosci. Meth., 5, 203-214.

Agnati, L.F., Fuxe, K., Zini, I., Benfenati, F., Hökfelt, T. and De Mey, J. (1982d). Principles for the morphological characterization of transmitter-identified nerve cell groups. J. Neurosci. Meth., 6, 157-167.

Agnati, L.F., Fuxe, K., Zini, I., Calza, L., Benfenati, F., Zoli, M., Hökfelt, T and Goldstein, M. (1982e). A new approach to quantitate the density and antigen contents of high densities of transmitter-identified terminals, immunocytochemical studies on different types of tyrosine hydroxylase immunoreactive nerve terminals in nucleus caudatus putamen of the rat. Neurosci. Lett., 32, 253-258.

Coons, A.H. (1958). Fluorescent antibody methods. In General Cytochemical Methods, Vol.1 (ed. J.F. Danielli). Academic Press, New York.

Dimova, R., Vaillet, J. and Seite, R. (1980). Study of the rat neostriatum using a combined Golgi-electron microscope technique and serial sections. Neuroscience, 5, 1581-1596.

Foster, G.A. and Johansson, O. (1985). Ultrastructural morphometric analysis of somatostatin-like immunoreactive neurones in the rat central nervous system after labelling with colloidal gold. Brain Res., submitted.

Fuxe, K., Agnati, L.F., Ganten, D., Lang, R.E., Calza, L., Poulsen, K. and Infantellina, F. (1982). Morphometrical evaluation of the coexistence of renin-like and oxytocin-like immunoreactivity in nerve cells of the paraventricular hypothalamic nucleus of the rat. Neurosci. Lett., 33, 19-24.

Graham, R.C. Jr and Karnovsky, M.J. (1966). The early stages of absorption of injected horseradish peroxidase in the proximal tubules of mouse kidney: ultrastructural cytochemistry by a new technique. J. Histochem. Cytochem., 14, 291-302.

Hartman, B.K., Zide, D. and Udenfriend, S. (1972). The use of dopamine-β-hydroxylase as a marker for the central noradrenergic nervous system in rat brain. Proc. natn. Acad. Sci. U.S.A., 69, 2722-2726.

Hökfelt, T., Fuxe, K., Goldstein, M. and Joh, T.H. (1973). Immunohistochemical localization of three catecholamine synthesizing enzymes: aspects on methodology. Histochemie, 33, 231-254.

Hökfelt, T., Johansson, O., Ljungdahl, Å., Lundberg, J.M. and Schultzberg, M. (1980a). Peptidergic neurones. Nature, 284, 515-521.

Hökfelt, T, Lundberg, J.M., Schultzberg, M., Johansson, O., Ljungdahl, Å. and Rehfeld, J. (1980b). Coexistence of peptides and putative transmitters in neurons. In Neural Peptides and Neuronal Communication (eds. E. Costa and M. Trabucchi). Raven Press, New York.

Hökfelt, T., Lundberg, J.M., Skirboll, L., Johansson, O., Schultzberg, M. and Vincent, S.R. (1982). Coexistence of classical transmitters and peptides in neurones. In Co-transmission (ed. A.C. Cuello). MacMillan Press, London.

IEEE (1983). Proceedings of the 1983 Computer Vision and Pattern Recognition Conference. IEEE Computer Society Press, Maryland.

Johansson, O. (1983). Peptide neurons in the central and peripheral nervous system. Light and electron microscopic studies. Doctoral Dissertation, Stockholm.

Johansson, O. and Backman, J. (1983). Enhancement of immunoperoxidase staining using osmium tetroxide. J. Neurosci. Meth., 7, 185-193.

Johansson, O., Foster, G.A. and Hökfelt, T. (1984). EM-immunocytochemistry of transmitter identified neurons. In Proceedings of the 1984 Scandinavian Society for Electron Microscopy Meeting (Abstr.).

Johansson, O. and Hallman, H. (1984). The use of interactive image analysis for demonstrating coexistence. Neurochem. Internat., in press.

Johansson, O., Hökfelt, T., Pernow, B., Jeffcoate, S.L., White, N., Steinbusch, H.W.M., Verhofstad, A.A.J., Emson, P.C. and Spindel, E. (1981). Immunohistochemical support for three putative transmitters in one neuron: Coexistence of 5-hydroxytryptamine, substance P- and thyrotropin releasing hormone-like immunoreactivity in medullary neurons projecting to the spinal cord. Neuroscience, 6, 1857-1881.

Johansson, O. and Vincent, S.R. (1984). Coexistence of somatostatin- and NPY-like immunoreactivity in the forebrain. In Proceedings of the VIIth International Congress of Histochemistry and Cytochemistry (Abstr.).

Lundberg, J.M. (1981). Evidence for coexistence of vasoactive intestinal polypeptide (VIP) and acetylcholine in neurons of cat exocrine glands. Morphological, biochemical and functional studies. Acta Physiol. Scand., Suppl. 496, 1-57.

Lundberg, J.M. and Hökfelt, T. (1983). Coexistence of peptides and classical transmitters. Trends in NeuroSciences, 6, 325-333.

Scherer-Singler, U., Kimura, H., Vincent, S.R. and McGeer, E.G. (1984). The NADPH-diaphorase technique: Methods description. J. Comp. Neurol., in preparation.

Sternberger, L.A. (1979). Immunocytochemistry, 2nd Edn. John Wiley and Sons, New York.

Vincent, S.R. and Johansson, O. (1983). Striatal neurons containing both somatostatin- and avian pancreatic polypeptide (APP)-like immunoreactivities and NADPH-diaphorase activity: A light and electron microscopic study. J. Comp. Neurol., 217, 264-270.

Vincent, S.R., Johansson, O., Hökfelt, T., Skirboll, L., Elde, R.P., Terenius, L., Kimmel, J. and Goldstein, M. (1983). NADPH-diaphorase: A selective histochemical marker for striatal neurons containing both somatostatin- and avian pancreatic polypeptide (APP)-like immunoreactivities. J. Comp. Neurol., 217, 252-263.

Vincent, S.R., Johansson, O., Skirboll, L. and Hökfelt, T. (1982a). Coexistence of somatostatin- and avian pancreatic polypeptide--like immunoreactivities in striatal neurons which are selectively stained for NADPH-diaphorase activity. In Regulatory Peptides: From Molecular Biology to Function (eds. E. Costa and M. Trabucchi). Raven Press, New York.

Vincent, S.R., Skirboll, L., Hökfelt, T., Johansson, O., Lundberg, J.M., Elde, R.P., Terenius, L. and Kimmel, J. (1982b). Coexistence of somatostatin- and avian pancreatic polypeptide (APP)-like immunoreactivity in some forebrain neurons. Neuroscience, 7, 439-446.

IMAGE ANALYSIS OF NEURONAL AND GLIAL MARKERS: FLUORESCENCE MICROSCOPICAL APPLICATIONS

LARS OLSON, HÅKAN BJÖRKLUND, MARIA ERIKSDOTTER-NILSSON, ANDREAS HENSCHEN and INGRID STRÖMBERG

Department of Histology, Karolinska Institutet, P.O. Box 60400, S-104 01 Stockholm, Sweden

ABSTRACT

The usefulness of computer-assisted image analysis with particular emphasis on fluorescence microscopy was evaluated and exemplified. Problems associated with image pick-up and transfer between microscope and computer are discussed. The importance of fully supported software programs adjusted to the needs of histology are emphasized. One such system, the IBAS (Kontron/Zeiss, West Germany) has been found suitable for use also by people with little or no background computer knowledge. Three examples where image analysis clearly adds a unique quantitative dimension to the evaluation of the results have been presented. (1) Nerve density measurements: A semi-automatic interactive program was used to evaluate the potentially neurotoxic effects of hexachlorophene and chlorhexidine, two disinfectant agents, using an intraocular screening model in which the density of the sympathetic autonomic ground plexus of the iris is studied by Falck-Hillarp fluorescence histochemistry applied to iris whole mounts. Pronounced neurotoxic effects were described. Dopaminergic nerve density measurements in striatum following neurotoxic drug treatments correlate well with other measurements of degree of denervation. (2) Transmitter release and diffusion: Experimentally induced unilateral parkinsonism in rats can be counteracted by intrastriatal implants of chromaffin tissue. These grafts work by releasing large quantities of catecholamines which diffuse through host neuropil. Image analysis was used to characterize in detail the diffusion of catecholamines using Falck-Hillarp fluorescence histochemistry. Linear scans of fluorescence intensity and imaging fluorescence gradients using false color look-up tables enables fast visual quantitative interpretation of the results. (3) Morphometry of smeared and sectioned astrocytes: A program was used that calculated area and perimeter of smeared astrocytes stained with an antiserum against glial fibrillary acidic protein, GFA. A study of astrocyte growth from adolescence to senescence revealed continuous growth of astrocytes throughout life. In this case the extreme complexity of astrocyte morphology necessitated

251

special interactive procedures to be used in which the experimenter
can "retouch" the digitized image prior to binary transformation.
Area and perimeter data of this kind could not have been obtained
without an image processing system. Astrocyte overgrowth in brain
tissue grafts to the anterior chamber of the eye and to the brain
were also described. It is concluded that image analysis is a powerful tool for the quantitative evaluation of microscopical images
with general usefulness in fluorescence histochemistry.

INTRODUCTION

　　Extraction of scientific information from microscopical images
is a tedious and complex procedure. Although statistically relevant
data can be obtained using semiquantitative evaluation scales,
coded specimens and non-parametrical statistical tests, recent developments of computer-assisted image analysis are now greatly improving the possibilities to characterize, sort and quantitatively
describe visual information. In the following we shall first briefly
discuss the four essential steps involved in computer-assisted image
analysis, mainly as applied to fluorescence microscopy, namely optimizing the signal-to-noise ratio in the fluorescence microscopical
specimen, image pick-up procedures, image processing, and data pre-
sentation. We will then exemplify the procedures with measurements
of nerve densities, receptor densities, transmitter diffusion through
tissue, and astrocyte morphology.

Image Pick-Up

　　When automated image analysis is to be used, particularly in
fluorescence microscopy, it is essential to optimize the histochemical techniques. A strong enough signal and a good enough signal-to-
noise ratio must be achieved. In Falck-Hillarp fluorescence histochemistry to visualize monoamines (Falck et al., 1962; Corrodi and
Jonsson, 1967), it is often possible to use regular high-resolution
black-and-white TV cameras (e.g. RCA TC 100 5XC). Epi-illumination
and a primary magnification of X 16 or larger may be required to
obtain a good signal-to-noise ratio. In weakly fluorescent specimens
and rapidly fading material such as the 5-hydroxytryptamine fluoro-
phore, image-intensified cameras may have to be used. For immunohistochemical techniques, rhodamine-labelled secondary antibodies
seem generally better suited than fluorescein-labelled seconary
antibodies, partly because of a slower fading. The fading problem
can be further diminished by the addition of anti-fading agents to
the mounting medium. It is, still, important to illuminate specimens
to be measured for approximately the same amount of time prior to
image pick-up.

　　There are several alternatives to the use of video cameras for
image pick-up such as flying laser spots, photomultipliers with
scanning stages, or light-sensitive diode array techniques. TV

cameras have the advantage of being sensitive and fast. In critical applications such as densitometry, it is important to choose a camera that does not self-adjust to light intensity changes inside or outside the measuring field and to test the linearity of the signal.

Image Analysis System

Developing the software needed for efficient and reliable processing of microscopical images fed directly from a video camera on-line to a computer is certainly not trivial. Thus, although powerful computers are now readily available, only research groups with a wealth of time and devoted programmers should consider this. Fortunately, systems are now available that enable non-experienced persons to apply advanced forms of image analysis with little training. Our own system, the IBAS (Kontron, Zeiss, West Germany) is one such fully software-supported system. A technical description is outside the scope of this chapter; suffice it to say that the image is digitized in a format that may be varied, but usually consists of 512 x 640 pixels, each with a grey-level depth of 8 bits, i.e. 256 levels. An array processor and a host computer handle image processing. The operator uses a cursor, a digitizing tablet and a keyboard to control image processing via a menu monitor. The image is presented on a high-resolution color monitor where it is also possible to interact via the cursor to change and measure directly in the image (exclude, fill in, other "retouch" procedures, perform interactive measurements of defined objects and interactive densitometry). A rich software, well suited for the needs in histology, including powerful filtering subroutines and other image enhancement procedures, boolean operations, scaling procedures, etc. prior to data extraction is provided. Of particular usefulness in fluorescence histochemistry is the possibilities to compensate efficiently for uneven illumination of the field of view. The software is user-friendly and enables easy construction of program strings designed for individual users. After automatic or interactive grey-level discrimination procedures, data extraction includes a large number of classical as well as newer morphometrical and densitometrical object- and field-associated parameters. Powerful menu-driven statistical programs can be applied before data are presented in the form of tables, histograms, and various types of two- and three-dimensional plots which may be photographed from the monitors or printed on a matrix printer.

NERVE DENSITY MEASUREMENTS

Neurotoxicity Screening Using Iris Whole Mounts

Neurotoxic effects of the two disinfectant agents hexachlorophene (HCP) and chlorhexidine (CHX) have been described (see Henschen and Olson, 1983, 1984). Until recently, however, little was known about the possible neurotoxic actions of HCP and CHX in the

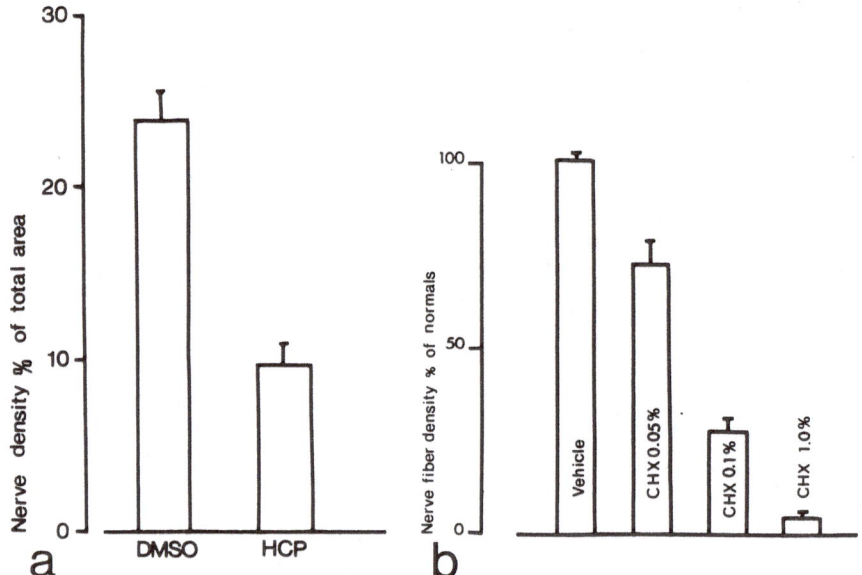

Fig. 1. Image analysis measurements of adrenergic nerve density in
irides after intraocular injection of disinfectant drugs. a. Effects of HCP (21 µg, 3 days). The area covered by adrenergic nerves,
expressed as % of total area, was measured in 4 different fields of
view in each iris and used to obtain mean individual values. The
means of 5 control and 6 HCP-treated irides are given. The difference is highly significant (p < 0.001). (From Henschen and Olson,
1983.) b. CHX-induced degeneration of adrenergic nerves as seen
2 days after treatment. The means of 8 control and 6 CHX-treated
irides are given. (From Henschen and Olson, 1984.)

autonomic nervous system. We have therefore used a recently devised
method for detecting neurotoxic actions on autonomic nerves (Björklund et al., 1981) to describe acute degeneration of adrenergic
nerve terminals following local drug application in rats. In these
experiments microliter amounts of drugs to be tested are injected
into the anterior chamber of the eye, where they have direct access to the nerve plexuses of the iris. The sympathetic adrenergic
plexus of the iris is then studied in whole-mounted stretch preparations using Falck-Hillarp fluorescence histochemistry. Both drugs
were shown to cause a marked dose-dependent degeneration of adrenergic nerves, followed, particularly after lower doses, by regeneration. In Fig. 1 the effects of HCP and different doses of CHX on
adrenergic nerve density as measured by the image analysis program
are shown. It is notable that degenerative changes are seen already
with a 0.05% solution of CHX, a concentration far below concentrations used clinically.

These experiments are important in two respects: firstly, they have demonstrated the reliability and validity of computer-based nerve density measurements. We have extensive experience of semiquantitative visual estimations of nerve densities in iris whole mounts (see Olson and Malmfors, 1970) and since such estimations correlate closely with measurements of uptake of labelled transmitter into the nerve fibers (Olson et al., 1968). the close correlation between computer-assisted image analysis of nerve densities and semiquantitative estimations found in the experiments with disinfectants demonstrate the validity of our approach. Secondly, once the soundness of the technique has been established, automated nerve density measurements can be used to screen with high precision other substances and lower doses to discover effects that might not be detected otherwise.

Density of Striatal Dopamine Terminals

Recently a similar nerve density program has been applied to a study of the neurotoxic effects of the meperidine derivative MPTP in the mouse brain. Immunohistochemical localization of tyrosine hydroxylase (TH) immunoreactivity was used as a marker of dopamine nerve terminals in striatum. Regional differences in the degree of degeneration within striatum were detected. The amount of degeneration detected in this way correlated well with biochemical and neurophysiological measurements of dopaminergic denervation (G. Jonsson, S. Johnson, R. Freedman, B. Hoffer, E. Sundström, I. Mefford, I. Strömberg, L. Olson, G. Gerhardt, G. Rose, unpubl. observations).

CATECHOLAMINE-RICH GRAFTS TO STRIATUM

Nigral Grafts: Densitometry of Receptor Autoradiographs

It is now well known that several of the behavioral consequences of 6-hydroxydopamine-induced experimental parkinsonism in rats can be counteracted by grafts of fetal substantia nigra neurons that provide striatum with a new dopamine innervation (Perlow et al., 1979; Björklund et al., 1980; see also Olson, 1984). One aspect of the function of nigral grafts is their ability to decrease receptor supersensitivity in reinnervated areas of striatum. Using image analysis, it has been demonstrated by in vitro receptor autoradiography that the degree of binding of tritiated spiroperidol is significantly lower in reinnervated areas (Freed et al., 1983). It should be noted that densitometry measurements put higher demands on the initial steps of the image analysis procedure than morphometry measurements. Thus stringent autoradiographic controls as well as stable video cameras are imperative.

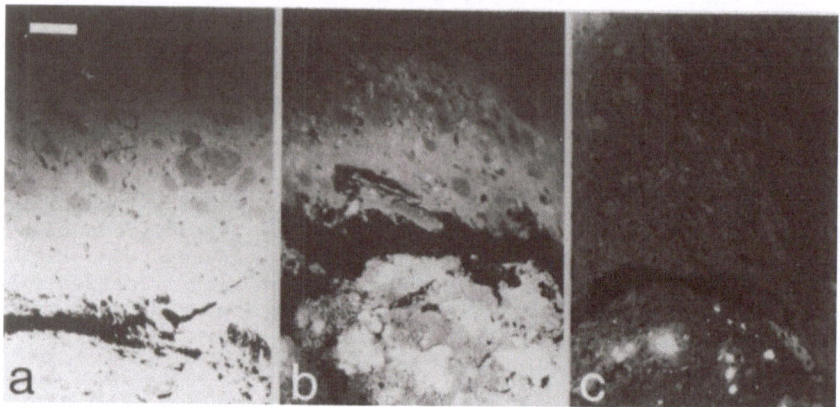

Fig. 2. Falck-Hillarp fluorescence histochemistry of grafts and the
host caudate different times after intrastriatal implantation of
adrenal medullary tissue. The blood-filled clefts between graft
(bottom) and host caudate (top) are dark. a: 2 min after grafting.
The chromaffin tissue is strongly fluorescent. The host caudate is
strongly fluorescent close to the graft with a gradient of decreas-
ing fluorescence with increasing distance from the graft. b: 100
min after grafting. In the graft, groups of chromaffin cells are
still strongly fluorescent, while the rest of the chromaffin tissue
is weakly fluorescent or non-fluorescent. The fluorescent halo in
host caudate is similar in size to the halo seen after 2 min, but
the fluorescence intensity of the halo is lower at 100 min. c: 400
min after grafting. Only occasional chromaffin cells are still
strongly fluorescent. The rest of the graft and the surrounding host
caudate is virtually non-fluorescent. Fluorescence microphotographs.
Calibration bar 100 microns. (From Strömberg et al., 1984.)

Adrenal Medullary Grafts: Catecholamine Diffusion Densitometry

        In an attempt to find a substitute for fetal substantia nigra,
we have tried adrenal medullary tissue and found that it too can
counteract the rotational behavior of unilaterally dopamine-denervated rats when implanted in close proximity to (Freed et al., 1981)
or into (Olson et al., 1984) the striatum. Intrastriatal chromaffin
tissue causes acute contralateral rotations in dopamine-denervated
animals that lasts for several hours (Herrera-Marschitz et al.,
1984). During this time they release large amounts of catecholamines
that diffuse through host striatal tissue (Strömberg et al., 1984).
This diffusion is illustrated in Fig. 2 and has been characterized
by image analysis. The spread of catecholamines from the site of implantation of the grafts was quantified using linear scans of fluorescence intensity and followed over time (Fig. 3). These measurements correlate well with the acute behavioral effects of the grafts
(Strömberg et al., 1984; Herrera-Marschitz et al., 1984). Moreover,

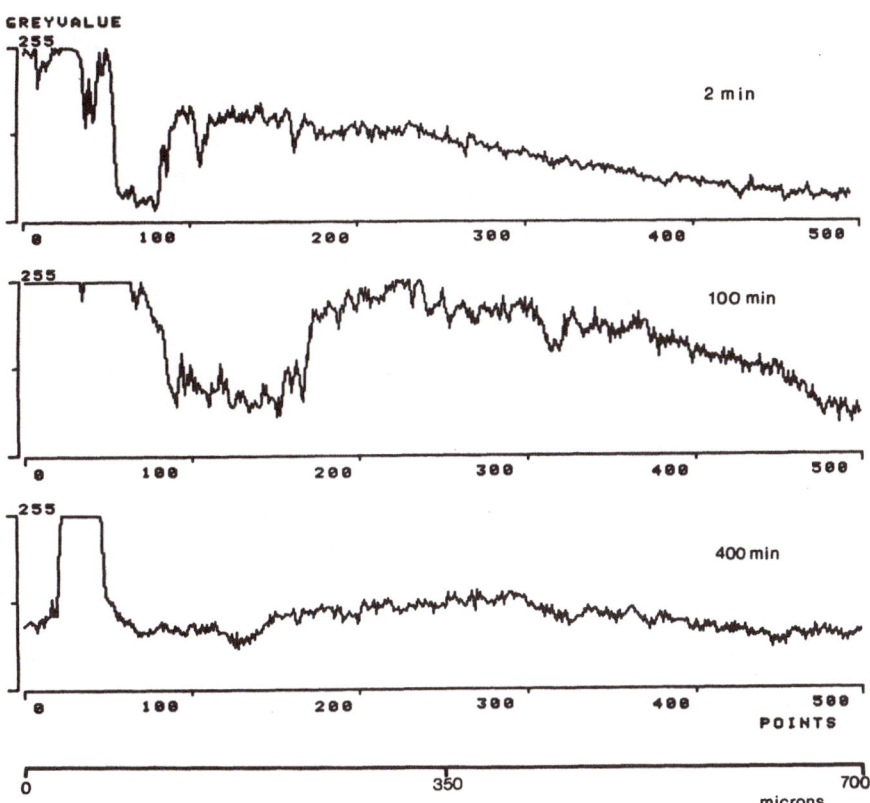

Fig. 3. Examples of linear scans of grey levels in digitized images
obtained from fluorescence microscopical images such as those in
Fig. 2. Scans were made 2, 100, and 400 min after grafting. To the
left in each panel the linear scan passes through chromaffin cells
having maximum (255) whiteness. The scan then passes through a non-
fluorescent blood-filled cleft between graft and host caudate. In
the upper two traces the fluorescence intensity is seen to decrease
with increasing distance from the graft. In the middle trace, a normalization of the distribution of grey levels was performed before
the scan. Therefore, the absolute grey levels of the middle trace
cannot be compared to the upper and lower traces. Below the lower
trace the distance in microns along the scanning lines is given.
(From Strömberg et al., 1984.)

false color look-up tables are useful in detecting and interpreting
the fluorescence intensity gradients.

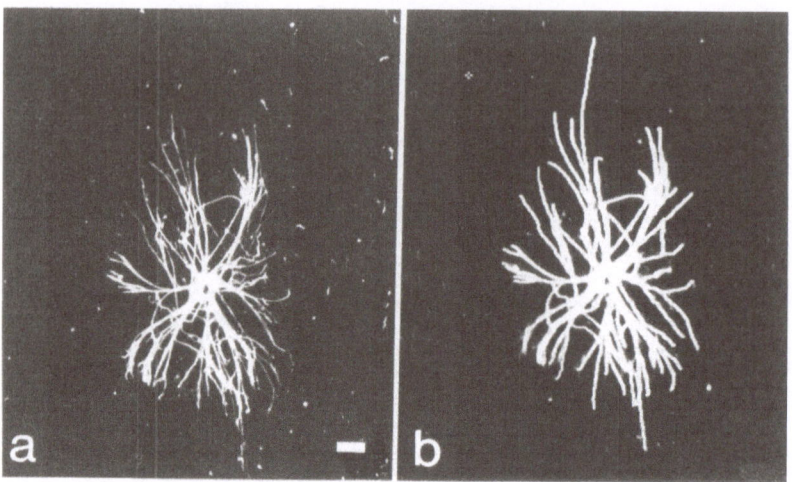

Fig. 4.   Astrocyte from the hippocampal formation of a 30-month-old
male Fisher 344 rat as visualized in a smear preparation using GFA
immunohistochemistry. a̲ demonstrates the cell as seen directly in
the fluorescence microscope. This is a typically aged astrocyte with
a large cell body and many long processes. b̲ is a photograph from
the video screen of the image analyzer showing the binary representation of the astrocyte as it appears after editing and discrimination procedures. Calibration bar 15 microns. (From Björklund et al.,
1985.)

ASTROCYTE MORPHOLOGY: GFA AND S-100 IMMUNOHISTOCHEMISTRY OF SECTIONS
AND SMEARS

Degree of Gliosis in Intraocular and Intracranial Grafts

     The astrocyte population of brain tissue is extremely sensitive
to disturbances during development and throughout life. Astrocytes
almost invariably react to various types of mechanical, chemical or
other types of trauma by hypertrophy, usually including an increased
intracellular content of glial fibrillary acidic protein (GFA) which
can be visualized immunohistochemically. It has been shown that
intra-ocular transplants of fetal cerebral cortex develop a relatively marked gliosis (Björklund and Dahl, 1982). Intraocular grafting experiments have shown that cortex cerebri is critically dependent upon contact with other brain areas for a more normal development (Björklund et al., 1983b; Palmer et al., 1983). A recently
described smear technique that allows visualization of whole individual GFA-positive astrocytes (Björklund et al., 1984b), was used
to analyze the gliosis reaction in grafts further (Björklund et al.,
1984a). The program in this case averaged five inputs from each
field of view to be measured. After grey-scale normalization, an
edit feature allowed "filling in" of weakly, and thus ill-detected

Image analysis of GFA -positive astrocytes
in grafted and lesioned cortex cerebri

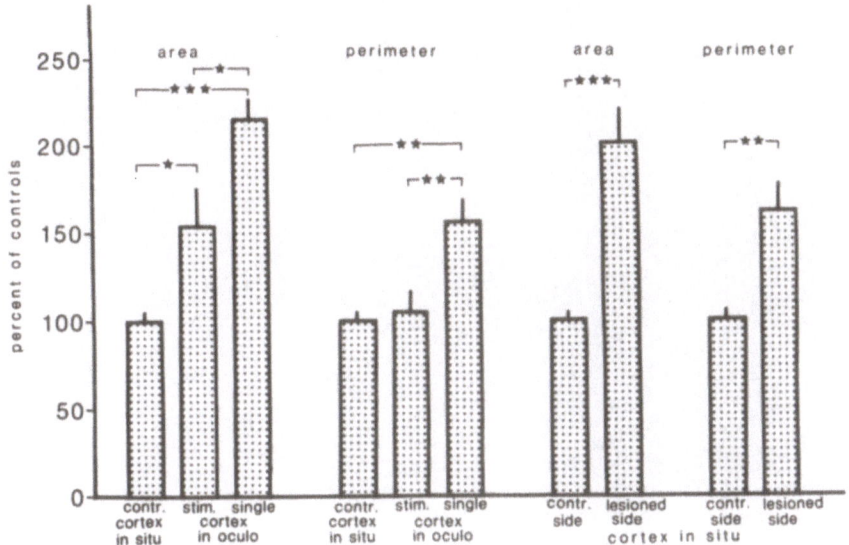

Fig. 5. Image analysis of GFA-positive astrocytes in grafted and lesioned cortex cerebri. Area and perimeter of smeared astrocytes ± S.E.M. are expressed as % of normal cortex cerebri values. * p < 0.005, ** p < 0.01, *** p < 0.001. (From Björklund et al., 1984.)

processes, as well as removal of interfering fluorescent structures. An interactive one-level discrimination procedure was then used to define objects to be measured. The program determined area and perimeter of each selected smeared astrocyte (Fig. 4). All measurements were made on coded slides. It was found that astrocytes in cortex cerebri grafts which were allowed to develop in contact, in the eye chamber, with previously grafted pieces of locus coeruleus were smaller and with lesser perimeters and thus more normal than astrocytes in isolated cortex grafts (Björklund et al., 1984a). However, also locus coeruleus-stimulated cortex grafts had a certain degree of gliosis, as evidenced by the increased size of astrocytes in this type of graft compared to normal cortex (Fig. 5). In the same study (Björklund et al., 1984a), lesions to cortex cerebri were found to cause increased sizes of GFA-positive astrocytes comparable to those seen in intraocular single cortex grafts (Fig. 5). Moreover, vimentin, an intermediate filament found in developing, but usually not in adult astrocytes, was found in a large population of astrocytes in grafted and lesioned cortex. Again, image analysis showed stimulated cortex grafts to be more normal, i.e. contain less vimentin, than non-stimulated grafts. In another study, intracranial

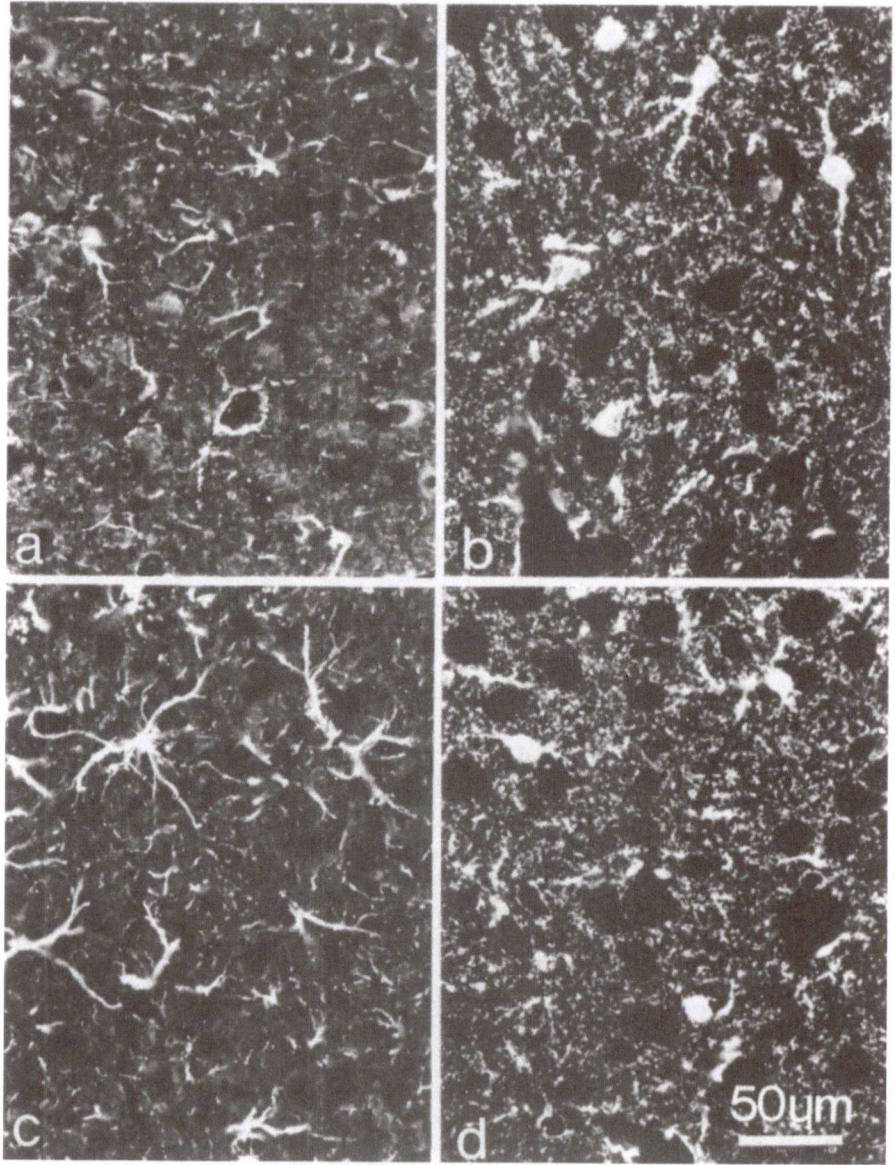

Fig. 6. Comparison of the amount and distribution of GFA-like immunoreactivity and S-100-like immunoreactivity in fetal cortex cerebri
pieces grafted to the cortical region and host cortex cerebri.
a: GFA-like immunoreactivity in normal parietal cortex. b: S-100-
like immunoreactivity in normal parietal cortex. c: GFA-like immunoreactivity in a graft. d: S-100-like immunoreactivity in a graft.

(Fig. 6, continued) The density of GFA-like immunoreactivity is higher in the graft than in normal cortex. Note also that star-shaped astrocytes are seen only in the graft. No difference in amount or distribution of S-100-like immunoreactivity can be detected between the graft and cortex in situ. Granular fluorescence as well as astrocytes with strongly fluorescent cell bodies are seen in both b and d. Fluorescence microphotographs X 340. (From Björklund et al., 1984.)

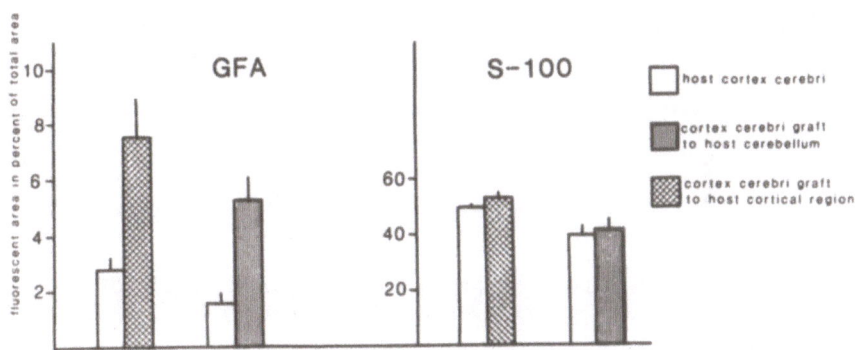

Fig. 7. Fluorescent area in % of total area as determined by computerized image analysis of sections immunohistochemically stained for GFA and S-100. Whereas the GFA antisera stain only approximately 2% of the total area in host cortex, 7.7 ± 1.3% (n=7) and 5.4 ± 0.8% (n=4) of the total area are stained in grafts to the cortical region and the cerebellar region, respectively. Using S-100 antibodies no differences in fluorescent area could be detected between grafts in either of the two locations. (From Björklund et al., 1984.)

grafts of cortex cerebri placed either in the cerebellar or in the cerebral cortex of immature hosts were studied (Björklund et al., 1983a). Image analysis of sections treated for GFA and S-100 immuno-histochemistry revealed that the total amount of GFA immunoreactivity was higher in both types of grafts than in host cortex cerebri, while there were no differences in amounts of S-100 immunoreactivity (Fig. 6 and 7). Since S-100 is considered to be present in all astrocytes, normal and reactive, it was suggested from these experiments that the amount of GFA had increased in individual astrocytes and thus hypertrophy rather than hyperplasia had occurred.

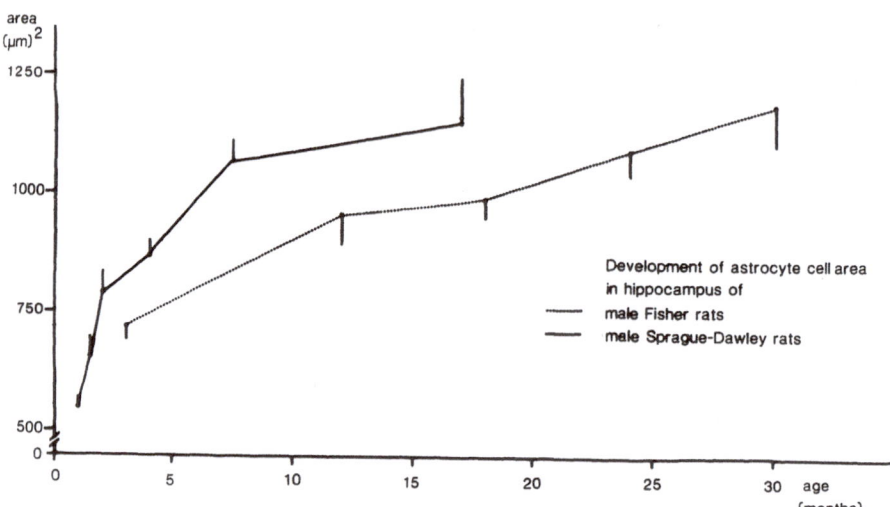

Fig. 8. Cell area of GFA-positive astrocytes in smears of the
hippocampus formation from male Sprague-Dawley (solid line) and
male Fisher 344 (dashed line) rats as determined by computerized
image analysis. Each point represents the mean ± S.E.M. of 7-10
smears taken from 4-5 animals. (From Björklund et al., 1985.)

Astrocyte Changes in Brain Cortices from Adolescence to Senescence

We have recently completed a study describing changes in size
and perimeter of individual astrocytes in different cortical regions
of rats ranging in age from 1-30 months (Björklund et al., 1985).
Using computerized image analysis, cell areas and the complex cell
perimeter was determined in smears of fresh tissues as described
above (Fig. 4). It was found that astrocytes in cerebellum and
hippocampus were significantly larger and had larger complex perimeters than astrocytes in cortex cerebri.

The lifetime development of astrocytes was followed in the
hippocampal formation. There was a rapid increase in cell area and
cell perimeter during the first few postnatal months followed by a
second slower, although highly significant, growth phase which continued through the life of the animal (Fig. 8). This continuous slow
astrocytic proliferation is an example of a change that could not
have been discovered without image analysis. It may have important
implications for our understanding of aging processes in the central
nervous system.

CONCLUDING REMARKS

In this chapter we have described how image analysis can be used to obtain quantitative information in studies of neurons and glial cells in the nervous system. Programs have been designed that measure nerve densities with high accuracy, that describe density distributions of transmitters and receptors and that generate morphometric data to characterize individual glial cells. In all cases, more precise information has been obtained than could have been obtained by semi-quantitative estimations. In some cases, notably astrocyte growth during aging, information has been obtained that cound not have been obtained without image analysis. Recent developments of histochemical and immunohistochemical techniques with improving signal-to-noise ratios together with the continuing rapid development of computer hardware and software makes it safe to conclude that automatic and semi-automatic computer-assisted image analysis is revolutionizing microscopical studies of the nervous system.

ACKNOWLEDGEMENTS

Supported by the Swedish Medical Research Council (14X-03185, 14P-5867), the Swedish Council for Planning and Coordination of Research, Magnus Bergvalls Stiftelse, Loo and Hans Ostermans Foundation, and Karolinska Institutets Fonder. For generous gifts of antibodies, we thank Dr. Doris Dahl, Boston, USA (GFA), Drs. Norman Weiner and William Tank, Denver, USA (TH), Drs. Kenneth Haglid and Lars Rosengren, Göteborg, Sweden (S-100), and Drs. Klaus Weber and Mary Osborn, Göttingen, West Germany (vimentin). We thank Ms. L. Hultgren, A. Hultgårdh, B. Standwerth and I. Engqvist.

REFERENCES

Björklund, A., Dunnett, S.B., Stenevi, U., Lewis, M.E. and Iversen, S.D. (1980). Reinnervation of the denervated striatum by substantia nigra transplants: functional consequences as revealed by pharmacological and sensorimotor testing. Brain Res. 199, 307-333.

Björklund, H. and Dahl, D. (1982). Glial disturbances in isolated neocortex: Evidence from immunohistochemistry of intraocular grafts. Develop. Neurosci.,5, 424-435.

Björklund, H., Dahl, D., Haglid, K., Rosengren, L. and Olson, L. (1983a). Astrocytic development in fetal parietal cortex grafted to cerebral and cerebellar cortex of immature rats. Develop. Brain Res., 9, 171-180.

Björklund, H., Dahl, D. and Olson, L. (1984a). Morphometry of GFA and vimentin positive astrocytes in grafted and lesioned cortex cerebri. Intern. J. Developm. Neurosci.,2, 181-192.

Björklund, H., Eriksdotter-Nilsson, M., Dahl, D. and Olson, L.
(1984b). Astrocytes in semars of CNS tissues as visualized by GFA
and vimentin immunofluorescence. Med. Biol., 62, 38-48.

Björklund, H., Eriksdotter-Nilsson, M., Dahl, D., Rose, G., Hoffer,
B. and Olson L. (1985). Image analysis of GFA-positive astrocytes
from adolescence to senescence. Exp. Brain Res. (in press).

Björklund, H., Hoffer, B., Olson, L. and Seiger, Å. (1981). Differential morphological changes in sympathetic nerve fibers elicited
by lead, cadmium and mercury. Environmental Res., 26, 69-80.

Björklund, H., Seiger, Å., Hoffer, B. and Olson, L. (1983b). Trophic
effects of brain areas on the developing cerebral cortex. I. Growth
and histological organization of intraocular grafts. Developm. Brain
Res., 6, 131-140.

Corrodi, H. and Jonsson, G. (1967). The formaldehyde fluorescence
method for the histochemical demonstration of biogenic monoamines.
A review on the methodology. J. Histochem. Cytochem., 15, 65-78.

Falck, B., Hillarp, N.-Å., Thieme, G. and Torp, A. (1962). Fluorescence of catecholamines and related compounds condensed with formaldehyde. J. Histochem. Cytochem., 10, 348-354.

Freed, W., Ko, G., Neihoff, D., Kuhar, M., Hoffer, B., Olson, L.,
Spoor, E., Morihisa, J. and Wyatt, R. (1983). Normalization of
spiroperidol binding in the denervated rat striatum by homologous
substantia nigra transplants. Science, 222, 937-939.

Freed, W., Morihisa, J., Spoor, E., Hoffer, B., Olson, L., Seiger,
Å. and Wyatt, R. (1981). Transplanted adrenal chromaffin cells in
rat brain reduce lesion-induced rotational behaviour. Nature (Lond.),
292, 351-352.

Henschen, A. and Olson, L. (1983). Hexachlorophene-induced degeneration of adrenergic nerves: Application of quantitative image analysis to Falck-Hillarp fluorescence histochemistry. Acta Neuropathol.
(Berl.), 59, 109-114.

Henschen, A. and Olson, L. (1984). Chlorhexidine-induced degeneration of adrenergic nerves. Acta Neuropathol. (Berl.), 63, 18-23.

Herrera-Marschitz, M., Strömberg, I., Olsson, D., Ungerstedt, U. and
Olson, L. (1984). Adrenal medullary implants in the dopamine-denervated rat striatum. II. Acute behavior as a function of graft amount
and location and its modulation by neuroleptics. Brain Res., 297
53-61.

Olson, L. (1984). On the use of transplants to counteract the symptoms of Parkinson's disease: Background, experimental models and
possible clinical applications. In Synaptic Plasticity and Remodel-

ling. (ed. C. Cotman). (in press).

Olson, L., Backlund, E.-O., Sedvall, G., Herrera-Marschitz, M., Ungerstedt, U., Strömberg, I., Hoffer, B. and Seiger, Å. (1984). Intrastriatal chromaffin grafts in experimental and clinical parkinsonism: first impressions. In Catecholamines: Neuropharmacology and Central Nervous System - Therapeutic Aspects. (eds. E. Usdin, A. Carlsson, A. Dahlström, J. Engel). (in press).

Olson, L., Hamberger, B., Jonsson, G. and Malmfors, T. (1968). Combined fluorescence histochemistry and ^{3}H-noradrenalin measurements of adrenergic nerves. Histochemie, 15, 38-45.

Olson, L. and Malmfors, T. (1970). Growth characteristics of adrenergic nerves in the adult rat. Fluorescence histochemical and ^{3}H-noradrenaline uptake studies using tissue transplants to the anterior chamber of the eye. Acta Physiol. Scand., Suppl., 348, 1-112.

Palmer, M., Björklund, H., Olson, L. and Hoffer, B. (1983). Trophic effects of brain areas on the developing cerebral cortex. II. Electrophysiology of intraocular grafts. Developm. Brain Res., 6, 141-148.

Perlow, M., Freed, W., Hoffer, B., Seiger, Å., Olson, L. and Wyatt, R. (1979). Brain grafts reduce motor abnormalities produced by destruction of nigrostriatal dopamine system. Science, 204 643-647.

Strömberg, I., Herrera-Marschitz, M., Hultgren, L., Ungerstedt, U. and Olson L. (1984). Adrenal medullary implants in the dopamine-denervated rat striatum. I. Acute catecholamine levels in grafts and host caudate as determined by HPLC-electrochemistry and fluorescence histochemical image analysis. Brain Res., 297, 41-51.

QUANTITATIVE IMMUNOCYTO-CHEMISTRY AND AMINE FLUORESCENCE HISTOCHEMISTRY

Chairman: C. OWMAN

THE USE OF QUANTITATIVE IMMUNOCYTO-CHEMISTRY TO STUDY THE REGULATION OF TYROSINE HYDROXYLASE IN NORADRENERGIC AND DOPAMINERGIC NEURAL SYSTEMS

R.H. BENNO*, L.W. TUCKER, T.H. JOH and D.J. REIS

Laboratory of Neurobiology, Cornell University Medical College, New York 10021, USA

Immunocytochemistry has now established itself as one of the most powerful techniques available for understanding the neurobiology of neuron-specific macromolecules in the brain. Using antibodies directed either against neurotransmitters, (Choy and Watkins, 1977; Steinbusch et al. 1978; Sternberger, 1979), the enzymes required for their biosynthesis (Pickel et al. 1975a, b; Sternberger, 1979), or even receptor associated molecules, (Strader, et. al. 1983) the distribution and biochemical anatomy of specific transmitter classes and their relationship to identifiable neuronal populations has been established.

However, in general, immunocytochemistry has been used <u>qualitatively</u> as a means of establishing the presence and/or localization of particular molecules within the brain. <u>Quantification</u> of the amounts of these substances on the other hand, has usually been performed by biochemical or immunochemical analysis whereby the amounts of these substances have been established in homogenates of particular brain regions. While the biochemical procedures may yield important information with respect to regional variations in the content of transmitters of their enzymes or may establish the dynamic changes which they undergo in response to lesions, environmental stimuli, drugs, during the performance of naturalistic behaviors, or in development, the values so obtained can only represent an average measurement of changes within populations of neurons or their terminals. It cannot determine whether such variations may, in fact, occur within a subpopulation of neurons within a nucleus otherwise considered to be homogeneous. It is evidently desireable to be able to quantitate the immunocytochemical procedures so that they can measure the amount of neurotransmitter-related substances in and around individual neurons. Ideally, such measurements would be truly quantitative yielding absolute values in terms of the number of molecules per cell. At present, no cytochemical technique can provide such information. Short of the ideal however, would be a semi-quantitative method in which the relative amounts of the substances could be accurately assessed.

*Present Address: William Paterson College of New Jersey

Supported by Grants: MH35638 and HL18974

Over the past several years, we have been attempting to develop a method for quantitative immunocytochemistry in brain (Benno, et al. 1982a, b). As a target molecule, we have studied the enzyme tyrosine hydroxylase, the enzyme catalyzing the first step in the biosynthesis of the catecholamine neurotransmitters in brain (Levitt, et al. 1965). We have used the peroxidase-antiperoxidase (PAP) technique of Sternberger (Sternberger, et al. 1970) to chart these changes.

The PAP method has great advantages over other techniques such as immunofluorescence, with respect to sensitivity, specificity, stability of the stain, compatibility with tissue preservation, and adaptability to electron microscopy (Mayersbach, 1959; Pickel et al. 1975a, b; Sternberger, 1979). We have, in a number of studies (Benno et. al 1982a, b; Reis et al. 1982), reviewed here, been able to show that the PAP method can be adopted for quantitative cytochemistry in that it can accurately detect changes in the amount of enzyme protein. This has been achieved by establishing the proper reaction conditions so as to make the method linear under changes in the amount of antigen in tissue. We have been able to validate the method so as to demonstrate that the magnitude of changes as detected by immunocytochemistry is closely parallel to that in tissue homogenates established by biochemistry. By developing this technique we have finally been able to examine the biology of TH within neurons of the noradrenergic locus ceruleus (LC). We have been able to show that the enzyme content varies over a three-fold range within different neurons of the LC, and that the variability relates to the topography of the nucleus. It has also allowed us to access the variations in the content of TH between dopaminergic neurons in the substantia nigra (SN) and noradrenergic neurons in the LC and differences in the response of neurons in these two populations to injections of resperpine. In addition, quantitative immunocytochemistry in conjunction with image processing has allowed us to study the time course of TH induction in the cell body relative to the neuropil and "track" the movement of TH from the cell body to its terminal field.

Development of A Computer-Assisted Method For Quantitative Immunocytochemistry of TH in Rat Brain

A.   Methods

i.   Animals and Tissue Preparation
The detailed methods of this study have been described elsewhere (Benno et al. 1982a, b) and will be summarized. Our studies were performed on male Sprague-Dawley rats adapted to the animal quarters for at least ten days so as to minimize stress effects on TH. The animals were anesthetized with pentobarbitol and perfused through the heart with 300 ml of 4% paraformaldehyde. The brains were removed, cut into half along the mid-saggital plane, and postfixed in picric acid (0.2% w/v) paraformaldehyde (2% w/v) for 30 min., placed in a phosphate buffer overnight, dehydrated through graded alcohols, cleared and embedded in paraffin. The half brains were sectioned at 5 um along the sagittal plane. This plane of section allows comparison of catecholamine neurons containing TH in the noradrenergic LC and the dopaminergic substantia nigra.

ii.   Immunostaining

The paraffin sections were hydrated, washed in Tris saline buffer and incubated with TH. The antibodies to TH were prepared in rabbits using TH purified from bovine adrenal medulla (Joh, et al. 1973). Staining was by the PAP method using diaminobenzidine (DAB) as the chromagen.

Images were analyzed an a Vidicon Based Image Analysis System (Bausch and Lomb Omnicon FAS II). Quantitative immunocytochemistry was performed on sections which were not counterstained. Background was accessed in an unstained region of the brain and values subtracted from those obtained in regions that were stained. Our image analysis system was able to resolve density differences on the order of 0.01 density units.

B. To establish the optimal conditions, we examined the intensity of staining of TH in the entire LC, examined in sagittal sections of the brain stem at 6.3x magnification (Fig. 1).

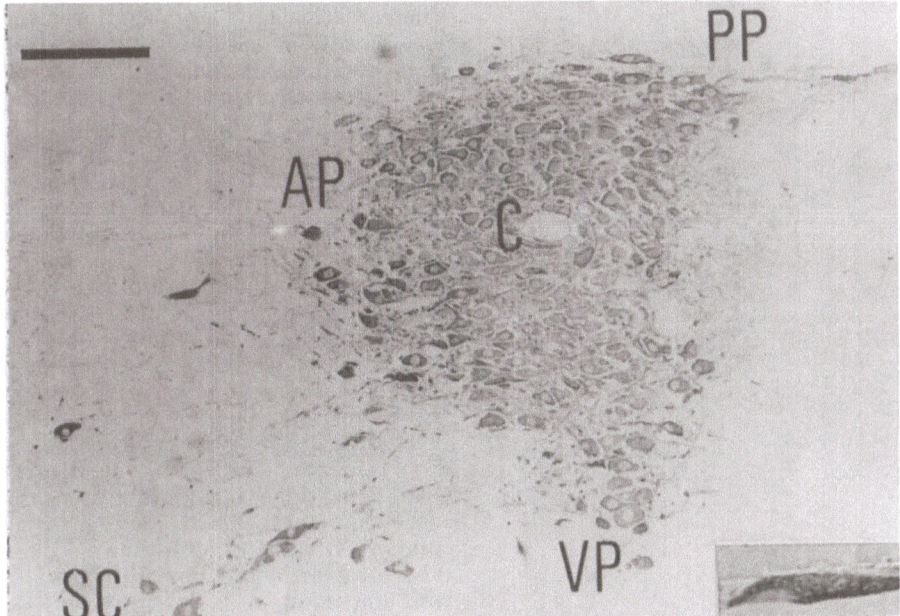

Fig. 1. The locus coeruleus of the rat brain showing specific staining for TH. A 5 µm sagittal TH stained paraffin section through the core region of the LC where the cells are tightly packed together. Abbreviations: AP, anterior pole; PP, posterior pole; VP, ventral pole; C, central core; SC, nucleus subcoeruleus; V, 4th ventricle. Bar = 150 µm. Inset: demonstration that the DAB reaction product is present throughout the full extent of the 5 µm paraffin section. This micrograph was made from a 5 um plastic embedded section which was cut from a 5 um paraffin section (e.g. Fig. 1) that had been re-embedded at 90° to the original plane of section.

It was possible to demonstrate that serial sections taken through the LC provided profiles of this nucleus which did not vary within animal with respect to average staining intensity.

We first examined the effects of independently varying three major variables on the intensity of staining through the LC: (a) the concentration of DAB; (b) the time of incubation in DAB and; (c) the TH antibody dilution. The variables were manipulated so that in the presence of a constant amount of antigen the intensity of staining was linear with respect to the amount of antigen. Establishment of the optimal reaction conditions was performed by holding two of the variables constant and manipulating the third.

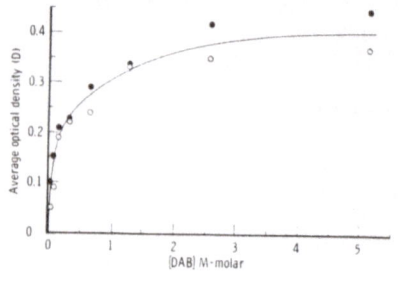

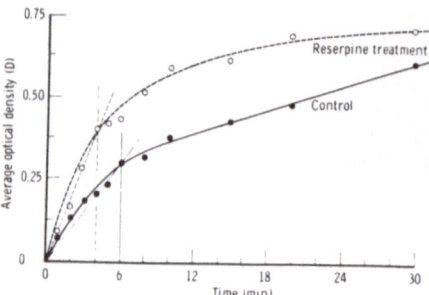

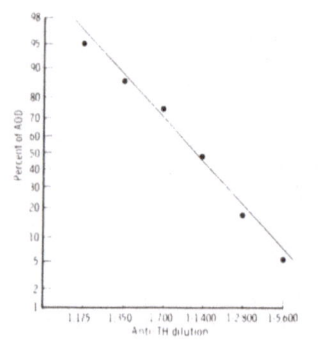

Fig. 2a. Analysis of the effect of DAB substrate concentration on the density of LC staining. Five-micron sagittal paraffin section through the core of the LC were immunostained with TH antibody (1:500) and analyzed for the effect of increasing DAB concentration on staining intensity. The reaction saturates at a DAB concentration of approximately 2.5 mM. Open circles, 3 min incubation; closed circles, 6 min incubation.

Fig. 2b. Analysis of the effect of increasing incubation time on staining intensity of TH in sagittal sections through the LC. The reaction is linear up to approximately 6 min in control animals and 4 min in reserpine treated animals. The line of best fit was determined from an analysis of the time point where there is a significant change in slope of a series of linear regression lines passing through three successive points on the graph. For the reserpine curve a significant change in slope occurs between the regression lines drawn through time points (3,4,5) and (4,5,6); thus at 4 min the reserpine curve begins to deviate from linearity.

Fig. 2c. Relationship of staining intensity (AOD) to dilution of TH. The line of regression (slope = -0.87) is the best fit for a series of LC sectiosn immunostained over a 32-fold dilution of antibody. Each value represents the averge of two sections. On the ordinate, 100% is the maximal value beyond which a further increase in antiserum concentration no longer yields an increase in AOD. Values on the Abscissa are expressed as the log of the TH antiserum dilution.

Figure 2 shows the effects of varying DAB substrate, incubation time, and antibody concentration on staining intensity. Keeping dilution and incubation time constant, staining intensity saturates at DAB concentrations of 2.5 mm (Fig. 2a). Using this concentration of DAB and constant dilution of antibody, the reaction is linear with incubation times up to 6 min. in the untreated rat (Fig. 2b). With optimal concentrations of DAB and incubation time at 6 min., the intensity varies with respect to antibody concentration (Fig. 2c). If one increases the amount of antigen by treating with reserpine (cf) the time of incubation through which the response is linear was shortened (Fig. 2b).

From these findings we concluded that optimal reaction conditions can be achieved by reacting tissues with saturated conditions of the substrate DAB and using a single concentration of antibody. It is then possible, by varying the incubation time, that the darkest and lightest elements within the tissue are within the linear range. While the optimal time varies with the tissue under inspection once established, the response is linear. The amount of incubation time necessary to keep elements within the linear range will vary with the experiment and must be established in pilot experiments.

By systematic analysis, we were able to demonstrate that variations in section thickness, penetration of DAB throughout the section, and fluctuations in the performance of the Vidicon system, did not influence our results. Moreover, we were able to demonstrate that by appropriate statistical analysis (Miller and Freund, 1977) using a sufficient number of animals, some between-animal comparison was permitted.

C.   Validation of the procedure

To validate the procedure we compared the effects upon TH content in the entire LC of administration of reserpine. Reserpine will increase the amount of TH in neurons of the LC up to 3-fold, reaching a maximum 72 hrs following administration of the drug (Reis et al. 1974; Ross et al. 1981). The increase is a consequence of an increase in the relative rate of biosynthesis of this enzyme (Ross et al., 1981).

As seen in Figure 3, when the time course and magnitude of changes of TH are compared, the magnitude of the changes and time course are roughly comparable with quantitative immunocytochemical procedures at all time points examined. The fact that the increase produced by reserpine is slightly less as detected by immunocytochemistry is not surprising, considering that cytochemical procedures require fixation of the tissue, unlike biochemical procedures in which fresh tissue is examined. Similarly, the greater variability in TH induction measured immunocytochemically is probably due to variations in perfusion and fixation of the tissue.

The present study has demonstrated that though the PAP reaction is a complex immunocytochemical procedure utilizing several separate antigen-antibody reactions (Sternberger, 1979), the reaction can be made linear. Linearity is achieved first by reacting tissues with saturated concentrations of the substrate. A saturating concentration of DAB not only sets necessary conditions so that the amount of product is proportional to the amount of enzyme by classic kinetics, but also avoids diffusion artifact resulting from variations in the penetration of DAB to peroxidase coupled to antigen in tissue (Van Duijn, et al. 1967) .

The amount of staining intensity was shown to vary sigmoidally as a function of antibody dilution. When these absorbance data were plotted as a

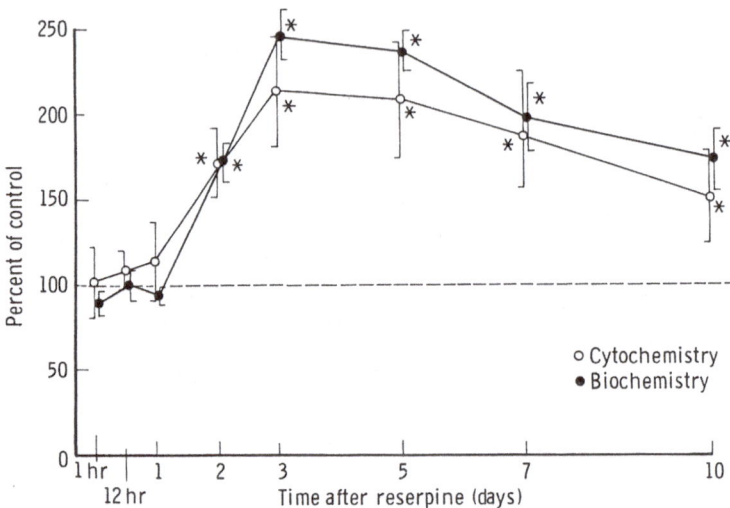

Fig. 3. Time course of changes in TH following reserpine (10 mg/kg) treatment. Six animals are used for each point. Values are % change $\pm$ SEM. $p \leq .05$

percentage of maximal absorbance index aganist the log of anitserum diultion a straight line was obtained (Fig. 2c). The relationship between relative absorbance index and antibody dilution suggests a direct relationship between antigen concentration and staining intensity. In addition, this relationship permits evaluation of enhancement or inhibition of staining as a result of experimental manipulations as previously described (Sternberger and Petrali, 1975).

For each biochemical or immunocytochemical reaction there is a critical time period of incubation during which the reaction conditions are linear and the amount of product formed varies linearly with respect to the amount of enzyme or antigen present. In this study, under control conditions, as long as the DAB reaction period did not exceed 6 minutes, the amount of DAB reaction product formed increased linearly with respect to the amount of the antigen present in the tissue.

However, following treatment with reserpine, which increased the amount of immunoreactive TH, linear reaction conditions were only present for 4 min. (In other words, if in reserpine-treated animals an incubation time of 6 min. were used the reaction would no longer be linear and the values so obtained would be highly underestimated.) It is therefore essential in the performance of quantitative immunocytochemistry that optimal reaction times be determined in pilot studies for whatever tissue is under examination.

That the PAP method can be used to detect changes of immunoreactive materials and tissues was validated in this study by demonstrating that the magnitude of the increase of TH within the LC produced by administration of reserpine was virtually comparable when determined biochemically and immunocytochemically. The close concordance between the results using the two methods moreover indicates that the competition for immunoreactive sites for the PAP complex, which may theoretically hinder quantitation particularly in the presence of large amounts of antigen, the so-called Bigbee effect (Bigbee, et al. 1977), does not hold true within this system. Thus this study demonstrates that it is possible using rigidly defined conditions to obtain relative quantitation of amounts of TH and changes in TH induced by drugs in the brain thereby allowing us to apply this method to biological problems.

Quantitative Immunocytochemistry at the Cellular Level

A.   Variations in the amount of TH enzyme protein within individual LC neurons

i.   Methods

By applying rigourous control of the reaction conditions for the PAP reaction we were thus able to demonstrate that the PAP method could be used quantitatively. We next sought to determine if the method could be used quantitively to examine TH at the level of the individual cell.

As before, studies were performed using male Sprague-Dawley rats perfused with paraformaldehyde. Similarly, tissues were processed and immunostained as described above. To localize individual cells in the LC relative to one another and to a fixed point in the brain it was necessary to define a reference point in the brain. The fixed reference point was obtained by placing the block in a sterotaxic holder and making two cuts perpendicular to the block face with their point of intersection lying immediately anterior to the substantia nigra. This point of intersection defined a fixed reference point so that the x, y, z co-ordinates of each indivdual LC neuron could accurately be determined.

In order to obtain linearity of the PAP reaction for indivdual cells it was necessary to adjust the reaction conditions so that individual LC cells were not saturated. These conditions were met by incubating the tissues in DAB for 4 minutes with a 1:700 TH antibody concentration. Under these reaction conditions the most darkly staining LC cells, those of the subceruleus are not saturated.

Image analysis procedures were used to perform morphometric and densitometric analysis of indivdual LC cells. The operator is able to select features by using a joystick controlled cursor which appears along with the image of the tissue section on the video monitor. Individual cells are defined by the operator at 40x magnification. The operator defines the boundary of the cell, which in these studies was arbitrarily defined by rounding off the cell at the base of the axon and dendrites. Cytoplasmic area was defined by outlining the nucleus of the cell and subtracting it from the total cellular area (Fig. 4). It was thus possible to determine relative size and density of the perikarya. Conversely it was possible to determine the area and staining intensity of the neuropil by outlining the perikarya and excluding them from the calculation.

Each neuron demonstrating a nucleolus in the plane of section was selected for analysis. This criterion prevents sampling error and potential variations in staining due to cutting through different portions of the cell. Individual neurons were analyzed and the following information was stored in a data base: a) density of staining, b) morphology, c) localization (topography

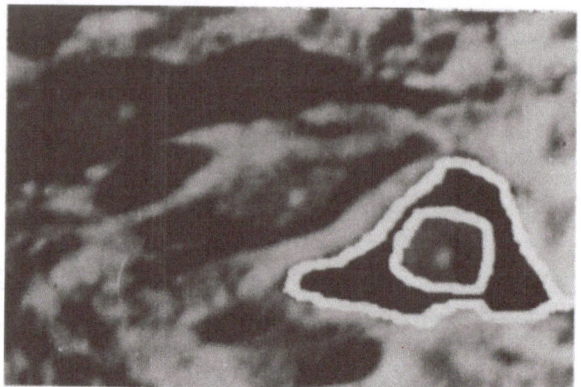

Fig. 4. Image analysis picture demonstrating how the operator can select features for analysis using the joystick; selection of a single cell with nucleus excluded

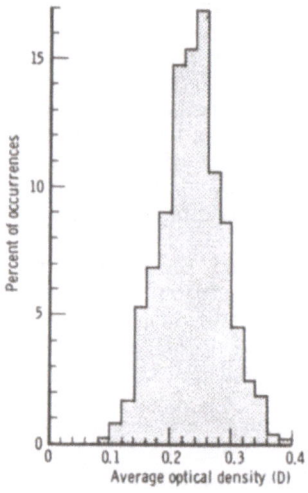

Fig. 5. Frequency distribution of staining intensity of TH–containing neurons throughout the LC. Data from 1,425 cells in five animals were grouped together and standardized to an overall mean of 0.220 ± 0.041 OD units.

within the LC), and d) packing density (the number of TH containing neurons found within a specified spherical radius of the target LC neuron).

      ii.   Variations in staining intensity

Inspection of sections of the LC stained for TH demonstrates that individual neurons vary in their intensity of staining, and hence immunoreactive TH. Computation of the average optical density of 1425 LC neurons obtained from 5 animals and standardized to an overall mean density and displayed as a density histogram (Fig 5) indicate that the distribution of staining density is unimodal over a 3-fold range with an average density of 0.220 ± 0.041 (optical density (O.D.) units $\pm$ S.D.). In individual animals, LC neurons varied over approximately a 2.4 - 3-fold range in staining intensity.

      iii.   Relationship of staining intensity to cellular morphometry, localization and packing density.

    Morphology. The realtionship between staining intensity and cellular morphology was assessed in 1425 LC neurons. We found little or no correlation between the average optical density and a) total cytoplsmic area ($r=0.20 \pm 0.06$); b) longest dimension:breadth ratio ($r=0.12 \pm 0.04$); and c) cytoplasmic area:nuclear area ratio ($r=0.10 \pm 0.08$). Therefore the amount of TH is apparently unrealted to size or shape of the neuron.

    Topography. We next sought to investigate the relationship between TH staining intensity and neuronal localization in the LC. We analyzed this relationship in two ways: a) using a distribution of LC cells based on conventional subdivisions of the LC (Fig. 1); and b) using a distribution of LC cells based on distances from the center of gravity of the LC, calculated from the equation: $r = ((x_i - x)^2 + (y_i - y)^2 + (z_i - z)^2)^{\frac{1}{2}}$ where x_i, y_i, and z_i are coordinates of the neuron and x, y, and z are coordinates for the center of gravity of the entire LC. Cells in the central "core" of the LC were found to be significantly lighter in staining intensity ($F = 5.69$, df= 16, $P \leq 0.01$) than cells in the remaining four subregions. Neurons in the other four subregions were found not to differ significantly in staining intensity from each other (Fig. 6a).

    When we analyzed the relationship between average optical density of individual LC cells in terms of their distance in spherical radii from the theoretical center of mass of the LC we found that cells located 0-300 μ from the LC center are significantly lighter in staining intensity ($F= 31.03$, df= 16, $P \leq 0.01$) than are cells further than 300 μ from the LC x, y, z center (Fig. 6b).

    Packing Density. It is obvious that cells in the center of the LC are more densely packed than are cells in the surrounding subregions (Fig. 1). Cells closest to the center of the LC in the central "core", were shown by analysis of variance to have fewer cells within 100 μ of themselves than do cells further from the center in regions defined as the "rim". The most densely packed neurons have significantly less TH enzyme protein ($F= 33.83$, df= 16, $P \leq 0.01$) than do cells having a lesser packing density of TH containing cells. Therefore, regions of the LC having a lower packing density, such as the ventral pole and the subceruleus contain more TH than do the tightly packed neurons in the central "core" of the LC. (Fig. 7a,b).

Comparison of the Amount and Regulation of TH enzyme protein in Noradrenergic and Dopaminergic Neurons

    Indirect biochemical evidence (Joh and Reis, 1975; Reis and Joh, 1977) suggests that the amount of TH enzyme protein is greater in individual DA neurons in the substantia nigra (SN) than in NE neurons in the LC.

    We therefore asked what happens to indivudal locus ceruleus and substantia nigra cells following a single dose of reserpine as described earlier. The amount of TH was therefore measured in the cell bodies of individual LC (noradrenergic) and SN (dopaminergic) neurons by computer assisted methods.

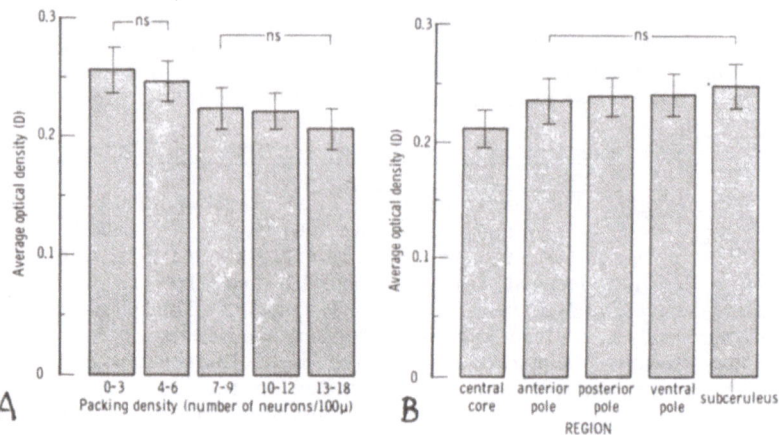

Fig. 6. Bar graphs showing the relationship between average optical density of TH staining in individual LC neurons as functions of (A) packing density, and (B) subregional localization of the neuron.

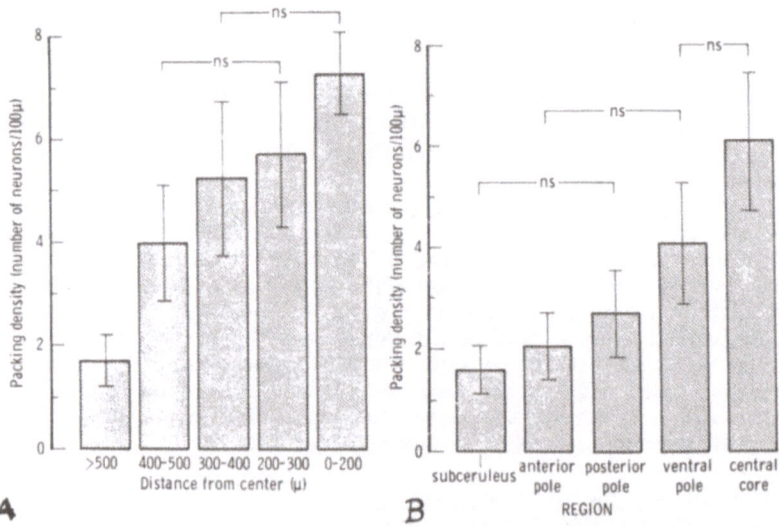

Fig. 7. Bar graphs showing the relationship between neuronal packing density (ie., the number of cells within a spherical radius of 100 μ of the target neuron) and (A) distance of the neuron from the center of gravity of the LC, and (B) subregion of the LC.

Since we had previously shown that there was a difference in the staining of "rim" and "core" neurons in the LC, a separate subanalysis was made to determine if there were any variations in the amount or time course of induction of TH enzyme protein in these two populations of LC cells.

Amount of TH in LC "Core and Rim" and SN Neurons

The average O.D. of "rim" cells in the LC was 0.140 ± 0.013 (mean $\pm$ sem) units, a significantly greater amount than in the "core" LC neurons $(0.108 + 0.011)$ units. The amount of TH in SN neurons (0.146 ± 0.017) units did not differ significantly from those of "rim" LC neurons, whereas dopaminergic SN neurons do contain significantly more TH than the noradrenergic cells that comprise the "core" of the LC. Analysis of variance of 900 LC and SN neurons in this study showed that there was a greater variability of staining in individual SN cells relative to either population of LC cells ($p \leq .05$).

Effects of Reserpine upon amount of TH within individual LC and SN neurons

Biochemical studies have shown that reserpine will increase the amount of TH enzyme protein in LC but not SN neurons (Reis and Joh, 1977). We therefore sought to determine if we could detect changes in TH in individual LC and/or SN neurons after a single injection of reserpine with animal survival times ranging from 1 to 10 days.

We have previously shown that three days following a single injection of reserpine, neurons of the SN show no changes in the amount of TH enzyme protein. (Reis et al. 1982). Analysis of SN neurons found in the plane of section in which LC nuerons were analyzed showed a similar lack of induction of TH at animal survival times from 1 to 10 days. Thus changes in TH enzyme protein in individual SN neurons will not be discussed further at this time.

Results in figure 8 compare percentage increases in TH in "rim" and "core" LC cells with TH induction in the LC as a whole over a time period of 1 to 10 days following reserpine administration. Three days following the administration of reserpine there is a maximal increase of TH of approximately 100% in the entire LC measured by regional immunocytochemistry. Similarly, individual neurons of the "rim" and "core" LC show maximal increases in TH at approximately 3 days following reserpine.

However, the induction within the entire LC differs from the induction in individual LC cells in several important ways: (a) the maximal increase in the perikarya of LC neurons is only 40% above controls whereas the entire LC increases by approximately 100%; (b) increases in levels of TH in individual cells are significant ($p \leq .05$) by day 1, but significant changes are not found in the entire LC until 2 days following reserpine; (c) induction of TH in perikarya remains close to maximal for up to 7 days following reserpine while levels in the entire LC are significantly less at 7 days than at the maximal 3 day point.

A possible explanation for these findings is that TH is made in the perikarya and shipped distally to terminal fields. Thus while TH levels are significantly increased by day 1 in the perikarya, the entire LC does not show significant increases until 2 days following reserpine, but shows a much larger

R.H. Benno, L.W. Tucker, T.H. Joh and D.J. Reis

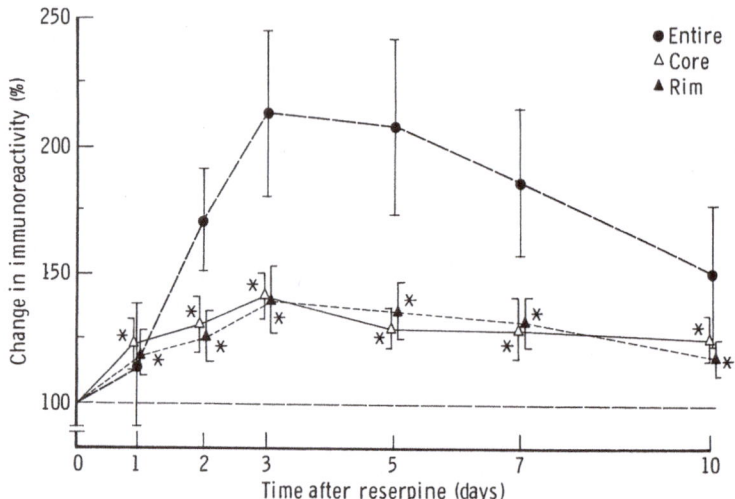

Fig. 8. Time course of changes in TH immunoreactivity in the entire LC and individual neurons in the core and rim of the LC. Each point represents either 50 core or rim neurons or 6 regional LC's. Data are % change $\pm$ SEM. Core induction vs. rim induction is not significant. Reserpine induction (core and rim) vs. control, significant p $\leq$ .05.

overall increase because the enzyme has been transported from the cell body into the processes where it is stored.

In order to test this hypothesis, we took advantage of the ability of the image processing system to separate perikarya from neuropil in the LC.

Results of this analysis comparing the amount of TH present in perikarya and neuropil in both control and reserpine treated animals is shown in Table 1.

Table 1

| | % Total area[1] | Control | | Reserpine | | |
		AOD	IOD[2]	AOD	% Change	IOD
Perikarya	24%	0.13	3.12	0.18	42	4.32
Neuropil	76%	0.08	6.08	0.16	100	12.16
Perikarya and neuropil	100%	—	9.20	—		16.48
Neuropil/ perikarya	—	—	1.94	—	—	2.81

[1]The relative areas of perikarya and neuropil were determined in sections taken through the LC by analyzing the area of neuropil separate from perikarya by image analysis procedures. Percent areas for control and reserpine treated animals were averaged together.

First, we determined the relative area contributed by both the neuropil and perikarya to the LC. The area of the LC occupied by perikarya was found to be approximately 24% while the neuropil occupied the remaining 76%. We then calculated the average optical density (A.O.D.) of individual LC cells (combining the rim and core cells) and separately the A.O.D. of the neuropil.

If one takes the A.O.D. of either the neuropil or perikarya and multiplies this density by the percentage of the LC occupied by that region, the product obtained represents the integrated optical density (I.O.D.) which is the total amount of enzyme contained within that region. The total amount of TH within the LC would therefore be the sum of the I.O.D. of both perikarya and neuropil. Using this compartmentalized method, the increase in TH, three days following reserpine treatment is 1.79-fold greater in reserpine treated animals than in controls (IOD (R)/IOD (C)). This increase in TH is approximately the same as was demonstrated in separate experiments when one views the LC as a whole at 6.3x as shown in Figure 3. One can also look at the data by examining the ratio of TH in the neuropil over the perikarya. In control animals, this ratio is 1.94, which indicates that normally there is approximately twice as much enzyme in the neuropil as in the perikarya. Three days following reserpine treatment, this ratio increases to 2.81, indicating that most of the TH accumulation three days following reserpine treatment is in the neuropil.

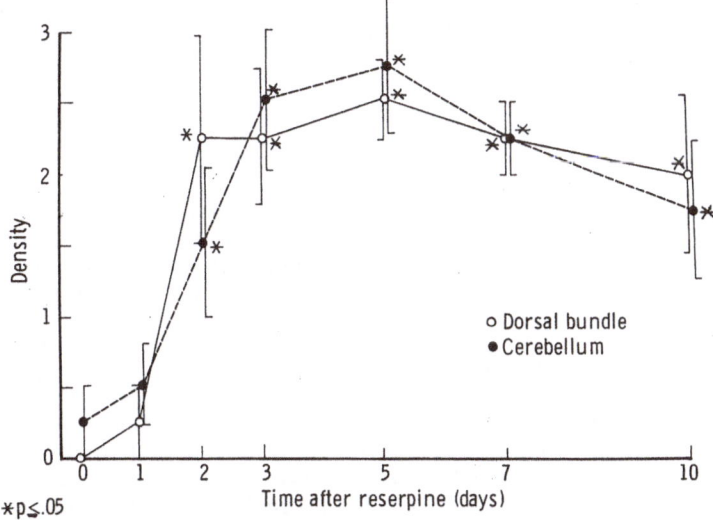

*p≤.05

Fig. 9. Time course of changes in TH immunoreactivity in dorsal noradrenergic bundle and pathway to cerebellum following reserpine. Density was scored as 0 to 4+ based on a semi-quantitative computer analysis. Four animals are represented at each point. Data expressed as mean ± SEM. Dorsal bundle vs. cerebellum not significant. Reserpine vs. control, p ≤ .05.

Further evidence for the hypothesis of shipment of TH from the perikarya to the terminal fields is demonstrated in figure 9. In this computer assisted semi-quantitative analysis the amount of TH in both the dorsal noradrenergic bundle (Maeda and Shimizu, 1972) and the LC projection into the cerebellum (Olson and Fuxe, 1971) was analyzed approximately 1 mm. away from the LC. Significant increases in TH were seen 2 days following reserpine in both fiber bundles (p ≤ .05). Density was scored as 0 to 4+ in this analysis. These semiquantitative values were obtained by determining the percentage of TH containing fibers exceeding a threshold value in a 1 mm. x

1 mm. square which contained the fiber bundle. The finding that significant increases in TH in these bundles occurs at 2 days following reserpine injection, the same time at which the entire LC shows an increase in TH, but 1 day after LC perikarya show increased levels of TH supports the hypothesis of TH synthesis in the cell body and subsequent shipment out to the terminal fields.

FORMULATION AND CONCLUSIONS

Variations in TH between single neurons of the LC

The use of this quantitative immunocytochemicl method in conjunction with computer assisted image processing has clearly demonstrated that neurons of the LC are heterogeneous with respect to the amount of TH enzyme they contain. The variation in TH is approximately three fold and correlates with the location of the cell in the LC and its topographic relationship to other TH containing cells in the LC. There is a lack of correlation between TH content and any morphometric feature of the neuron including cytoplasmic area and neuronal shape. Cells in the "rim" portion of the LC (which includes the anterior and posterior and ventral poles as well as the subceruleus) contain significantly more TH than do those cells in the central "core" of the LC. These "rim" cells were also shown to have a less extensive packing density than the more lightly stained "core" neurons.

There are several possible mechanisms that might account for the relationship between the amount of TH protein and the location of the cell in the LC. First, the more heavily stained TH containing neurons might represent a subpopulation of neurons having a more extensive projection path than do the lighter stained neurons. Thus the more darkly stained "rim" neurons might represent a population with projections terminating both in the spinal cord and forebrain, while the lighter staining "core" neurons might possess only a single projection field, such as to the hippocampus. In support of this hypothesis are numerous studies demonstrating variations in LC neuronal location and projection paths of these topographically distinct cells (Guyenet, 1980; Foote, 1980; Mason, 1979). Although correlations between the LC neurons having more extensive projection pathways and variations in amount of TH enzyme protein have not been made, the possibility exists that the extent of projection for an individual neuron may dictate the amount of protein that is made by that cell.

A second possible mechanism is that variations in the amount of TH enzyme protein might reflect differences in the firing rates of individual neurons. In peripheral sympathetic neurons it has previously been demonstrated that the amount and activity of TH enzyme protein is directly related to the discharge rate of the cell (Zigmond and Yehezkel, 1977; Zigmond, et al. 1980). Thus the lighter staining cells of the "core" LC might be firing at a slower rate than are the "rim" neurons. This slower firing frequency might relate in turn to the less extensive projection pathway proposed for these "core" neurons above.

Likewise, the control of firing frequency of these LC neurons might be related to the packing density of the individual neurons. Thus, diminished levels of TH in "core" neurons might be due to powerful feedback inhibition of the firing of these cells caused by the intimate contact of these "core" neurons with the extensive network of axons and dendrites in the center of the LC. Feedback inhibition of LC neurons has previously been demonstrated and such activity may account for the enzyme differences in the LC cell population (Aghajanian, et al. 1977; and Koda, et al., 1980). Further support of the "rim" versus "core" concept of variation in staining intensity comes from studies of topographical specificity of firing frequency of LC neurons in the behaving rat (Aston-Jones and Bloom, 1981). Neurons within 50 μ of the edge of the compact portion of the LC were shown to fire consistently faster than those neurons located closer to the center of the LC. Thus if LC neurons behave in a similar fashion to peripheral sympathetic neurons we would expect the faster firing "rim" neurons to have more TH enzyme protein as demonstrated in the present study.

Another possible explanation for the variation in TH amount in individual LC cells is the presence of a second neurotransmitter co-existing in a subpopulation of LC cells. The polypeptide, avain pancreatic polypeptide, APP, has been shown to be co-localized within approximately 40% of the TH containing LC cells (Hunt, et al. 1981). Although the relationship between the amount of TH enzyme protein and the presence of immunocytochemically detectable APP in the subpopulation of LC cells is unknown, this heterogeneity of co-localization of neurotransmitters might relate to differences in staining intensity and function of LC cells.

Differences in the Regulation of TH in Noradrenergic and Dopaminergic Neurons

In the present study the quantitative method has been shown to parallel biochemical studies of the regulation of TH by reserpine. In addition, we have been able to assess the regulation of TH at the level of the individual neuron, a degree of resolution not possible by biochemical methods. Thus while confirming previous studies showing selective increases in TH in NE but not DA neurons, this method has allowed us to distinguish two populations of LC cells and analyze diferences in the induction of TH in these populations 1 to 10 days after injection of reserpine. TH accumulation in both "rim" and "core" neurons was significant relative to controls at all time points analyzed, but the amount of increase in these two populations relative to each other was not significantly different at any time.

In addition, image processing procedures have allowed us to look at the accumulation of TH within the perikarya of the neuron and compare these changes with those occurring in the neuropil. We have shown that an increase in TH can be detected in both "rim" and "core" LC cells as early as 1 day following reserpine, whereas changes in the neuropil cannot be detected until day 2. This delayed increase in TH accumulation in the neuropil closely parallels changes in TH detected by the regional (6.3x) analysis of TH as well as by biochemical dissection.

The delayed increase in TH in the neuropil versus the perikarya might be evidence for the fact that reserpine increases TH amount by turning on protein synthesis in the perikarya and subsequently shipping the newly systhesized enzyme from the cell body to the terminal field. The close concordance between the amount of TH induction in cell bodies followed by a subsequent increase in neuropil and terminals over a 10 day time course lends support to this concept of perikaryal synthesis and subsequent shipment of TH to the terminals. Biochemical studies of increases in TH enzyme activity in terminal fields at varying survival times following reserpine administration (Zigmond, 1979), are in agreement with our immunocytochemical data. In addition, image processing procedures have allowed us to look at the accumulation of TH within fibers of the dorsal noradrenergic bundle and the tract projecting from the locus ceruleus to the cerebellum. Analysis of these fiber tracts approximately 1 mm. away from the LC demonstrated an increase in TH occurring 2 days after reserpine administration. This increase in TH within the axons projecting away from the LC parallels the time course of TH induction in the neuropil but occurs a day later than increases in TH in the perikarya. Thus the analysis of increases in TH in the fiber bundles leading away from the LC supports the hypothesis of TH synthesis in the perikarya and subsequent shipping out to the terminal fields.

SUMMARY

In summary, these studies have demonstrated the following:
(1) By maintaining rigorous control over staining conditions, the PAP method can be quantified.
(2) The increase in amount of TH present in the LC following reserpine induction measured by quantitative immunocytochemistry closely parallels biochemical data.
(3) TH staining in the LC is heterogeneous
(4) Heterogeneity in staining is a function of location and/or packing density in the LC and is not a funciton of cell morphometry.
(5) In the LC more TH enzyme protein is present in the neuropil than in the perikarya; following reserpine treatment, the neuropil/perikarya staining ratio is even greater.
(6) Changes in immunoreactive TH can be detected earlier in individual cells by immunocytochemistry than by biochemical dissection, which looks at changes in the LC as an whole (e.g., perikarya and neuropil).
(7) TH was shown to be made in the perikarya and shipped out distally into dendrites and terminal fields.

The technique of quantitative immunocytochemistry promises to one day provide us with a powerful method for analyzing the regulation of neurotransmitters and neuropeptides within individual neurons and relating differences in these neurons to specific parameters such as cell localization and projection pathway. This new information of how cells are regulated at the cellular level may provide us with valuable information as to how pharmacological and physiological manipulation or even disease states may effect the behavior of an organism by acting at the cellular level.

REFERENCES

Aghajanian, G.K., Cederbaum, J.M., and Wang, R.Y., Evidence for norepinephrine-mediated collateral inhibition of locus coeruleus neurons, Brain Res. 136 (1977) 570–577.
Aston–Jones, G. and Bloom F.E., Activity of norepinepherine–containing locus coeruleus neurons in behaving rats anticipates fluctuations in the sleep-waking cycle, J. Neurosci., 1, #8 (1981), 876–886.

Benno, R.H., Tucker, L.W., Joh, T.H., and Reis, D.J., Quantitative immunocytochemistry of tyrosine hydroxylase in rat brain: I. Development of a computer assisted method using the peroxidase-antiperoxidase technique, Brain Res., 246 (1982a) 225-236.

Benno, R.H., Tucker, L.W., Joh, T.H. and Reis, D.J., Quantitative immunocytochemistry of tryosine hydroxylase in rat brain: II. Variations in the amount of tryosine hydroxylase among individual neurons of the rat locus ceruleus in relation to neuronal morphology and topography. Brain Res. 246 (1982b) 237-247.

Bigbee, J.W., Kosek, J.C. and Eng, L.F., Effects of primary antiserum dilution on staining of antigen rich tissues with the peroxidase antiperoxidase technique. J. Histochem. Cytochem., 25 (1977) 443-447.

Choy, V.J. and Watkins, W.B., Immunocytochemical study of the rat hypothalamo-neurohypophysial system. II. Distribution of neurophysin, vasopressin and oxytocin in the normal and osmotically stimulated rat. Cell Tissue Res. , 180 (1977) 467-480.

Foote, S.L., Laughlin, S.E., Cohen, P.S., Bloom, F.E., and Livingston, R.B. Accurate three-dimensional reconstruction of neuronal distributions in brain: reconstruction of the rat nucleus locus ceruleus, J. Neurosci. Methods, 3 (1980) 159-173.

Grzanna, R. and Molliver, M.E., The locus ceruleus in the rat: an immunohistochemical delineation, Neuroscience, 5 (1980) 21-40.

Guyenet, P.G., The cerulospinal noradrenergic neurons: anatomical and electrophysiological studies in the rat, Brain Res., 189 (1980) 21-33.

Hunt, P.C., Emson, R. Gilbert, R., Goldstein, M., and Kimmell, J.R., Presence of Avian pancreatic polypeptide-like immunoreactivity in catecholamine and methionine-enkaphalin-containing neurones within the central nervous system; Neuroscience Letters, 21 (1981) 125-130.

Joh, T. H. and Reis, D.J., Different forms of tyrosine hydroxylase in central dopaminergic and noradrenergic neurons, sympathetic ganglia and adrenal medulla. Brain Res., 85 (1975) 146-151.

Koda, L.Y., Aston-Jones, G., and Bloom, F.E., Small granular vesicles in the locus ceruleus may indicate dendritic release of norepinephrine, (Abstract) Society for Neuroscience, 6 (1980) 352.

Levitt, M., Spector, S., Sjoerdsma, A. and Undenfriend, S., Elucidation of the rate limiting step in norepinephrine biosynthesis in the perfused guinea-pig heart, J. Pharmacol. Exp. Ther., 148 (1965) 1-8.

Maeda, T. and Shimizu, N., Ascending projections from the locus ceruleus and other aminergic neurons to the prosencephalic area of the rat, Brain Res. 36 (1972) 19-35.

Mason, S.T. and Fibiger, H.C., Regional topography within noradrenergic locus ceruleus as revealed by retrograde transport of horseradish peroxidase. J. Comp. Neurol., 187 (1979) 703-724.

Mayersbach, H., Unspecific interactions between serum and tissue sections in the fluorescent-antibody technique for tracing antigens in tissues. J. Histochem. Cytochem., 7 (1959) 427.

Miller, I. and Freund, J.E., Probability and Statistics for Engineers, Prentice Hall, New Jersey (1977)

Olson. L. and Fuxe, K., On the projections from the locus ceruleus noradrenaline neurons: The cerebellar innervation, Brain Res., 20 (1971) 165-171.

Pickel, V.M., Joh, T.H. and Reis, D.J., Immunohistochemical localization of tyrosine hydroxylase in brain by light and electron microscopy. Brain Res., 85 (1975a) 295-300.

Pickel, V.M., Joh, T.H. and Reis, D.J., Ultrastructural localization of tyrosine hydroxylase in noradrenergic neurons of brain. Proc. Nat. Acad. Sci., U.S.A., 72: (1975b) 659-663.

Reis, D.J., Benno, R.H. Tucker, L.W., and Joh, T.H., Quantitative immunocytochemistry of tyrosine hydroxylase in brain. In: Palay, S. and Chan-Palay, V. (eds) Cytochemical methods in neuroanatomy, New York, Alan R. Liss, 205-228.

Reis, D.J. and Joh, T.H., Long term regulation of brain tyrosine hydroxylase. In: Structure and Function of Monoamine Enzymes, Usdin, E., Weiner, N., and Youdim, M.B.H. (Eds.) Marcel Dekker, New York, 1977, pp. 119-124.

Reis, D.J., Joh, T.H., Ross, R.A. and Pickel, V.M., Reserpine selectively increases tyrosine hydroxylase and dopamine-B-hydroxylase enzyme protein in central noradrenergic neurons. Brain Res., 81 (1974) 380-386.

Ross, R.A., Joh, T.H. and Reis, D.J., Increase in the relative rate of synthesis of dopamine-B-hydroxylase in the nucleus locus ceruleus elicited by reserpine. J. Neurochem. 31 (1981) 1491-1500.

Steinbusch, H.W.M., Verhofstad, A.A., and Joosten, H.W., Localization of serotonin in the central nervous system by immunohistochemistry: description of a specific and sensitive technique and some applications, Neuroscience, 3 (9) (1978), 811-819.

Sternberger, L.A., Immunocytochemistry (second edition) Prentice Hall, 1979.

Sternberger, L.A., Hardy, P.H., Jr., Cuculis, J.J. and Meyer, H.G., The unlabeled antibody method of immunohistochemistry. Preparation and properties of soluble antigen-antibody complex (horseradish peroxidase-anti-horseradish peroxidase and its use in identification of spirochetes. J. Histochem. Cytochem., 18 (1970) 315-339.

Sternberger, L.A. and Petrali, J.P., Quantitative immunocytochemistry of pituitary receptors for luteinizing hormone-releasing hormone. Cell Tiss. Res. 162 (1975) 141-176.

Strader, C.D., Pickel, V.M., Joh, T.H., Strohsacker, M.W., Shorr, R.G.L., Lefkowitz, R.J., and Caron, M.G., Antibodies to beta-adrenergic receptor: attenuation of catecholamine-sensitive adenylate cyclase and demonstration of postsynaptic receptor localization in brain, Proc. Natl. Acad. Sci. 80 (1983) 1840-1844.

van Duijn, P., Pasco, E., and van der Ploeg, M., Theoretical and experimental aspects of enzyme determination in a cytochemical model system of polyacrylamide films containing alkaline phosphatase. J. Histochem. Cytochem., 15 (1967) 631-695.

Zigmond, R.E., and Yehezkel, B.A., Electrical stimulation of preganglionic nerve increases tyrosine hydroxylase activity in sympathetic ganglia. Proc. Natl. Acad. Sci. 74 (1977) 3078-3080.

Zigmond, R.E., Chalazonitis, A. and Joh, T.H., Preganglionic nerve stimulation increases the amount of tyrosine hydroxylase in the rat superior cervical ganglion, Neuroscience Lett., 20 (1980) 61-67.

Zigmond, R.E., Tyrosine hydroxylase activity in noradrenergic neurons of the locus ceruleus after resperpine administration: sequential increase in cell bodies and nerve terminals, Brain Res. 32 (1979) 23-29.

QUANTITATIVE CYTOCHEMICAL STUDIES ON CATECHOLAMINERGIC AND PEPTIDERGIC NERVE TERMINALS

J. SCHIPPER*

Department of Pharmacology, Free University, Van der Boechorststraat 7, 1081 BT Amsterdam, The Netherlands

FORMALDEHYDE INDUCED FLUORESCENCE OF CATECHOLAMINERGIC NERVE TERMINALS

Quantitative aspects

Catecholamines (CA) and indolamines (IA) can be visualized in histochemical preparations by the use of the formaldehyde induced fluorescence (FIF) method (Falck and Hillarp, 1962). In addition to morphological information on the localization, size and shape of monoaminergic neurones, this method can also provide information on the local concentration of CA and IA.

Theoretically, the FIF intensity is linearly related to the monoamine concentration in the preparation (for extensive discussion see Schipper and Tilders, 1982). This has been confirmed in studies on nonbiological models (Lichtensteiger, 1970; Einarsson, 1975; Schipper and Tilders, 1982). Also in a number of biological preparations that contain high concentrations of monoamines, such as mastcells (Enerback, 1975), pineal gland (Tilders et al. 1974), iris (Schipper et al. 1980c) and caudate nucleus (Einarsson, 1975), a good correlation was found between the FIF intensity and the monoamine concentration. Although quantitative cytochemistry does not allow interpretation in terms of absolute amounts of monoamines, the FIF intensity can be used as an index for changes in local monoamine content.

The advantage of quantitative FIF measurements compared to microchemical methods is the possibility to determine changes in amine content in specific brain structures that are too small to be dissected for microchemical analysis. For example, microfluorimetric measurements on FIF have revealed the existence of two dopamine systems in the median eminence (Lofstrom et al. 1976) and also two intermingling dopamine systems in the striatum with different turnover rates (Fuxe et al. 1978).

* present adress: Dept. Pharmacology, Duphar B.V.,
        P.O. Box 2, 1380 AA Weesp, The Netherlands.

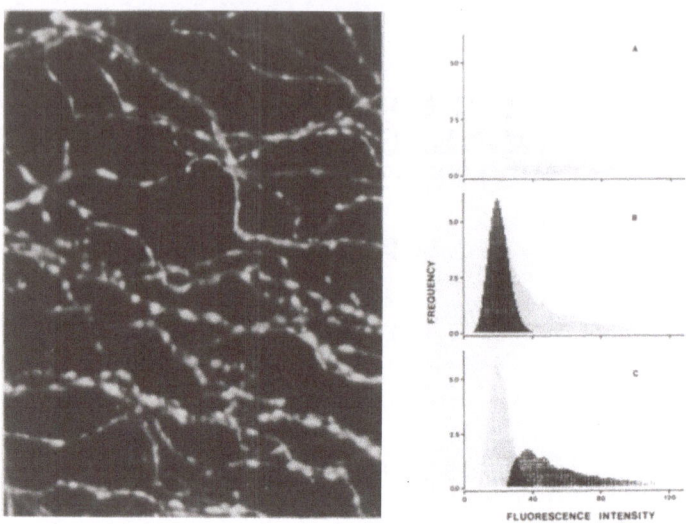

Fig. 1 <u>Left panel</u>: Microphotograph of formaldehyde induced fluorescence in noradrenergic nerve terminals in the dilatato-
ry muscle of the rat iris (whole mount preparation). Note
the differences in dimensions and in fluorescence intensity of various varicosities.
<u>Right panel</u>: A: Fluorescence histogram obtained after
scanning a 50 x 50 μm area in an iris preparation. B: calculated extraneuronal histogram (dark hatched). C: nerve
fiber histogram (dark hatched) obtained by substraction of
the extraneuronal histogram from the total histogram.
Abscissa: fluorescence intensity (arbitrary units). Ordinate: frequency expressed as percentage of total number of
measurements (10000). Data from Schipper et al. (1980a).

Most quantitative fluorescence data have been obtained with
"large field measurements", which involve measurements of the
total fluorescence of large areas (100 – 1000 μm^2) containing a
large number of varicosities. Much more detailed information can
be obtained by using scanning microfluorometry, which enables FIF
intensity measurements in structures as small as individual varicosities. Within a terminal network, strong differences in FIF
intensity between varicosities can be observed (see Fig. 1).
The functional relevance of these differences is unknown. In
order to study the implications of this heterogeneity we have
used the noradrenaline (NA) containing symapthetic terminal axons
in the dilatator muscle of the iris as a model. With the aid of a
computer controlled fluorescence scanning microscope, large areas
(50 x 50μm) in iris preparations have been scanned with a 0.5 μm

measuring spot (for details see Schipper et al. 1978). The fluorescence data (10000/scan) have been transformed into histograms (see Fig. 1) and subjected to a mathematical analysis in order to separate the measurements on nerve fibers from those on extraneuronal tissue (i.e. the underlying smooth muscle). Briefly, the mathematical histogram analysis is based upon the characteristics of (extraneuronal) histograms obtained in NA depleted iris preparations (denervation, reserpine). Using these characteristics, a curve fitting program has been applied to the low intensity part of the histograms that represent measurements on extraneuronal tissue only. Based on this, the remainder of the extraneuronal histogram can be calculated. After subtraction from the original histogram, the nerve fiber histogram is obtained (see Fig. 1). In this way, quantification of the nerve fiber (neuronal) fluorescence is independent of variation in the extraneuronal (smooth muscle) fluorescence (Schipper et al. 1980a).

Differences in NA turnover within a neuronal network

In an attempt to study the functional implications of the large variation in FIF intensity (=NA content) of varicosities, we have analyzed the rate of decline in FIF intensity after administration of a CA synthesis inhibitor (α-methyl-p-tyrosine; α-mpt) as parameter for NA turnover. Results of such studies show that the half life ($T\frac{1}{2}$) of disappearance of FIF in the total population of varicosities (4,2 hr) is in agreement with the chemically determined $T\frac{1}{2}$ of disappearance of NA in the iris (4,3 hr). Assuming that measurements of the highest fluorescence intensities in iris preparations of control rats and of α-mpt treated rats represent identical populations of varicosities, we have found pronounced differences in NA turnover in varicosities with different FIF intensity (see Table 1). It is interesting to note that the observed differences are even larger if our assumption is not correct.

The most intensely fluorescent varicosities appear to have the highest turnover rate. However, in these varicosities, the fraction of FIF that dissappears per unit of time after blockade of biosynthesis is the lowest and results in the longest $T\frac{1}{2}$ values. Since it is generally assumed that the turnover of NA is possitively correlated with neuronal activity and thus with transmitter release, we conclude that the most strongly fluorescent varicosities release the largest amount of NA. This will cause high concentrations of NA in the immediate surroundings of the strongest fluorescent varicosities. As a result, a high portion of presynaptic alpha-2 receptors will be activated, which inhibit the release of NA from that varicosity. This will be reflected in a decrease of the fraction of NA that is utilized per unit of time and thus in a longer $T\frac{1}{2}$.

The observed differences in turnover indicate that the turnover of CA may be determined by presynaptic mechanisms that operate at the level of individual varicosities. Furthermore, the functional heterogeneity observed here between varicosities of

Table 1. Turnover of NA in varicosities that exhibit different FIF intensities.

Fluorescence	FIF°	$T\frac{1}{2}$	Turnover rate
Strong	9.2	6.3	1.01
Medium	3.8	4.7	0.56
Low	1.6	3.9	0.28

Scanning microfluorometry was performed in iris preparations of sham- and α-methyl-p-tyrosine treated rats. Meaurements on neuronal FIF has been divided into 20 intensity classes and data of 3 classes are presented above. FIF° represents the total FIF intensity per class expressed as % of total neuronal FIF (=NA content) of untreated rats. $T\frac{1}{2}$ represents the half life (in hours) of dissapearance of FIF after blockade of CA biosynthesis. The turnover rate is calculated with the formula: turnover rate = ln 2 / $T\frac{1}{2}$ x FIF° and is expressed as % of total FIF intensity (=NA content) per hour. Data from Schipper et al. (1980c).

the same neuronal plexus, indicate that different "CA pools" found by neurochemical methods may in fact represent different populations of varicosities rather than different storage compartments within a varicosity.

Extraneuronal FIF: an index for in vivo NA release

In microfluorimetric studies on the iris, the intensity of the extraneuronal fluorescence appeared to be an interesting parameter. After surgical or pharmacological depletion of neuronal CA stores in the iris, we found a decrease of about 40% in the extraneuronal fluorescence. This decrease suggested the presence of CA in an extraneuronal compartment. Spectral analysis of extraneuronal FIF confirmed the presence of CA outside the nerve fibers (Schipper et al. 1979a). Since extraneuronal and neuronal FIF reacted differently to various experimental conditions, it is highly unlikely that an experimental artefact (e.g. diffusion) was involved.

Blockade of the impulse flow by extirpation of the superior cervical ganglion resulted in a rapid decrease in extraneuronal FIF. On the other hand, stimulation of NA release by administration of amphetamine resulted in an increased extraneuronal FIF intensity. Based on these and other pharmacological manipulations, we concluded that the extraneuronal FIF in the iris reflects the in vivo release of NA from the sympathetic nerve fibers (Schipper et al. 1979a, 1979b, 1980b).

Accordingly, a good correlation was obtained between the increase in extraneuronal FIF and the postsynaptic response of the
iris (mydriasis) after treatment with indirectly acting sympatho-
mimetic drugs (see Fig. 2). Similarly, the decrease in pupil diameter after reserpine or after blockade of the impulse flow coin-
cided with a decrease in extraneuronal FIF.

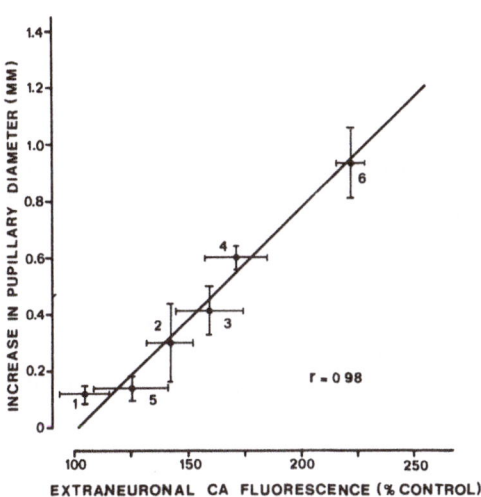

Fig. 2 Relationship between pupillary diameter in vivo and extra-
       neuronal CA fluorescence in histochemical preparations of
       this iris of rats after treatment with (1) nialamide (100
       mg/kg, 6,5 hr); (2) amphetamine (10 mg/kg, 2 hr); (3) am-
       phetamine (10 mg/kg, 1 hr); (4) amphetamine (10 mg/kg, 20
       min); (5) cocaine (10 mg/kg, 30 min); (6) cocaine (10
       mg/kg, 2 hr). Data from Schipper et al. (1979).

These data support the view that extraneuronal FIF reflects
NA that is available for interaction with catecholamine receptors
in the iris. Interestingly, the concentration of NA responsible
for this extraneuronal FIF falls well within the range of NA concentrations that are necessary to evoke a contraction of the
dilatory muscle of the iris in vitro (Schipper et al. 1980b).
This indicates that the extraneuronal concentration of NA is sufficiently high to explain the sympathetic tone in the dilatory
muscle of the iris of intact rats. These observations fit into
the concept of nonsynaptic neurotransmission, in which CA are
released in an extended extracellular space to reach the effector
cells.
   Further support for this concept comes from electronmicrosco-
pical studies which indicate that most of the sympathetic varicosities lack the "synaptic membrane complex" which serves as a
morphological criterium for the presence of a synapse (Ehinger et
al. 1980; Katzman et al. 1977). Usually, the majority of the

varicosities is localized at a considerable distance from the nearest postsynaptic membrane. Similar observations have been reported for brain tissue (Descarries et al. 1975). Whether extraneuonal FIF occurs in the brain as a result of nonsynaptic neurotransmission remains to be established. If this is the case, extraneuronal FIF may be an interesting parameter for measuring neuronal activity in the CNS.

IMMUNOFLUORESCENCE OF PEPTIDERGIC NERVE TERMINALS

Quantitative aspects

The localization of peptide containing neurones has been extensively studied during the last decade by use of various immunocytochemical techniques. Like the FIF method, the immunofluorescence method has the potential of providing quantitative information on the local concentration of antigens. So far however, quantitative immunocytochemistry of peptides has been performed only to a limited extent (Benno et al. 1982; Sternberger and Petrali, 1975), which might be due to the many pitfalls involved.

Within certain limits the immunofluorescence intensity is expected to be proportional to the concentration of the antigen in the preparation. In order to study this relationship at the microscopic level, we have developed a nonbiological model. This model consists of a gelatine gel to which a substance (antigen) in any desired concentration can be added (for details see Schipper et al. 1983).

The gels are fixed, sectioned and stained identically to tissue samples. Due to the homogenous distribution of immunofluorescence, these gels are suitable for quantitative microfluorimetric studies. In addition to studies on the quantitative aspects of immunocytochemistry, we have also used this model for studying the specificity of immunocytochemical procedures and for selection of antisera (Schipper et al. 1983; 1984).

As illustrated in Figure 3, a linear relationship is found between the concentration of serotonin added to the gels and the immunofluorescence intensity after staining with a serotonin antiserum. A similar result was found in substance P containing models stained with a substance P antiserum, although in a higher concentration range. These results indicate that within certain concentration limits, the immunofluorescence intensity can be used for obtaining quantitative information on the local concentration of peptides and that changes in immunofluorescence intensity may reflect changes in the concentration of antigens.

For quantitative purposes, standardization of the immunocytochemical procedure is of upmost importance. Especially the fixation procedure is critical. For reasons of standardization we prefer (if possible) immersion fixation instead of perfusion fixation of brain tissue. Furthermore, variations in tissue thickness of cryostat section should be limited by use of a high quality, electrically driven cryostat. Incubation of the sections

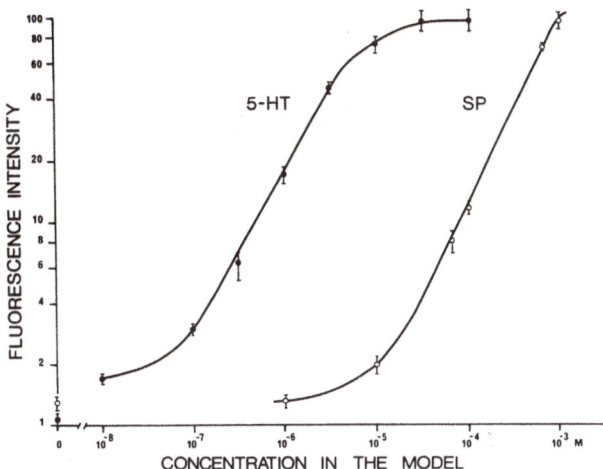

Fig. 3 The relationship between the concentration of serotonin (5-HT) and substance P (SP) incorporated in the model and the immunofluorescence intensity after staining of cyostat sections with 5-HT and SP antiserum respectively. Both immunofluorescence intensity and concentrations in the model are plotted on logarithmic scales. The fluorescence intensity is expressed in arbitrary units and data represent mean and SE of 18 measurements in 3 different sections.

should be performed under standardized conditions (indubation-time, antibody dilution, wash procedure) and all sections of one experiment should be treated simultaneously (for details see Schipper et al. 1983, 1984). With such precautions for standardization, quantification of immunofluorescence intensity in brain tissue is meaningfull as will be demonstrated by some studies on Corticotrophin Releasing Factor (CRF).

Functional aspects of CRF immunofluorescence

CRF was the first hypophysiotropic factor that was postulated, but its structure was only recently elucidated (Vale et al. 1982). By using antibodies raised against the 41 amino acid containing CRF peptide isolated from ovine hypothalami, immunoreactive (CRF$_i$) nerve fibers have been demonstrated in immunocytochemical preparations of brain tissues of various species including rat (Bugnon et al. 1982; Tilders et al. 1982; Schipper et al. 1984).

A dense plexus of CRF$_i$ fibers in the rat are localized in the external zone of the median eminence (see Fig. 4).

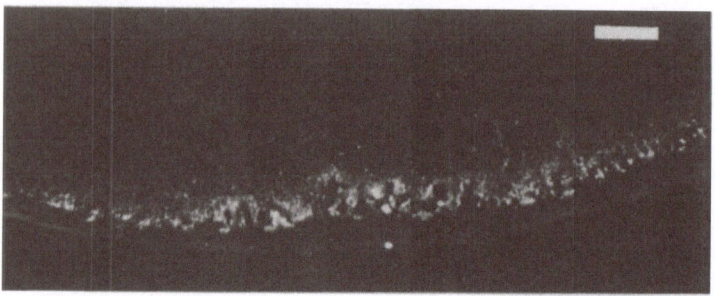

Fig. 4 Microphotograph of CRF immunofluorescence in the median
eminence of the rat. Bar represents 100 μm.

These fibers are in close proximity to the capillary loops, suggesting that CRF$_i$ material may gain access to the portal blood
and to affect pituitary function.

In order to establish the involvement of CRF$_i$ material in
the median eminence in the control of the pituitary – adrenal
system, the intensity of the CRF immunofluorescence was studied
in hypophysectomized and adrenalectomized rats. One week after
the removal of the pituitary or of the adrenals, the CRF$_i$ in
the median eminence was reduced to 48% and 39% respectively (see
table 2).

The alterations in CRF$_i$ under these conditions probably
reflect changes in CRF content due to changes in synthesis and/or
release of the peptide induced by removal of negative feedback
control (Smelik 1977). This is further supported by the observation that treatment of the adrenalectomized and hypophysectomized
rats with dexamethasone in doses that effectively reduce plasma
ACTH and corticosterone levels, largely prevented the decrease in
CRF$_i$ in the median eminence (see Table 2). These results indicate that the immunofluorescence intensity can be indicative for
the functional activity of these neurones.

Turnover of CRF

In experiments using the axonal transport inhibitor colchicine, we observed an increase in the fluorescence intensity of the
cell bodies, but on the other hand a decrease in fluorescence
intensity of the CRF$_i$ nerve terminals in the median eminence.
Microfluorimetric measurements on the median eminence of rats at
different times after administration of colchicine showed that
the decrease in CRF$_i$ was time-dependent. The decrease in CRF$_i$
corresponded to a monoexponential decline with a half-life of 15
hr (see Fig. 5).

Table 2. Effects of one week adrenalectomy, hypophysectomy
and dexamethasone (2 mg/kg s.c. daily for 8 days)
treatment on CRF immunofluorescence in median eminence of rats.

Treatment	CRF fluorescence intensity (% control)	
	−	+ dexamethasone
Control	100 ± 4	92 ± 4
Hypophysectomy	48 ± 6 *	73 ± 2 °
Adrenalectomy	39 ± 3 *	72 ± 8 °

Fluorescence intensity is expressed as percentage of that of
sham-operated controls. Data represent the mean and SE (5
rats per group). * p<0.01 compared to controls; ° p<0.01 compared to non-dexamethasone treatment (Student t-test).

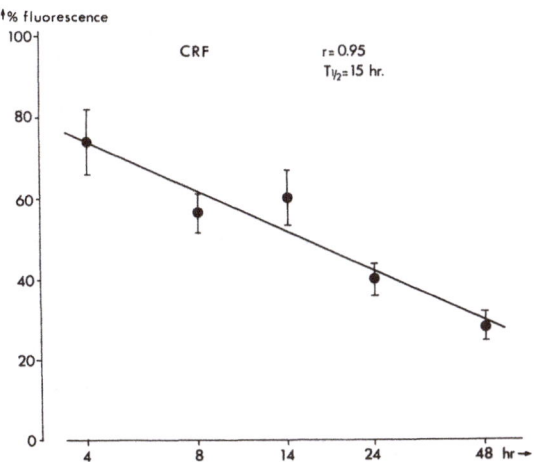

Fig. 5 The decrease in CRF immunofluorescence in the median eminence of rats on different times after an intracerebroven-
tricular injection of colchicine (50 µg). Fluorescence
intensity is expressed as percentage of that of untreated
controls. Data represent the mean and SE (5 rats per goup).

Blockade of the axonal transport induces a deprivation of newly synthesized CRF in the nerve terminals, since peptide synthesis does not occur at the nerve terminal level. Therefore, the reduction in the content of CRF in the nerve terminals can be considered as a measure for turnover of the peptide, which is analogous with the α-methyl-p-tyrosine model for studies on the turnover of CA. It is likely that the decline in CRF_i in the median eminence is not a general phenomenon, since other peptidergic neurones are affected differently. For instance, 48 hr after colchicine, the vasopressin immunofluorescence in the posterior lobe of the pituitary was slightly reduced only by 10-20%. This result is in agreement with the data obtained by Parish et al. (1981), who found after pulse labeling a vasopressin turnover rate of 5% per day in the posterior lobe.

The immunofluorescence intensity of other peptidergic nerve terminals (e.g. substance P) does not seem to be affected 48 hr after colchicine. This indicates that CRF neurons display a relatively high turnover rate compared to other peptide containing neuronal systems. The decline in CRF_i can be diminished by the administration of dexamethasone (see Table 3). This can be interpreted as a reduction in CRF turnover caused by the inhibition of CRF release induced by dexamethasone (Smelik 1977).

These results suggest that colchicine can be used to obtain information on the turnover of certain neuropeptides, which undoubtedly will be a powerful tool in establishing the functional activity of peptidergic neuronal systems. However, further experiments are necessary to evaluate whether colchicine disturbs the normal turnover rate. We have already some indication that this

Table 3. Effects of colchicine (50 μg, icv) and dexamethasone (2 mg/kg s.c. daily for 2 or 3 days) on CRF immunofluorescence in the rat median eminence.

Treatment	CRF fluorescence intensity (% control)	
	−	+ dexamethasone
Control	100 ± 6	96 ± 5
Colchicine 24 hr	49 ± 3 *	75 ± 3 °
Colchicine 48 hr	28 ± 3 *	52 ± 3 °

Fluorescence intensity is expressed as percentage of that of sham-operated controls. Data represent the mean and SE (5 rats per group). Dexamethasone treatment was started 24 hr before colchicine injection. * $p < 0.01$ compared to controls; ° $p < 0.01$ compared to non-dexamethasone treatment (Student t-test).

might be the case, since colchicine treated rats have slightly higher basal plasma corticosteron levels (Schipper, unpublished), suggesting an increased CRF release under these conditions.

SUMMARY

Changes in the local concentration of monoamines in brain structures after physiological or pharmacological manipulations can be determined by measuring the FIF intensity in histochemical preparations. The use of this microscopical method enables studies on structures as small as individual varicosities. Analysis of neuronal FIF in the rat iris after CA biosynthesis inhibition has revealed pronounces differences in NA turnover in different populations of varicosities. These differences indicates a functional heterogeneity of varicosities within a nerve plexus, which might be due to local regulatory mechanisms on synthesis and release of NA.

In the same preparation, quantification of FIF has revealed the presence of CA fluorescence outside the nerve fibers. This extraneuronal CA fluorescence reflects NA released in situ from the sympathetic nerve fibers and can be used as a histochemical parameter for monitoring neuronal activity.

In additional to these quantitative studies on CA histofluorescence, the quantitative aspects of peptide immunofluorescence has been studied by use of a nonbiological model. Within certain limits, the immunofluorescence intensity is linearly related to the antigen concentration in the preparation.

Studies on the CRF immunofluorescence in the median eminence of rats have revealed that manipulations of the pituitary-adrenal system (adrenalectomy, hypophysectomy, dexamethasone) induce changes in the CRF immunofluorescence intensity. The quantitative immunofluorescence data support the view that CRF neurons play a central role in the control of pituitary activity.

After administration of an axonal transport inhibitor, a monoexponential decline in the CRF immunofluorescence has been observed in the median eminence. Inhibition of CRF release by dexamethasone results in a slower decline of CRF immunofluorescence. It is tempting to speculate that the decline of immunofluorescence intensity after blockade of axonal transport can be used as a parameter for the turnover rate of peptides. The data presented here indicate that quantitative evaluation of cytochemical staining represent a tool to study changes in activity of monoamine and peptide containing neurones.

REFERENCES

Benno, R.H., Tucker, L.W., Joh, T.H. and Reis, D.J. (1982). Quantitative immunocytochemistry of tyrosine hydroxylase in rat brain. Brain Res. 246, 225-236.

Bugnon, C., Fellman, D., Gauget, A. and Cardor, J. (1982). Onto-geny of corticoliberin neuroglandular system in rat brain. Nature (Lond.) 298, 159–161.

Descarries, L., Beaudet, A. and Watkins, K.C. (1975). Sertotonin nerve terminals in adult rat neocortex. Brain Res. 100, 563–588.

Ehinger, B., Falck, B. and Sporrong, B. (1970). Possible axo-axonal synapses between peripheral adrenergic and cholinergic nerve terminals. Z. Zellforsch. 107, 308–321.

Einarsson, P., Hallmann, H. and Jonsson, G. (1975). Quantitative microfluorimetry of formaldehyde induced fluorescence of dopamine in the caudate nucleus. Med. Biol. 53, 15–24.

Enerbäck, L. and Jarlstedt J. (1975). A cytofluorimetric and radiochemical analysis of the uptake and turnover of 5-hydroxy-tryptamine in mast cells. J. Histochem. 23, 128–135.

Falck, B., Hillarp, N.A., Thieme, G. and Torp, A. (1962). Fluo-rescence of catecholamines and related compounds condensed with formaldehyde. J. Histochem. Cytochem. 10, 348–354.

Fuxe, K., Fredholm, B.B., Agnati, L.F. and Corrodi, H. (1978). Dopamine receptors and ergot drugs. Evidence that an engolene derivative is a differential agonist at subcortical limbic dopa-mine receptors. Brain Res. 46, 295–311.

Katzman, R., Broida, R. and Raine, C.S. (1977). Reinnervation, myelination and organization of iris tissue implanted into the rat midbrain-an ultrastructural study. Brain Res. 138, 423–433.

Lichtensteiger, W. (1970). Katecholaminhaltige Neurone in der neuroendokrinen Steuerung. Prog. Histochem. Cytochem. 1, 185–276.

Löfström, A., Jonsson, G., Wiesel, F.A. and Fuxe, K. (1976). Microfluorimetric quantitation of cathecholamine fluorescence in rat median eminence II turnover changes in hormonal states. J. Histochem. Cytochem. 24, 430–442.

Schipper, J. and Tilders, F.J.H. (1979 a). On the presence of extraneuronal catecholamine in the iris of the rat: a scanning microfluorimetric study. Neurosci. Lett. 12, 229–234.

Schipper, J. and Tilders, F.J.H. (1982). Quantification of for-maldehyde induced fluorescence and its application in neurobiolo-gy. Brain Res. Bull. 9, 69–80.

Schipper, J. and Tilders, F.J.H. (1983). A new technique for studying specificity of immunocytochemical procedures; specifici-ty of serotonin immunostaining. J. Histochem. Cytochem. 31, 12–18.

Schipper, J., Tilders, F.J.H., Groot Wassink, R., Boleij, H.F. and Ploem, J.S. (1980a). Microfluorimetric scanning of sympathetic nerve fibres: quantitation of extraneuronal and neuronal fluorescence with aid of histogram analysis. J. Histochem. Cytochem. 28, 124-132.

Schipper, J., Tilders, F.J.H. and Mulder, A.H. (1980b). Extraneuronal catecholamine in the iris of the rat: a consequence of nonsynaptic neurotransmission? Neuroscience 5, 745-751.

Schipper, J., Tilders, F.J.H. and Ploem, J.S. (1978). Microfluorimetric scanning of sympathetic nerve fibres: an improved method to quantitate biogenic amines. J. Histochem. Cytochem. 26, 1055-1066.

Schipper, J., Tilders, F.J.H. and Ploem, J.S. (1979b). Extraneuronal catecholamine as an index for sympathetic activity: A scanning microfluorimetric study on the iris of the rat. J. Pharmacol. Exp. Ther. 211, 265-270.

Schipper, J., Tilders, F.J.H. and Ploem, J.S. (1980c). A scanning microfluorimetric study on sympathetic nerve fibres: intraneuronal differences in noradrenaline turnover. Brain Res. 190, 459-471.

Schipper, J., Werkman, T.R. and Tilders, F.J.H. (1984). Quantitative immunocytochemistry of corticotropin Releasing Factor. Studies on nonbiological models and on hypothalamic tissues of rats after hypophysectomy, adrenalectomy and dexamethasone treatment. Brain Res. 293, 111-118.

Smelik, P.G., (1977). Some aspects of corticosteroid feedback actions. Ann. N.Y. Acad. Sci. 297, 580-590.

Sternberger, L.A. and Petrali, J.P. (1975). Quantitative immunocytochemistry of pituitary receptors for luteinizing hormone-releasing hormone. Cell Tiss. Res. 162, 141-146.

Tilders, F.J.H., Ploem, J.S. and Smelik, P.G. (1974). Quantitative microfluorimetric studies on formaldehyde induced fluorescence of 5-hydroxytryptamine in the pineal gland of the rat. J. Histochem. Cytochem. 22, 967-975.

Tilders, F.J.H., Schipper, J., Lowry, P.J. and Vermes, I. (1982). Effects of hypothalamus lesions on the presence of CRF immunoreactive nerve terminals in the median eminence and on the pituitary-adrenal response to stress. Regulatory Peptides 5, 77-84.

Vale, W., Spiess, J., Rivier, C. and Rivier, J. (1981). Characterization of a 41-residue ovine hypothalamic peptide that stimulates secretion of corticotropin and β-endorphin. Science 213, 1394-1397.

QUANTITATION OF NERVE TERMINAL NETWORKS OF TRANSMITTER–IDENTIFIED NEURONS AFTER SELECTIVE NEUROTOXIC LESIONS

G. JONSSON, H. HALLMAN and J. LUTHMAN

Department of Histology, Karolinska Institutet, P.O. Box 60400, S–104 01 Stockholm, Sweden

ABSTRACT

Examples are presented on the use of computer–assisted image analysis for the quantitation of nerve terminal density of certain transmitter–identified (noradrenaline, serotonin, substance P) neuron systems in CNS and PNS demonstrated by histochemical and immunocytochemical techniques. The effects of selective neurotoxic lesions on these neurotransmitter systems and their behavior during conditions of regrowth was investigated. The results obtained with image analysis were compared with measurements of relevant neurochemical parameters for the respective neuron types under identical experimental conditions. The results from the neurochemical analyses were in general in good agreement with those obtained by quantitative image analysis. It is concluded that image analysis is a very powerful and useful tool for quantitation of transmitter–identified nerve terminal networks under conditions of dynamic changes (degeneration–regrowth) in terminal density.

INTRODUCTION

Lesion techniques have long been used to explore the structural organization and function of the nervous system. During the last decade there have been great improvements of these techniques with the introduction of target–directed chemical neurotoxins capable of selectively lesioning particular classes of neurons defined with respect to their chemical messenger (transmitter), e.g. 6–hydroxy-dopamine (6–OH–DA), which can induce a selective degeneration of catecholamine neurons (see Jonsson, 1983). These chemical neurotoxins are increasingly used for the elucidation of neuronal structure–function relationships as well as for creating lesion models for studies of degeneration–regrowth properties. For a proper evaluation of such lesions in acute and chronic stages, it is necessary to use specific and quantitative neurochemical and/or histochemical techniques in order to be able to evaluate the extent of the alter-

ation of the lesioned neuronal system. The most precise information
on this point can generally be obtained by quantitative morphological techniques for neuron-specific marker (e.g. transmitters or
transmitter synthesizing enzymes). This approach has in the past
generally been time-consuming and therefore not always feasible as an
evaluation instrument   However, this situation has now changed considerably due to the introduction of computer-assisted image analysis systems making it possible to analyze microscopical pictures of
nervous tissue at high speed and precision. The aim of the present
article is to present data where we have used image analysis to
quantitate alterations in neuronal networks of transmitter-identified (noradrenaline = NA; 5-hydroxytryptamine = 5-HT; substance P
= SP) central and peripheral neuron systems demonstrated by specific
histochemical and immunocytochemical techniques after selective
neurotoxic lesions. 6-OH-DA was used to produce degeneration of sympathetic adrenergic nerves, while neonatal lesions of 5-hydroxy-
tryptamine (5-HT) and substance P neurons in the brain and spinal
cord were induced by 5,7-dihydroxytryptamine (5,7-HT) and capsaicin
respectively (see Jonsson, 1981, 1983). Data on the effect of exogenous administration of the monosialoganglioside GM$_1$ in these
lesion models will also be presented in view of the increasing evidence indicating that gangliosides may promote sprouting-regenera-
tion processes and functional recovery after lesions of nerve tissue
(see Rapport and Gorio, 1981; Gorio et al., 1983; Agnati et al.,
1983). In order to get information on the precision and reliability
of the image analysis technique used for quantitation of nerve terminal density of the lesioned transmitter systems studied, measurements of relevant neurochemical parameters for the various systems
were also conducted.

MATERIALS AND METHODS

     For neonatal lesions, newborn rats of both sexes (Sprague-
Dawley) were injected with 5,7-HT (50 mg/kg s.c.) or capsaicin
(50 mg/kg s.c.) within 6-8 h after birth. The pups thereafter received 4-12 daily injections of GM$_1$ (30 or 20 mg/kg s.c., Fidia).
The regional CNS dissection for sampling of tissue for chemical
5-HT assay was carried out as described previously (Jonsson et al.,
1982). Chemical sympathectomy was produced in adult mice (N.M.R.I.,
25 g) by injecting 6-OH-DA (25 mg/kg i.v.). In these experiments
the first GM$_1$ dose (20 mg/kg i.p.) was administered 1 h before
6-OH-DA.

Falck-Hillarp Fluorescence Histochemistry

     Mouse irides were dissected out and prepared as whole mounts
(Malmfors, 1965), dried and exposed to gaseous formaldehyde of optimum humidity at +80°C for 1 h. The processed whole mounts were taken
directly for fluorescence microscopical analysis (Falck et al.,
1962; Corrodi and Jonsson, 1967; Fuxe et al., 1970).

5-HT and SP Immunocytochemistry

The rats were anesthetized and perfused with cold fixative
(buffered 4% paraformaldehyde). The brain and spinal cord were dissected out, post-fixed, frozen and cut on a cryostat. The sections
(10 μm) were processed for 5-HT or SP immunocytochemistry according
to the indirect immunofluorescence technique of Coons (1958) as described previously (Steinbusch et al., 1978; Ljungdahl et al., 1978;
Cuello et al., 1979).

Quantitation of NA, 5-HT and SP Nerve Density

Mouse irides processed according to the Falck-Hillarp technique
for demonstration of NA nerves and rat CNS sections stained immunocytochemically for 5-HT and SP were analyzed employing an interactive computer-assisted image analysis system (IBAS, Kontron/Zeiss)
for determination of NA, 5-HT and SP nerve terminal density. An
image intensified high-resolution black-and-white TV camera (Siemens
K5b M21005) was attached to the fluorescence microscope with proper
filter settings used to visualize the NA, 5-HT and SP neurons. Both
trans- and epi-illumination and a primary magnification of x10, x16
or x25 (Neofluar, Zeiss) were used. After digitizing the fluorescence microscopical image to be analyzed, correction for uneven
background fluorescence, contrast enhancement and editing steps
were carried out. Thereafter a standardized measuring field was
defined and a gray-level threshold chosen to select for fluorescent
nerves excluding virtually all background fluorescence. The binary
picture was generated and the computer calculated the area covered
by fluorescent nerves in % of the measuring field after editing
(see Fig. 4). Assuming that the fluorescence morphology of the nerve
terminal types analyzed is relatively uniform, the percentage values
recorded are a measure of the relative nerve density per unit area
in the section or whole mount.

^{3}H-NA Uptake in Vitro

Isolated mouse irides were incubated in vitro in Krebs-Ringer
bicarbonate buffer (pH 7.4) containing 0.05 μM ^{3}H-NA for 10 min at
+37ºC. Radioactivity taken up and retained in the irides was after
solubilization determined by liquid scintillation spectrometry
according to Olson et al. (1968). ^{3}H-NA uptake was expressed as
nCi/iris and corrected for 'extra-neuronal uptake' determined by
performing incubations in 0.5 μM desipramine (a NA uptake blocker).
The radioactivity values obtained can be considered as ^{3}H-NA uptake
in NA nerve terminals.

Chemical 5-HT Assay

Endogenous 5-HT in CNS tissue samples was determined by liquid

chromatography with electrochemical detection (LCEC) according to
Ponzio and Jonsson (1979). The values were corrected for recovery
and expressed as ng/g wet weight of the tissue.

Neurotoxins Used

     5,7-HT creatinine sulphate (Regis); 6-OH-DA HCl (Hässle); cap-
saicin (8-methyl-N-vanillyl-6-nonenamide; Sigma).

RESULTS

Chemical Sympathectomy of Mouse Iris

     The sympathetic adrenergic nerves innervating mouse iris con-
stitute a dense network of evenly distributed varicose nerve fibers
which are very distinctly visualized by Falck-Hillarp fluorescence
histochemistry (see Fig. 1). This is also illustrated by the very
high signal-to-noise ratio found when performing image analysis of
this network of NA-containing fibers (Fig. 1; see also Fig. 2). It
is therefore clear that this tissue preparation is an ideal one for
nerve density quantitation with the interactive image-analysis system used, allowing close comparison between original and computer-
generated pictures. The rationale for the presently used technique
is to generate a binary picture of the network that is as identical
as possible to the original fluorescence microscopical image and
calculate the fraction of the area that is covered by fluorescent
nerve terminals (see Fig. 4). Providing that the fluorescence morphology of the nerve terminals is reasonably uniform, the percentage
area measured is a measure of the relative nerve density per unit
area in the histological specimen analyzed. The area covered by
fluorescent NA nerve terminals in normal irides measured in this
way contributes about 20-30% of the total measuring field (see Fig.
1). For each animal 4-8 measurements were made and averaged to represent the NA nerve density value for the animal in question.

     Administration of the catecholamine neurotoxin 6-OH-DA (25 mg/
kg i.v.) produces a very rapid (within a few hours) and selective
degeneration of NA nerve terminals in iris as reflected by an almost
complete disappearance of the peaks representing NA fibers 16 h
after 6-OH-DA (Fig. 2A). The recorded mean NA nerve density value
was also found to be reduced by more than 90% at this time interval
after 6-OH-DA (Fig. 3). The ganglionic NA cell bodies remain intact
after the 6-OH-DA dose used (Jonsson and Sachs, 1970) making it
possible for regeneration of the adrenergic nerves which takes about
2-3 months to become complete (Jonsson and Sachs, 1970, 1972). In
the present study it was found that there was a gradual recovery of
the NA nerve plexus and the NA nerve density was restored to about
50% of control value 4 weeks after 6-OH-DA (Fig. 3). The regenerat-
ing NA nerve terminals had a similar fluorescence morphology compared to normal NA terminals (see Fig. 2B), although they displayed

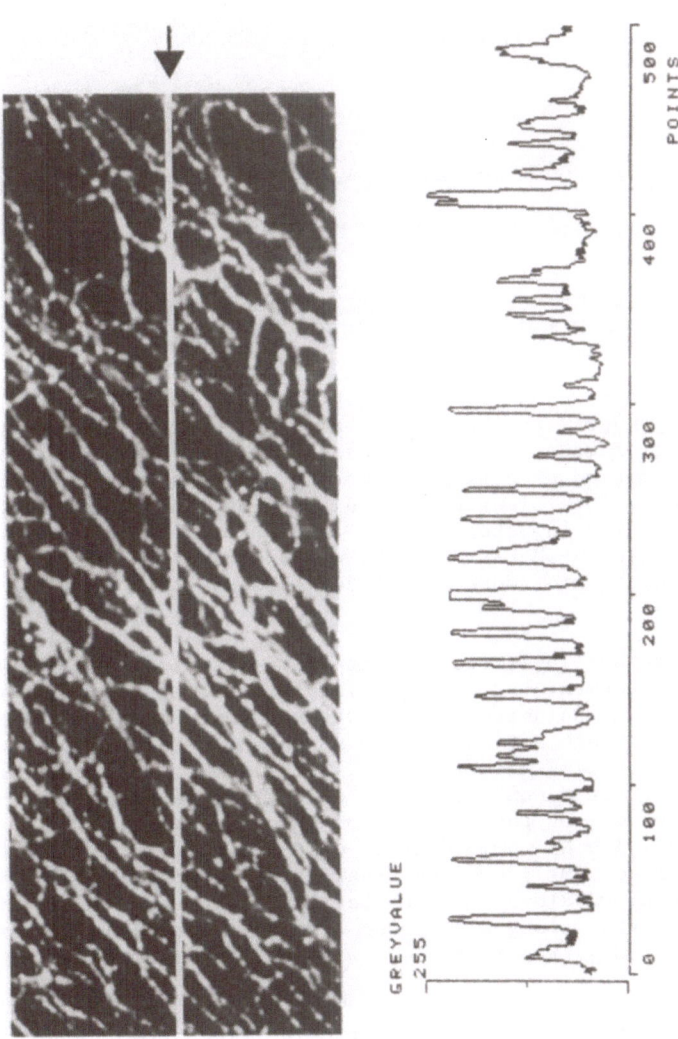

Fig. 1. The diagram shows gray levels (0–255) along the straight line (→) across the digitized image of NA-containing sympathetic adrenergic nerves in mouse iris demonstrated by fluorescence microscopy according to Falck-Hillarp. The peaks represent NA fibers crossed by the scanning line. Total length of the scanning line = 670 μm).

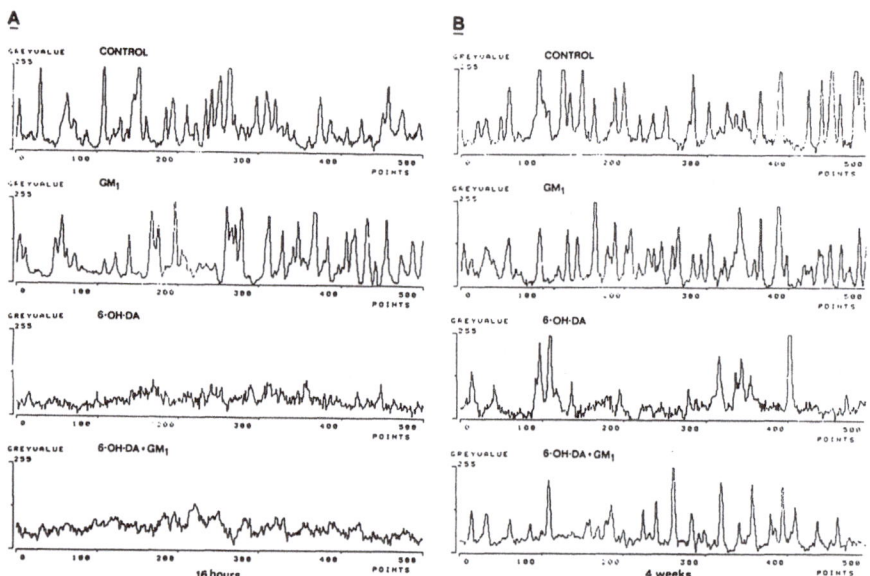

Fig. 2.   Effect of 6-OH-DA (25 mg/kg i.v.) and/or GM₁ (20 mg/kg i.p.) treatment on gray levels (0-255) along a straight line across digitized images of mouse irides processed for fluorescence micro-scopical visualization of adrenergic nerves according to the Falck-Hillarp technique. Total length of each scanning line = 640 μm. The first dose of GM₁ was injected 1 h before 6-OH-DA. A: The mice were sacrified 16 h after 6-OH-DA. B: GM₁ was administered once daily for one week and the mice were sacrificed 4 weeks after 6-OH-DA.

in general a somewhat lower fluorescence intensity indicating lower NA levels. GM₁ treatment had no significant effect on the 6-OH-DA induced reduction in NA nerve density when measured 16 h or 2 weeks after 6-OH-DA, although a tendency for an enhanced recovery was noted at the latter time interval (Fig. 3). However, measurements 4 weeks after 6-OH-DA demonstrated a clearcut and significant en-hancing effect of GM₁ on the recovery of NA nerve density after 6-OH-DA (Fig. 3). The NA nerve density was thus found to be about 70% of control after GM₁ + 6-OH-DA treatment as compared to 50% of control after 6-OH-DA alone. This effect of GM₁ was also noted when recording gray levels along a scanning line in digitized images of irides (Fig. 2B). GM₁ treatment alone was ineffective to alter NA nerve density.

Measurement of ³H-NA uptake in vitro in irides from identically 6-OH-DA and/or GM₁ treated mice displayed almost exactly the same results as those obtained by measuring NA nerve density by image analysis (Fig. 3). Previous studies have shown that ³H-NA uptake (initial rate) is a sensitive and reliable parameter for monitoring

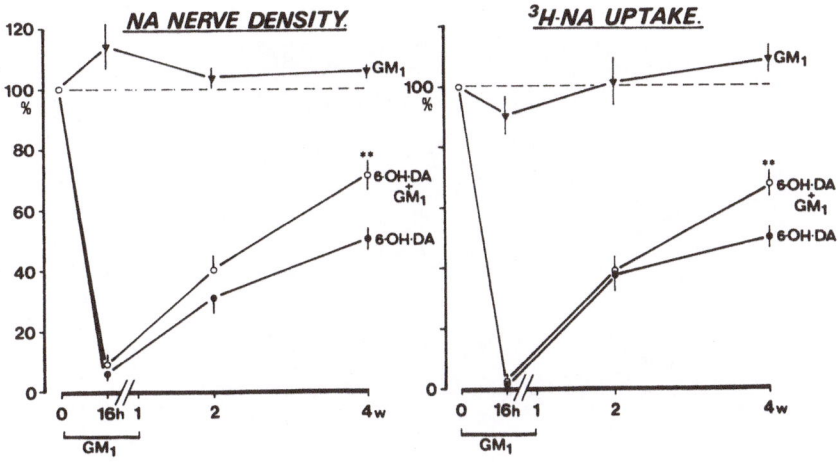

Fig. 3. Effect of 6-OH-DA (25 mg/kg i.v.) and/or GM_1 (20 mg/kg
i.p.) on NA nerve density and ^{3}H-NA uptake in mouse irides various
periods of time after the 6-OH-DA administration (16 h, 2 weeks and
4 weeks). NA nerve density was determined in irides processed for
Falck-Hillarp fluorescence histochemistry using image analysis,
while ^{3}H-NA uptake was measured by liquid scintillation spectrometry
after in vitro incubation in 0.01 μM ^{3}H-NA for 10 min. Each point
represents the mean ± S.E.M. (n = 4-6), expressed as % of control.
The first GM_1 injection was given 1 h before 6-OH-DA. Only one GM_1
injection was given to mice sacrificed 16 h after 6-OH-DA, while
mice sacrificed 2 or 4 weeks after 6-OH-DA received daily injections
of GM_1 for 1 week. ** = 0.001 > p > 0.01 (Student's t-test).

NA nerve density (see Jonsson, 1983). The only minor discrepancy was
noted at the time interval 16 h after 6-OH-DA where the ^{3}H-NA uptake
values indicated a more marked reduction of NA nerve density than
that found by image analysis. The exact reason for this is unclear
but might be related to the particular sensitivity properties of
the TV camera used, which tends to increase its sensitivity when the
light intensity of the measuring field is reduced. This property of
a non-linear response is reflected in Fig. 2A where there are signs
of an increase in measured background fluorescence intensity after
6-OH-DA induced denervation as compared to control. It will therefore be difficult to choose a proper discriminatory threshold (gray-
level definition) selecting only for the few NA fibers left after
6-OH-DA without including some areas with high background. Nevertheless, in view of the almost identical results obtained with the two
methods used to monitor NA nerve density, it can be concluded that
the present image analysis technique is a very sensitive and precise
method to measure NA nerve density in mouse iris under dynamic conditions of alterations of its adrenergic nerve plexus.

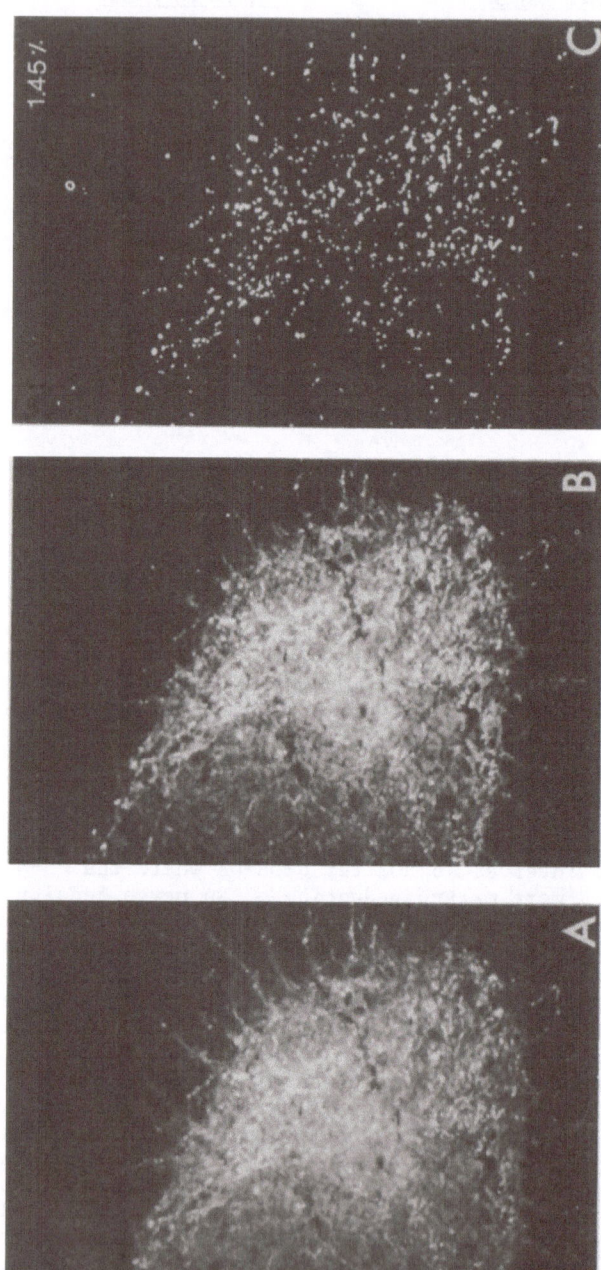

Fig. 4. Illustration of the various steps in the measurement of the density of 5-HT nerve terminals visualized by fluorescence immunohistochemistry in the anterior horn of rat thoracic spinal cord using interactive computer-assisted image analysis. A: Digitized TV image. B: After contrast enhancement. C: Binary picture after correction for uneven background fluorescence, with display of the percentage figure denoting the area covered by fluorescent 5-HT terminals (X 160).

Neonatal Lesion of 5-HT Neurons in Rat CNS

The central 5-HT neurons form a widespread neuronal system innervating many brain regions to variable degrees. The present immunocytochemical technique gives a distinct visualization of central 5-HT neurons with a good signal-to-noise ratio giving possibilities for reliable nerve density measurements employing image analysis, as demonstrated in Fig. 4. The present 5-HT nerve density measurements from selected CNS regions were obtained by recording 12 measurements per region (sampled from 3 sections) which were averaged to represent one observation (one animal). In control rats it was observed that the average nerve density values (computed as the area covered by fluorescent 5-HT nerve terminals in % of the whole measuring field) varied from region to region, being 2.9% in frontal and 1.6% in occipital cortex. The density value for dorsal hippocampus was 1.7% and for lumbar spinal cord (anterior horn) 9.5%, while for piriform cortex 7.7% and for the amgdaloid region 10.8%. Endogenous 5-HT levels were recorded in some of these regions and it was found that there was a very high correlation (r = 0.996) between 5-HT levels recorded by chemical assay in frontal and occipital cortex, hippocampus and spinal cord (corrected for 5-HT concentration in the gray matter) and the 5-HT nerve density values in these regions obtained by image analysis (Fig. 5). This strongly suggests that the endogenous 5-HT level is a reasonably reliable index of 5-HT nerve terminal density, which also has been found in other studies (see Jonsson and Hallman, 1982). It should be noted, however, that spinal cord is a 'drop-out', showing higher 5-HT levels than in the other regions analyzed. This is in agreement with experience from Falck-Hillarp fluorescence histochemistry where 5-HT nerve terminals in the spinal cord are more readily visualized than in most other regions, indicating high endogenous 5-HT levels in the terminals of this region. This suggests that when using endogenous 5-HT levels as an index for 5-HT nerve terminal density, comparisons between different regions cannot be made without reservation and that this parameter is more suitable for comparisons between different experimental conditions of the same region. It should, however, be pointed out that alterations of endogenous 5-HT levels as well as of other monoamines can occur without changes in nerve density. For a proper evaluation it is necessary to use both neurochemical and quantitative morphological techniques. Systemic administration of the serotonin neurotoxin 5,7-HT (50 mg/kg s.c.) to newborn rats has previously been shown to produce a heterogenous alteration of the postnatal development of central 5-HT neurons, leading to among other things a marked and permanent degeneration of distant 5-HT nerve terminal projections in e.g. cerebral cortex, hippocampus and spinal cord (Sachs and Jonsson, 1975; Jonsson et al., 1978). The 5-HT perikarya remain largely intact after this 5,7-HT treatment. The data presented in Fig. 6 show that neonatal 5,7-HT administration produced marked reductions of 5-HT nerve density recorded by image-analysis in frontal cortex (-74%), occipital cortex (-66%), hippocampus (-84%) and lumbar spinal cord (-80%). No alterations were found in piriform cortex and the amygdaloid region studied.

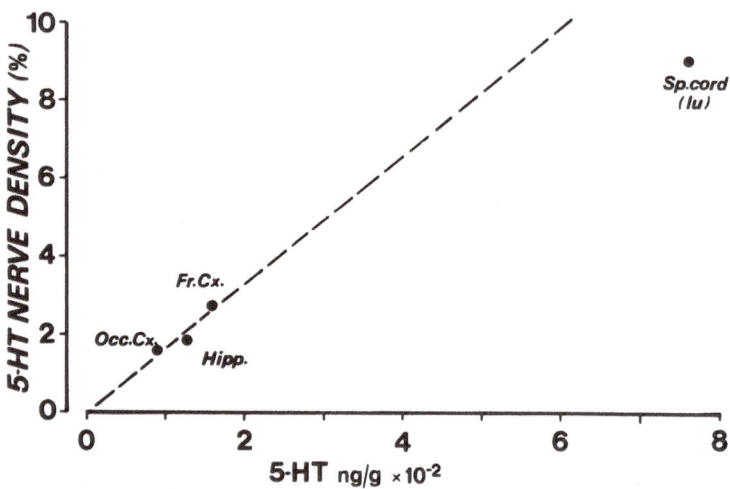

Fig. 5. Relation between endogenous 5-HT concentration (ng/g) measured by chemical 5-HT assay and 5-HT nerve density estimated by
image analysis in sections processed for 5-HT immunohistochemistry
in various regions of rat CNS. Occ.Cx. = occipital cortex; Hipp.
= hippocampus; Fr.Cx. = frontal cortex; Sp.cord = spinal cord
(lumbar). Data from Jonsson et al. (1984).

These reductions correlated fairly well with those obtained by chemical 5-HT assay of the different regions (cf. Sachs and Jonsson,
1975; Jonsson et al., 1984). After combined 5,7-HT + GM$_1$ treatment
it was found that the 5-HT nerve terminal density was markedly
higher in frontal and occipital cortex as well as hippocampus and
spinal cord than after 5,7-HT alone (Fig. 6). The counteracting
effect of GM$_1$ in frontal cortex was quantitatively very similar to
data obtained by chemical 5-HT assay, although the effects in occipital cortex, hippocampus and spinal cord were more pronounced than
those found when measuring 5-HT levels in these regions (see Jonsson
et al., 1984). This latter discrepancy is most likely related to
the fact that 5-HT levels were measured in relatively large pieces
of tissue, leading to a 'dilution effect' masking relatively large
differences locally, although it cannot be excluded that endogenous
5-HT levels under the present experimental conditions do not exactly
reflect 5-HT nerve density. Although the present results do not
give a definitive answer as to how precisely it is possible to measure 5-HT nerve terminal density using image-analysis, it seems
reasonable to conclude that the present technique is adequate for
this purpose.

EFFECT OF GM₁ ON 5,7-HT INDUCED ALTERATION OF 5-HT NERVE DENSITY.

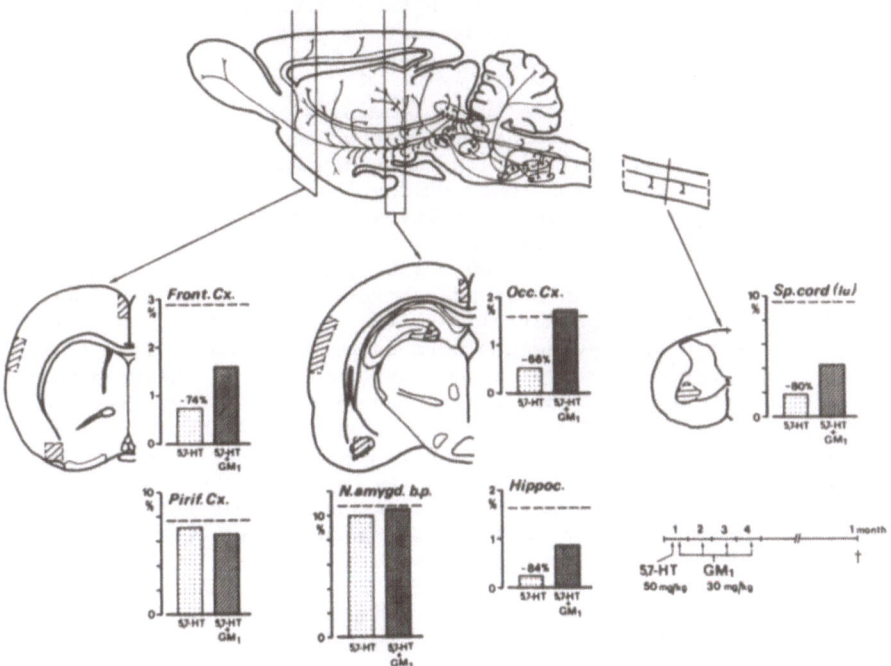

Fig. 6. Effect of GM₁ on the 5,7-HT induced alteration of 5-HT
nerve terminal density in various CNS regions of one-month-old rats.
5,7-HT (50 mg/kg s.c.) was injected at birth and thereafter GM₁
(4 x 30 mg/kg s.c., 24 h interval) or saline was administered as
indicated. 5-HT nerve terminal density was estimated in sections
(levels indicated) processed for 5-HT immunohistochemistry using
interactive image analysis. Nerve density is expressed as the area
covered by fluorescent 5-HT nerve terminals in % of the measuring
field. Each bar represents the mean of data obtained from three
rats. The figure above bars denoted 5,7-HT indicates the percentage
decrease of 5-HT nerve density in % of control value (dashed line).
Hatched areas in the schematic cross-sections of the CNS indicate
the areas from where the 5-HT nerve density measurements were recorded. Data from Jonsson et al. (1984). Front.Cx. = frontal cortex;
Occ.Cx. = occipital cortex; Pirif.Cx. = piriform cortex; N.amygd.
b.p. = nuc. amygdaloideus basalis posterior; Hippoc. = hippocampus;
Sp.Cord = lumbar spinal cord.

Neonatal Lesion of SP Neurons in Rat Spinal Cord

Neonatal treatment with capsaicin (50 mg/kg s.c.), the pungent
factor of red pepper, has been shown to cause degeneration of a distinct population of primary sensory neurons involved in mediation

of chemogenic pain (Janscó et al., 1977). This population consists
of several neuron types defined with respect to neuropeptide content and one type of these neurons contains the neuropeptide SP with
the central branch innervating the superficial layers of the dorsal
horn of the spinal cord (see Janscó et al., 1981). In the present
study the effect of capsaicin on SP terminals in the dorsal spinal
cord visualized by fluorescence immunocytochemistry was investigated
by quantitative image analysis. The immunocytochemical technique
used demonstrates with high contrast the SP nerve terminals in the
spinal cord. SP nerve terminal density was measured in a defined
area enclosing the superficial layers of the dorsal horn (A; measuring field approx. rectangular 500 x 130 µ) as well as in a small
area containing SP terminals in the dorsomedial part of the lateral
funiculus adjacent to the dorsal horn (B; see Fig. 7). From each
animal and area 6-8 measurements were recorded (sampled from 4 sections) and averaged to represent one observation. The nerve density
values recorded from the superficial layers (area A) were about 20%
of the whole measuring field, while this value was about 15% for
area B. Neonatal capsaicin treatment caused a reduction of SP nerve
terminal density to about 50% of control value in the superficial
layers, while a small but not significant effect was noted in area
B (Fig. 7). This former reduction corresponds fairly well with previously reported capsaicin-induced reductions of SP levels in the
dorsal horn of the spinal cord recorded by radioimmunoassay (Nagy
et al., 1980; Brodin, unpubl. data). This would indicate that the
present image-analysis technique is a reliable method for SP nerve
terminal quantitation. It was furthermore observed that GM_1 treatment had a significant antagonizing effect on the capsaicin-induced
reduction of SP nerve terminal density in the superficial layers,
while no effect was noted after GM_1 treatment alone. None of the
treatments produced any significant alterations in SP nerve density
in area B (Fig. 7). Although more methodological work is needed,
the present results strongly indicate that the present image-analysis technique may prove to be a sensitive and reliable technique
for the quantitation of the density of SP nerve terminals demonstrated by immunocytochemistry.

DISCUSSION

The results presented show that the interactive image analysis
technique used is very useful and powerful for quantitation of
transmitter-identified nerve terminal networks in experimental
studies where the terminal density undergoes alterations due to
degeneration and regrowth. The limitations of this technique appears
mainly to be confined to the histochemical and immunocytochemical
procedures used the demonstrate the neurotransmitter systems in
terms of sensitivity (detectability), reproducibility and specificity as well as quality of the sections and other microscopical
preparations used. Needless to say, the microscopical demonstration
of the particular neuron system to be studied with high-contrast is
one of the most important factors for a proper quantitation of nerve

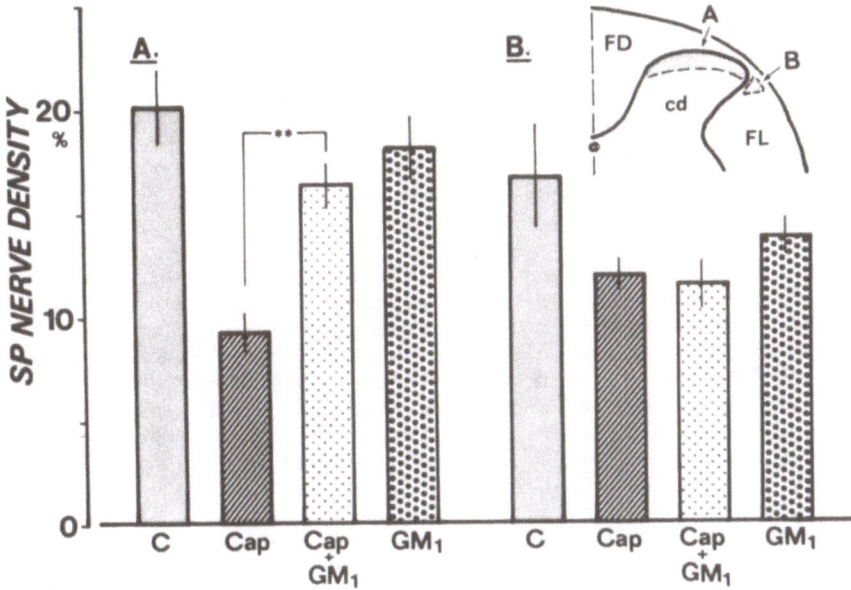

Fig. 7. Effect of GM$_1$ on the capsaicin (Cap.) induced alteration of SP nerve terminal density in the lumbar spinal cord in adult rat (2 months old). Capsaicin (50 mg/kg s.c.) was injected at birth and thereafter GM$_1$ (12 x 20 mg/kg s.c.; 24 h interval) or saline was administered. The first GM$_1$ injection was made 1 h after capsaicin. Quantitation of SP nerve terminal density in cross-sections processed for SP immunohistochemistry was made by interactive image analysis. The dotted areas A (superficial layers of the dorsal horn) and B (dorsomedial part of lateral funiculus adjacent to the dorsal horn) in the schematic cross-section of the dorsal aspect of the lumbar spinal cord indicate the areas from where the SP nerve density measurements were recorded. Nerve density is expressed as the area covered by fluorescent SP nerve terminals in % of the measuring field. Each bar represents the mean ± S.E.M. of 4-6 determinations. FD = funiculus dorsalis; FL = funiculus lateralis; cd = dorsal horn. ** = 0.001 < p < 0.01 (Student's t-test).

density. Although the present image analysis system is very advanced and allows sophisticated measurements of morphological parameters, it would improve significantly with the availability of a TV camera with a better response in terms of linearity than the one used, especially in situations of measurements of very low nerve terminal densities. Another problem that may arise using the present technique analyzing images originating from fluorescence microscopy is in situations of quantitation of very dense terminal networks with very high fluorescence intensity in the single terminals. In such cases the binary picture generated tends to exaggerate the area covered by fluorescent fibers leading to an overestimation of the nerve terminal density. This problem can be solved or at least partly eliminated by the use of thinner sections and/or reducing the intensity of the excitation light of the fluorescence microscope by a suitable filter. Although interactive computer-assisted image analysis appears to be a very powerful tool for morphometric quantitation, it still seems advisable, at least with respect to the applications presented here, to conduct analyses of relevant neurochemical parameters in parallel with the quantitation by image analysis.

Finally, the experiments with GM_1 should be commented upon briefly. In all lesion models investigated it was found that GM_1 had a counteracting effect of the neurotoxin-induced lesion in the chronic stage. The data so far available indicate that GM_1 does not have any detectable effects on the primary neurotoxic action of the neurotoxins used as analyzed in the acute stage. The results obtained are therefore compatible with the view that exogenous GM_1 has a regrowth stimulatory effect on lesioned NA, 5-HT and SP neurons. It is also possible that the present results are related at least partly to protective actions of GM_1 against retrograde degeneration of the lesioned neurons which might occur following the initial neurotoxin-induced nerve terminal damage.

ACKNOWLEDGEMENTS

The present studies have been supported by grants from the Swedish MRC (04X-2295), Karolinska Institutet, Expressen's Prenatal Research Foundation, Socialstyrelsen, Bergvall's and Jeansson's Foundations. The skillful technical assistance of Ms. B. Käller, Ms. E. Lindqvist and Ms. B. Drevinger is gratefully acknowledged. The authors are also very grateful to Ms. Ida Engqvist for expert secretarial help.

REFERENCES

Agnati, L.F., Fuxe, K., Calza, L., Benefenati, F., Cavicchioli, L., Toffano, G. and Goldstein, M. (1983). Gangliosides increase the survival of lesioned nigral dopamine neurons and favour the recovery of dopaminergic synaptic function in striatum of rats by collateral sprouting. Acta Physiol. Scand. 119, 347-363.

Coons, A.H. (1958). Fluorescent antibody methods. In General Cytochemical Methods, Vol. I. (ed. J.F. Canielli). Academic Press, New York.

Corrodi, K. and Jonsson, G. (1967). The formaldehyde fluorescence method for the histochemical demonstration of biogenic monoamines. A review on the methodology. J. Histochem. Cytochem. 15, 65-78.

Cuello, A.C., Galfre, G. and Milstein, C. (1979). Detection of substance P in the central nervous system by a monoclonal antibody.
Proc. Natl. Acad. Sci. 76, 3532-3536.

Falck, B., Hillarp, N.-Å., Thieme, G. and Torp, A. (1962). Fluorescence of catecholamines and related compounds condensed with formaldehyde. J. Histochem. Cytochem. 10, 348-354.

Fuxe, K., Hökfelt, T., Jonsson, G. and Ungerstedt, U. (1970). Fluor-
escence microscopy in neuroanatomy. In Contemporary Research in
Neuroanatomy. (eds. W.J.K. Nauta and S.O.E. Ebbesson). Springer-
Verlag, Berlin.

Gorio, A., Marini, P. and Zanoni, R. (1983). Muscle reinnervation.
III. Motoneuron sprouting capacity, enhancement by exogenous gangliosides. Neuroscience 8, 417-429.

Janscó, G., Hökfelt, T., Lundberg, J.M., Kiraly, E , Halász, N.,
Nilsson, G., Terenius, L., Rehfeld, J., Steinbusch, H., Verhofstad,
A., Elde, R., Said, S. and Brown, M. (1981). Immunohistochemical
studies on the effect of capsaicin on spinal and medullary peptide
and monoamine neurons using antisera to substance P, gastrin/CCK,
VIP, enkephalin, neurotensin and 5-hydroxytryptamine. J. Neurocytol.
10, 963-980.

Janscó, G., Kiraly, E. and Jancsó-Gábor (1977). Pharmacologically
induced selective degeneration of chemosensitive primary sensory
neurons. Nature 270, 741-743.

Jonsson, G. (1981). Lesion methods in neurobiology. In Techniques in
Neuroanatomical Research. (eds. Ch. Heym and W.G. Forssman).
Springer-Verlag, Berlin.

Jonsson, G. (1983). Chemical lesioning techniques: Monoamine neuro-
toxins. In Handbook of Chemical Neuroanatomy. Vol. 1. Methods in
Chemical Neuroanatomy. (eds. A. Björklund and T. Hökfelt). Elsevier,
Amsterdam.

Jonsson, G., Gorio, A., Hallman, H., Janigro, D., Kojima, H. and
Zanoni, R. (1984). Effect of GM1 ganglioside on neonatally neuro-
toxin induced degeneration of serotonin neurons in the rat brain.
Develop. Brain Res. (in press).

Jonsson, G. and Hallman, H. (1982). Response of central monoamine
neurons following an early neurotoxic lesion. Biblthca Anat. 23,
76-92.

Jonsson, G., Hallman, H., Pollase, T. and Sachs, Ch. (1978).
Developmental plasticity of central serotonin neurons following 5,7-
dihydroxytryptamine treatment. Ann. N.Y. Acad. Sci. 305, 328-345.

Jonsson, G., Hallman, H. and Sundström, E. (1982). Effects of the noradrenaline neurotoxin DSP4 on the postnatal development of central noradrenaline neurons in the rat. Neuroscience 7, 2895-2907.

Jonsson, G. and Sacns, Ch. (1970). Effects of 6-hydroxydopamine on the uptake and storage of noradrenaline in sympathetic adrenergic neurons. Eur. J. Pharmacol. 9, 141-155.

Jonsson, G. and Sachs, Ch. (1972). Neurochemical properties of adrenergic nerves regenerated after 6-hydroxydopamine. J. Neurochem. 19, 2577-2585.

Ljungdahl, Å., Hökfelt, T. and Nilsson, G. (1979). Distribution of substance P-like immunoreactivity in the central nervous system of the rat. I. Cell bodies and nerve terminals. Neuroscience 3, 861-943.

Malmfors, T. (1965). Studies on adrenergic nerves. Acta Physiol. Scand. 64 Suppl., 248.

Nagy, J.I., Vincent, S.R., Staines, W.M.A., Fibiger, H.C., Reisine, T.D. and Yamamura, H.I. (1980). Neurotoxic action of capsaicin on spinal substance P neurons. Brain Res. 186, 435-444.

Olson, L., Hamberger, B., Jonsson, G. and Malmfors, T. (1968). Combined fluorescence histochemistry and ^{3}H-noradrenaline measurements of adrenergic nerves. Histochemie 15, 38-45.

Ponzio, F. and Jonsson, G. (1979). A rapid and simple method for the determination of picogram levels of serotonin in brain tissue using liquid chromatography with electrochemical detection. J. Neurosci. 32, 129-132.

Rapport, M.M. and Gorio, A. (eds.) (1981). Gangliosides in Neurological and Neuromuscular Function, Development and Repair. Raven Press, New York.

Sachs, Ch. and Jonsson, G. (1975). 5,7-Dihydroxytryptamine induced changes in the postnatal development of central 5-hydroxytryptamine neurons. Med. Biol. 53, 156-164.

Steinbush, H.W.M., Verhofstad, A.A.J. and Joosten, H.W. (1978). Localization of serotonin in the central nervous system by immunohistochemistry: Description of a specific and sensitive technique and some applications. Neuroscience 3, 811-819.

CYTOFLUORIMETRIC SCANNING: A QUANTITATIVE METHOD TO STUDY AXONAL TRANSPORT

ANNICA DAHLSTRÖM and PÄR-ANDERS LARSSON

Institute of Neurobiology, University of Göteborg, S–400 33 Göteborg, Sweden

INTRODUCTION

The neuron is a specialized cell, which has one or more long processes. One of these, the axon, may assume an impressive length (up to 1 m) relative to the perikaryon (~ 100 µm). The perikaryon is the metabolic center of the cell, where macro-molecules and organelles are manufactured, products that are exported into the axon, to undergo axonal transport towards the nerve ending. Retrograde axonal transport also occurs, and this bidirectional phenomenon, intra-axonal transport, is of vital importance for the function of the neuron (cf. Grafstein and Forman 1980, Dahlström 1983). It has therefore been the subject of a number of investigations that have dealt with e.g. the influence of various experimental procedures on amount of transported material or rate of transport.

If an endogenous substance is to be studied, a technique must be applied which results in arrest of axonal transport, e.g. a cold block or ligature. The transported organelles then accumulate on either side of the crush region in a manner depending on amount, rate and direction of transport.

Using microscopical techniques the number of axons with accumulated material, the distance the accumulations reach from the crush as well as the character of the axon type involved can be registered. However, in order to obtain quantitative data biochemical estimation methods must be employed, using homogenates of nerve.

The situation is examplified in Fig. 1, symbolizing two consecutive sections from a ligated peripheral nerve, treated for indirect immunofluorescence. One section was incubated with an antiserum to cholinergic vesicles, and the immunoreactive material has accumulated in thick myelinated axons reaching a

considerable distance (5 mm) on either side of the crush region.
The lower section was incubated with an antiserum to substance P
(SP), and very thin axons with material accumulated over 0.5–1 mm
proximal to the crush are demonstrated. This can be clearly
discerned in the fluorescence microscope. However, in order to
get a sufficiently good quantitative resolution of these two
accumulation patterns, the nerve would need to be chopped up in
very short segments before assay, and probably many nerves would
have to be pooled.

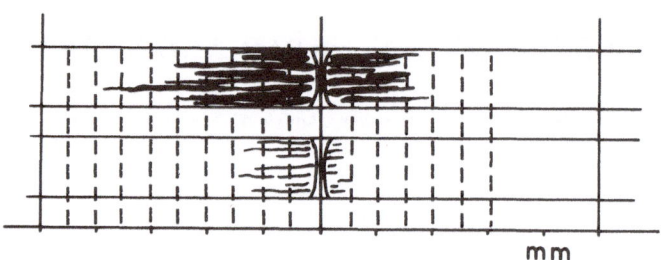

Fig. 1. Schematic illustration of the different patterns
of accumulation between cholinergic vesicle-like
material in thick myelinated axons (top) and SP-like
material in thin axons (bottom) as viewed in a fluorescence microscope. As indicated by the hatched lines
0.5 mm sections would have to be assayed to give accumulation profiles with sufficient resolution.

We have therefore developed a method, based on morphological techniques, which can supply quantitative data as well as
morphological details.

METHODOLOGY

Equipment

The technique is in its present form based on fluorescence
microscopy. Any substance that is fluorescent, or can be made to
fluoresce, can be studied. With the modern immunofluorescence
technique any substance that can be identified by appropriate
antisera can thus be measured.

The basic instrument in our equipment is a Leitz MPV II
fluorescence microscope with incident light, equipped with a
scanning system (Fig. 2). The fluorescent light passes via a
narrow slit (measuring diaphragm) into a PM-tube, and the emitted
light is amplified in a Leitz amplifier and fed into a recorder
with integrator (see Fig. 3). The recorded curve shows the

intensity and extent of the accumulations and the area under the accumulation curve is proportional to the amount of fluorescence, which can be expressed in arbitrary units.

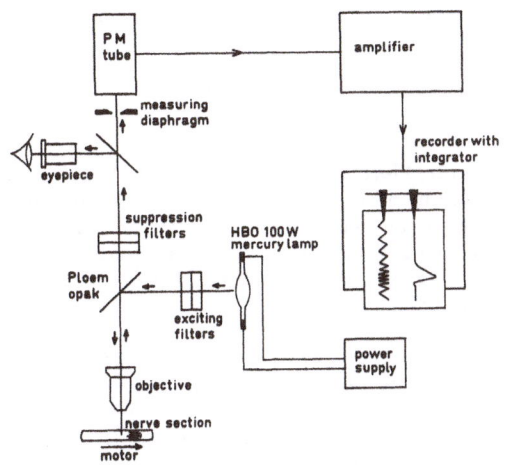

Fig. 2. Schematic representation of the technical equipment for cytofluorimetric scanning.

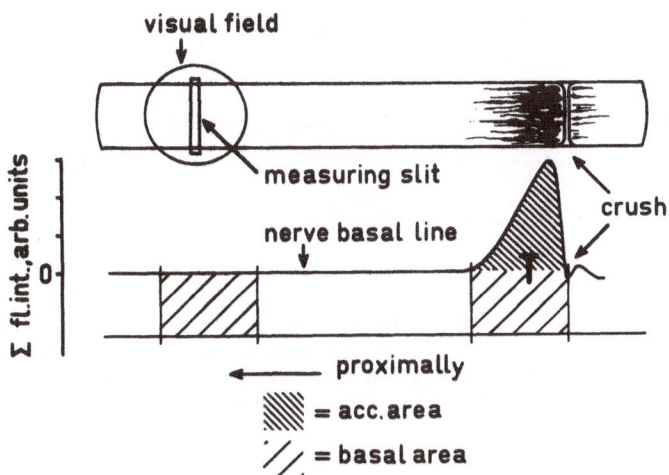

Fig. 3. Cytofluorimetric scanning of a longitudinal section of a crushed nerve (top). The section is passed under the measuring slit at a constant speed, and the fluorescence intensity is recorded as a curve (bottom). The background or basal fluorescence intensity (coursely hatched area) is subtracted from the accumulation peak (T), yielding the accumulation area (finely hatched area), which is proportional to the amount of fluorescent material transported towards the crush.

The scanning system consists of a motor driven cross-table
with a control unit giving a scanning speed of up to 40 μm/sec,
an adjustable square measuring diaphragm, which we usually set at
1000x100 μm. Before scanning the section is oriented and adjusted
in position using transmitted light microscopy, to minimize
fading. All details on the equipment are given by Larsson et al.
(1984).

Preparation of Tissue

In principle all substances that can be made to fluoresce
can be assayed by scanning. With immunofluorescence techniques
the number of substances that can be studied is limited only by
the access of specific antibodies. Also, catecholamines and
serotonin can be studied using the formaldehyde-induced
fluorescence method of Hillarp-Falck.

When the tissue is prepared there are important points to
consider at 3 stages: In the living organism, it is essential
that all procedures to interrupt the axonal transport cause a
lesion extending perpendicular to the length of nerve (or axon
bundle) and that the lesion area is broad enough not to cause
"contamination" from proximal into distal regions and vice versa,
as indicated in Fig. 4.

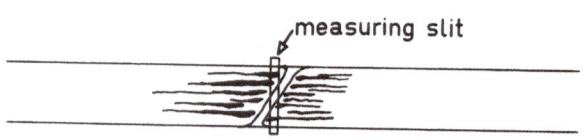

Fig. 4. If the crush operation is placed obliquely the
measuring slit will allow fluorescence from segments
both proximal and distal to the crush to be registered
simultaneously, giving incorrect results.

When the **tissue is dissected** from the unfixed (for FIF-
method) or perfusion fixed (for immunofluorescence) material care
must be taken to strip off as much of the connective tissue
sheaths as possible while not disturbing the organization of the
crush region (or cooled region). The sheath should be removed
since it may contain cells (monocytes, mast cells, macrophages)
in addition to connective tissue fibres that may disturb the
fluorescence signal.

When freezing the tissue for FIF (in liquid propane at
-150°C) or for cryostat sectioning, the specimen must be kept
absolutely straight. For the FIF-method, thin metal plates with
an engraved straight groove to harbour the nerve, have been used
during the freezing and freeze drying. Perfusion fixed nerves
have been placed in a block of agarose gel where a groove has
been produced by molding the block with a steel rod of appropriate diameter. After solidification the steel rod is carefully removed, leaving a straight furrow for placement of the
nerve.

The sectioning has to be done carefully, cracks or folds
are deletorious, as well as dust.

RESULTS

Adrenergic Nerves

Rat sciatic nerves were crush operated 1-12 h before
sacrifice and the nerves were after freeze drying exposed to
formaldehyde vapour, sectioned and mounted as described earlier
(e.g. Corrodi and Jonsson 1967). In longitudinal sections the
noradrenaline (NA) accumulations increased with time as
demonstrated earlier with fluorescence microscopy (Dahlström
1965) and biochemical methods (Dahlström and Häggendal 1966). The
registered fluorescence under the scanning curves also increased
with time (Fig. 5). When the results of the scanning were
compared with biochemical data, a very good agreement was noted
(Fig. 6A). Thus, accumulated fluorescence in the sections
increased with time in parallel with the increase in NA.

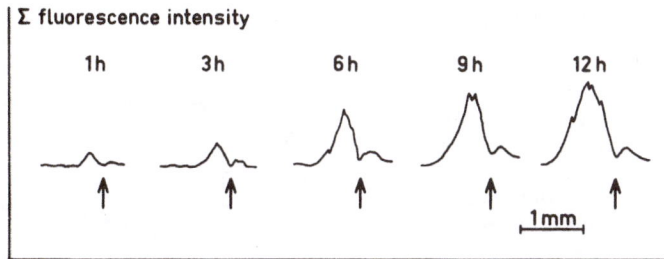

Fig. 5. The accumulation profiles of NA in sections of
rat sciatic nerves, crush operated 1-12 h before sacrifice. The abscissa indicates in arbitrary units the area
under the accumulation curves. Site of crush indicated
by arrow. Proximal is to the left.

322   A. Dahlström and P.–A. Larsson

    In perfusion fixed rats the accumulations of material with
dopamine-β-hydroxylase   (DβH)-like   immunoreactivity   were
followed with time. In this case the inclination of the
accumulation curve as measured with cytofluorimetric scanning of
immunoreactive material was steeper (Fig. 6B) than the curve
obtained with assay of enzymatic activity of DβH (Häggendal
1980). DβH-molecules with low or no enzyme activity (cf. Nagatsu
et al. 1976) will be detected by immunofluorescence and therefore
the scanning data probably represent total DβH transported.

    Consecutive sections from the perfusion fixed nerves were
incubated with antisera to **tyrosine hydroxylase** (TH). The
accumulation curve demonstrated a slower increase with time after
crushing than was the case for NA and DβH. Comparison with an
accumulation curve of TH-enzymatic activity revealed an excellent
agreement.

Fig. 6. A) Time course
curves for the accumulation
of NA proximal to a nerve
crush (3-12 h) assayed with
a biochemical method (From
Dahlström and Häggendal
1966; •---•) as compared
to the cytofluorimetrically
registered data (•——•).

B) The accumulation curve
for DβH-enzymatic activity
(From Häggendal 1980; •---•)
compared with the accu-
mulation curve for immunoreactive DβH as registered
by cytofluorimetric scanning
of nerve sections incubated
with an antiserum to DβH.
Note that the slope is
steeper when immunoreactive
DβH was measured (•——•).

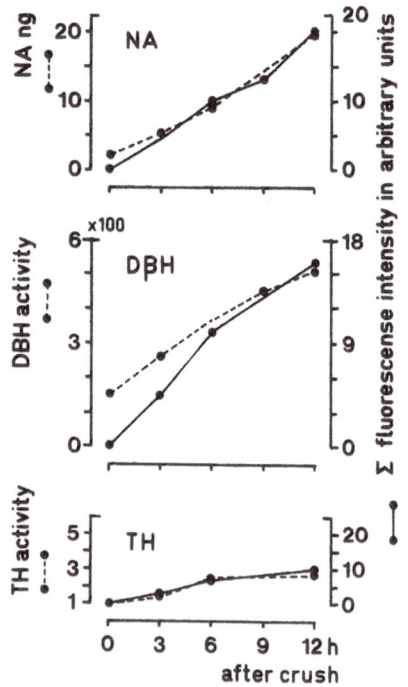

C) The accumulation of TH in
crush operated sciatic nerve
of rat. Biochemical data,
demonstrating increase in TH-
enzymatic activity (•---• , from Wooten and Coyle 1973) show a
slow accumulation, similar to the curve obtained with cytofluorimetric scanning of sections incubated with antisera to TH (•——•).
(n = 5-11.)

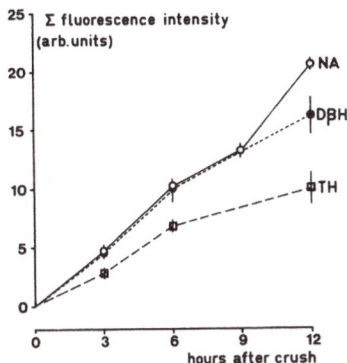

Fig. 7. The accumulations of NA, DβH—immunoreative material, and TH—immunoreactive material in rat crushed sciatic nerve, measured with cytofluorimetric scanning. NA and DβH follow the same inclination up to 9 h postoperatively, where an increased loading of NA in the accumulated granules is seen as a steeper increase in NA than in DβH. (From Larsson et al. 1984.)

When cytofluorimetric data on NA, DβH and TH were compared (Fig. 7), NA and DβH accumulated in parallel up to 9 h, which is expected since they are localized in the same organelle. Between 9 and 12 h postoperatively NA increased more than DβH, as demonstrated earlier by biochemical assays (Dahlström et al. 1975). TH increased more slowly and appeared not to move together with the amine granules (containing NA and DβH). Thus, the cytofluorimetric scanning data gave similar results as the biochemical studies, but with less number of animals involved.

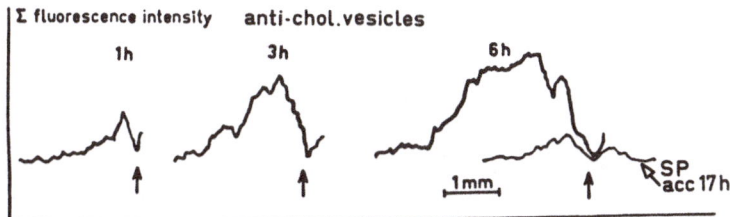

Fig. 8. Scanning graphs of the accumulation of cholinergic vesicle—like material proximal to a crush in rat sciatic nerves. The rats were sympathectomized 7 days prior to the crush operation to remove postganglionic sympathetic fibres. Proximal is to the left. Bars indicate mm along the nerve.

In the right graph of a section from a 6 h crush operated nerve a SP scanning graph is indicated. This section was from a nerve crushed 17 h before sacrifice. The difference between the accumulation patterns of the two substances can be clearly discerned.

Cholinergic and Peptide Containing Axons

        Using an antiserum against whole synaptic cholinergic
vesicles, purified from the electric organ of Narcine Brasi-
liensis (see Carlson and Kelly 1980) immunoreactive material was
demonstrated to accumulate in crushed sciatic nerves in a pattern
indicated in Fig. 1. Thick myelinated axons on either side of the
crush contained strongly immunoreactive material extending
several mm away from the crush region. Scanning profiles from 1,
3 or 6 h crushed nerves proximal to the crush area are shown in
Fig. 8. Based on such scanning profiles an accumulation curve for
the proximodistal transport of immunoreactive axonal cholinergic
vesicles was prepared (Fig. 9).

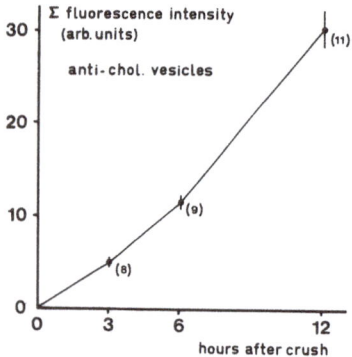

Fig. 9. Time course curve for
the accumulation of cholinergic vesicle-like material proximal to a crush (3–12 h) in rat
sciatic nerve. The curve is
based on cytofluorimetric measurements. The abscissa represents fluorescence of the
area under the accumulation
curves, expressed in arbitrary
units. (n = sections from 3
nerves at each time after
crushing.)

        Using an antiserum to SP immunoreactive SP-like material
accumulated in this axons. Proximal to the crush the accumulations reached up to 1 mm, while distally the accumulations could
be traced only a fraction of a mm. The microscopic picture is
shown schematically in Fig. 1. The scanning profile of a 17 h
crushed nerve is inserted in Fig. 8, where the difference in
accumulation pattern between cholinergic vesicle-like material
and SP-like material is clearly shown. The SP-like material
accumulated during 17 h extends proximally over a much shorter
distance than the accumulation of cholinergic vesicle-like
material having accumulated only during 6 h.

Advantages of the Technique

        Resolution of accumulation curves: In the scanning graphs
of Fig. 8 the length of the proximal accumulations of cholinergic
vesicle-like material can be seen to extend far more proximally

than the accumulation of SP-like material, occurring in thin
axons, during 17 h after crushing. This picture can be compared
to the schematical drawing in Fig. 1. The scanning procedure can
clearly demonstrate the two accumulated substances with excellent
resolution.

Simultaneous morphological registration: As demonstrated
in Fig. 10, the scanned nerve sections also can be photographed.
Here two crushes were placed on the sciatic nerve 6 h before
sacrifice, 3.5 mm apart. The accumulations were of about the same
magnitude proximal and distal to the isolated segment, and a
small fraction of material had also moved distally in the
isolated segment and accumulated proximally to the distal crush.

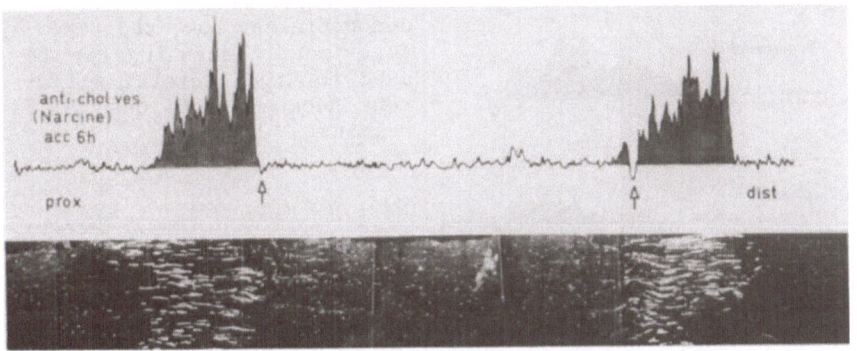

Fig. 10. Scanning graph of a section from a sciatic
nerve which was crush operated at two sites (arrows), 3
mm apart, 6 h before sacrifice. Incubated with anticholinergic vesicle antiserum. The same section was also
photographed and shown in the montage at the bottom of
the figure.

Several different substances can be assayed in one nerve:
If longitudinal sections are collected for immunofluorescence
studies, consecutive sections can be incubated with various
antisera. A sciatic nerve, for instance, is about 1 mm in
diameter. The middle third (~300 μ) can be used for longitudinal
scanning and considered to be a representative fullwidth section.
If 10 μ cryostat sections are performed, 30 sections can theoretically be collected. In Fig. 11 four consecutive sections were
collected and incubated with antisera against TH, VIP, somatostatin or SP. After a certain experimental procedure each animal
can thus be analyzed for many different substances. This will
reduce the biological variation in the experiment, it is

economical (the price of experimental animals has increased drastically), and it is time saving.

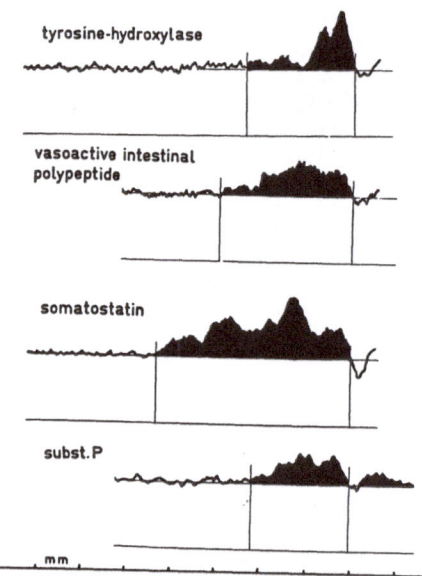

Fig. 11. Set of scanning graphs of four consecutive sections of a sciatic nerve which was crush operated 18 h before sacrifice. The four sections have been incubated with antisera against TH, VIP, SOM or SP. The different accumulation patterns are obvious. This figure demonstrates how the cytofluorimetric scanning can be used to study several different substances in one nerve segment.

Flexibility: Scanning can be performed either along the length of the axon as shown in Fig. 3, or with a broader slit across the axon, i.e. perpendicular to the direction indicated in Fig. 3. Then the individual axons or axon bundles can be registered as individual peaks on the graph. If the nerve is cross-sectioned instead of sectioned longitudinally all axons in the nerve are seen and measured, and the variation between sections in distribution density of various axon types can be avoided. Single axons can be scanned, thus the system may be useful in tissue culture experiments. Accumulations in axons in the CNS can be measured provided appropriate angles of sectioning are used.

Is the Technique Quantitative or Semiquantitative?

In all likelihood, the scanning technique is quantitative for the FIF-method. Several investigators have demonstrated (Jonsson 1969, Schipper et al. 1979) that there is a linear relationship between fluorescence intensity and concentration of amine up to a certain MA level. This has been used for cytofluorimetric assays of MA by e.g. Jonsson (1971) and Löfström et al. (1976). Also the comparison between the accumulation curves obtained with scanning and with biochemistry revealed an excellent agreement. However, so far our data are expressed in arbitrary units. At present we are developing a reference model which will enable a translation from arbitrary units into absolute NA values.

For the immunofluorescence samples there are several factors to consider: 1) The incubations are performed on glass-slide adhered sections, 10 μ thick. The antibodies have most likely penetrated the section, since Triton-X-100 is always included in the incubation step. Therefore, when the proper antibody dilution has been determined by the chess-board procedure it is likely that the antigenic sites all through the section are labelled.

2) The FITC-fluorophore, used in most experiments, have a tendency to fade. Therefore the sections are viewed in weak transmission light microscopy for selection and positioning prior to scanning. When scanning has started, using the incident activation light, each area of the section is illuminated for the same period of time, due to the controlled speed scanning. Thus, fading would be equal along the nerve section. In the later experiments we have included an antifading substance, para-phenylene diamine (Johnson and Gloria de Nogueira Aranjo 1978), which in our hands is efficiently preventing the fading of the FITC-fluorescence.

3) The comparisons between immunofluorescence scanning data and biochemical data carried out so far (Figs. 6B, C) indicate that immunofluorescence is in good agreement with enzymatic measurements of DβH and TH. We are at present carrying out experiments using an ELISA technique to quantitate the amount of immunoreactive molecules in sections, consecutive to sections which are scanned. This procedure will tell us if there is a reliable proportionality between scanning-registered fluorescence and actual amounts of material in the sections.

4) In order to standardize our cytofluorimetric measurements of immunofluorescence, gel slabs with various concentrations of antigens will be prepared. These gel slabs should then be included in the series of procedures to enable comparison of fluorescence intensity. This will of course be possible only for antigens available.

In conclusion: We have developed a cytofluorimetric scanning method to study axonal transport phenomena. The method can give quantitative as well as morphological data on tissue sections. Many different substances can be studied with immuno-fluorescence techniques in one single biological sample, thus reducing biological variations and costs. The method can be applied to peripheral nerves, CNS tissue and probably also to tissue cultures, since single axons or bundles of axons can be scanned. Available comparative data so far indicate that the method can be used for quantitation, not only semiquantitation, but there are yet several control experiments and standardization procedures to be worked out. The principles for the method should be possible to apply also on material viewed in a light microscope, e.g. by measuring absorbtion or extinction instead of fluorescence.

ACKNOWLEDGEMENTS

This work was supported by the Swedish Medical Research Council (2207), Magnus Bergvall´s Foundation, Lars Hierta´s Foundation, Torsten and Ragnar Söderberg´s Foundation and the Medical Faculty, University of Göteborg. For generous supply of TH and D H antisera we are grateful to Prof. M. Goldstein, New York, and for supply of the Narcine anticholinergic vesicle antiserum we thank Prof. R. Kelly, San Francisco.

REFERENCES

Carlson, S.S. and Kelly, R.B. (1980). An antiserum specific for cholinergic synaptic vesicles from electric organ. J. Cell Biol. <u>87</u>, 98.

Corrodi, H. and Jonsson, G. (1967). The formaldehyde fluorescence method for the histochemical demonstration of biogenic amines. A review of the methodology. J. Histochem. Cytochem. <u>15</u>, 65.

Dahlström, A. (1965). Observations on the accumulation of noradrenaline in the proximal and distal parts of peripheral adrenergic nerves after compression. J. Anat. (Lond.) <u>99</u>, 677.

Dahlström, A. (1983). Presence, metabolism, and axonal transport of transmitters in peripheral mammalian axons. In <u>Handbook of Neurochemistry</u> (ed. A. Lajtha). Plenum Press, New York, <u>5</u>, 405.

Dahlström, A. and Häggendal, J. (1966). Studies on the transport and life-span of amine storage granules in a peripheral adrenergic neuron system. Acta Physiol. Scand. <u>67</u>, 278.

Dahlström, A., Häggendal, J. and Larsson, P.–A. (1975). On the noradrenaline loading in axonal amine storage granules in rat crushed sciatic nerves. Acta Physiol. Scand. <u>94</u>, 451.

Grafstein, B. and Forman, D.S. (1980). Intracellular transport in neurons. Physiol. Review 60, 1167.

Häggendal, J. (1980). Axonal transport of dopamine-β-hydroxylase to rat salivary glands: studies on enzymatic activity. J. Neural Transm. 47, 163.

Johnson, G.D. and Gloria de Nogueira Aranjo, M. (1981). A simple method of reducing the fading of immunofluorescence during microscopy. J. Immunol. Meth. 43, 349.

Jonsson, G. (1969). Microfluorimetric studies on the formaldehyde-induced fluorescence of noradrenaline in adrenergic nerves of the rat iris. J. Histochem. Cytochem. 17, 714.

Jonsson, G. (1971). Quantitation of fluorescence of biogenic amines. Prog. Histochem. Cytochem. 2, 299.

Larsson, P.-A., Goldstein, M. and Dahlström, A. (1984). A new methodological approach for studying axonal transport. J. Histochem. Cytochem. 32, 7.

Löfström, A., Jonsson, G. and Fuxe, K. (1976). Microfluorimetric quantitation of catecholamine fluorescence in rat median eminence. I. Aspects on the distribution of dopamine and noradrenaline nerve terminals. J. Histochem. Cytochem. 24, 415.

Nagatsu, I., Kondo, Y., Kato, T. and Nagatsu, T. (1976). Retrograde axoplasmic transport of inactive dopamine- -hydroxylase in sciatic nerves. Brain Res. 116, 277.

Schipper, J., Tilders, F.J.H. and Ploem, J.S. (1978). Microfluorimetric scanning of sympathetic nerve fibres. An improved method to quantitate formaldehyde-induced fluorescence of biogenic amines. J. Histochem. Cytochem. 26, 1057.

Wooten, G.F. and Coyle, J.T. (1973). Axonal transport of catecholamine synthesizing and metabolizing enzymes. J. Neurochem. 20, 1361.

QUANTITATIVE MICROFLUORIMETRY AND SEMIQUANTITATIVE IMMUNOCYTOCHEMISTRY AS TOOLS IN THE ANALYSIS OF TRANSMITTER IDENTIFIED NEURONS

KJELL FUXE[1], LUIGI F. AGNATI[2], KURT ANDERSSON[3], MICHELE ZOLI[2], FABIO BENFENATI[2], PETER ENEROTH[4] and CLAUDIO CUELLO[5]

[1]Department of Histology, Karolinska Institutet, Stockholm, Sweden
[2]Department of Human Physiology, University of Modena, Modena, Italy
[3]Hormone Laboratory, Karolinska Hospital, Stockholm, Sweden
[4]Department of Anatomy, University of Bologna, Bologna, Italy
[5]Department of Pharmacology and Human Anatomy, Oxford University, Oxford, UK

INTRODUCTION

In our laboratories we have in previous studies been able to develop methods for the determination of dopamine steady-state levels (nmol/g) and dopamine turnover rates (nmol/g x min^{-1}) in discrete dopamine nerve terminal systems of the tuberculum olfactorium and the nucleus caudatus putamen (Agnati et al. 1979), using the tyrosine hydroxylase inhibition method for the studies on dopamine turnover rate. Microfluorimetrical quantitation of catecholamine fluorescence has also been made in the discrete catecholamine nerve terminal systems of the median eminence (Löfström et al. 1976a). Using the tyrosine hydroxylase inhibition method it was possible to determine rate constants and half-lives of CA in dopamine and noradrenaline nerve terminal systems of the median eminence (Löfström et al. 1976b). In all these studies an apparent monophasic decline of the catecholamine stores was obtained following injection of the tyrosine hydroxylase inhibitor α-methyl-dl-p-tyrosine methyl ester, suggesting the existence of mainly one pool of catecholamines.

This work has now been continued (see Andersson et al. 1984) by determining if the tyrosine hydroxylase inhition method in combination with quantitative microfluorimetry could also be used to determine half-lives and turnover rates in absolute amounts also in other types of hypothalamic, preoptic and telencephalic dopamine and noradrenaline nerve terminal systems. This has been made possible by comparing the results obtained with biochemical methodologies (high pressure liquid chromatography) with those obtained with quantitative microfluorimetrical measurements using the tyrosine hydroxylase inhibition method. This comparison between results obtained with histochemical and biochemical methodologies have also been extended to include a study of increases of catecholamine levels in discrete catecholamine nerve terminal systems following monoamine oxidase inhibition in view of the possible development of a concentration-dependent quenching of

catecholamine fluorescence as the catecholamine levels are increased.

In the present paper we have also introduced a new type of
semiquantitative method for the evaluation of immunoreactivity in
structures based on the assumption that the monoclonal antibody
antigen reaction is controlled by the law of mass action (see also
Fuxe et al. 1983). In this analysis monoclonal antibodies against
substance P have been employed (Cuello et al. 1979) and the
results obtained in the substantia nigra using immunocytochemistry
have been compared with those obtained by means of radioimmunoassay determination of substance P.

MATERIAL AND METHODS

Male specific pathogen free Sprague Dawley rats have been
used. The rats were given food pellets and water ad libitum and
were kept under standardized lighting conditions (lights on at
6.00 a.m. and off at 8.00 p.m.). In the experiments on the development of the semiquantitative method for evaluation of immunoreactivity in neuronal structures also lesioned rats were used.
Thus, in order to study changes in substance P immunoreactivity in
the substantia nigra partial unilateral hemitransections were
performed of the connections between substantia nigra and striatum
in order to lesion the strio-nigral substance P pathways (Kanazawa
et al. 1977). The operation performed has been described previously
(Agnati et al. 1983). Briefly, a 4 mm wide knife was inserted in
the coronal plane close to the midline 1 mm caudal to the bregma
and lowered at an angle of 70° to the horizontal plane to reach
the ventral border of the brain at a König & Klippel level of
A3200. This level is located just in front of the substantia
nigra.

Experiments using quantitative microfluorimetry

The tyrosine hydroxylase inhibitor α-methyl-dl-p-tyrosine
methyl ester (H44/68, 250 mg/kg, i.p., Sigma Chemical Comp.) was
given 30, 60 and 120 minutes before decapitation. Some rats were
given saline alone in order to study also the levels of catecholamines in discrete catecholamine nerve terminal systems of the
forebrain, the hypothalamus and the preoptic region.

In the histochemical analysis the di- and telencephalon were
rapidly dissected out and taken to the standard Falck-Hillarp
procedure for the cellular localization of catecholamines (see
Fuxe & Jonsson 1973; Löfström et al. 1976). In the quantitative
microfluorimetrical analysis a Zeiss fluorescence microscope
equipped with MPV-systems was used. The sections were exposed to
ultraviolet light from a mercury lamp (Osram HB 100W/2 lamp)

(epiilumination). On its way to the section the ultraviolet light passed an interference filter (BP 400–440, Zeiss) which served as an excitatory filter and was then reflected on a dichroic mirror (FT 460, Zeiss) and passed through a fluorite oil immersion objective (Plan-Neofluar 25/0.8 1mm). The circular measuring field used was 40 μm in diameter and before entering the photomultiplier the emitted light passed the dichroic mirror and a barrier filter (LP 470, Zeiss). The emitted light then reached a photomultiplier and the electrical signal created was digitalized by means of a digital voltmeter (modified Opti-lab, GUNA- konsult HB, Stockholm, Sweden). Coded sections were used, and from each area two sections per animal were analyzed and in each section three fluorescence intensity measurements were recorded in the region under study. Tissue blank values were obtained by making three measurement per section of general tissue background fluorescence intensity. The coefficients of variation (CV) within and between experiments were between 5–10%. The catecholamine fluorescence values could be converted into nmol/g of tissue by the use of catecholamine standards (Agnati et al. 1979; Andersson et al. 1984). The catecholamine fluorescence values in the forebrain represents dopamine fluorescence and thus dopamine nerve terminal systems were analyzed (see Fuxe 1965). However, the catecholamine fluorescence values obtained within the hypothalamus and the preoptic area represents noradrenaline fluorescence. Thus, the hypothalamic and preoptic dopamine and adrenaline nerve terminal networks cannot be demonstrated with the standard Falck–Hillarp procedure, with the exception of the dopamine nerve terminal networks of the median eminence. It is also well known that the adrenaline nerve terminal networks are not visible in view of the low fluorescence yield of adrenaline in the formaldehyde condensation reaction (see Fuxe et al. 1970).

Biochemical analysis of catecholamines was performed by the use of high pressure liquid chromatography in combination with electrochemical detection (Jonsson et al., 1980; Keller et al., 1976). As an internal standard was used α–methyl dopamine. The noradrenaline and dopamine levels were determined in the nucleus interstitialis striae terminalis pars ventralis (Nist vent.) and within a periventricular and paraventricular hypothalamic region (PV + PA). The areas were punched from frozen brain tissue using a circle punch (diameter 1 mm) for Nist vent. and a rectangular punch (2 x 1 mm) centered around the third ventricle for the peri- and paraventricular area.

The monoamine oxidase inhibitor used in the experiments was Catron (Lakeside, USA) (10 mg/kg, i.p.). It was given 30, 60, 120 and 240 minutes before decapitation. Peri- and paraventricular catecholamine nerve terminal systems were analyzed histochemically and biochemically. This was true also for the noradrenaline nerve terminal in the Nist vent..

The catecholamine fluorescence concentration relationship has been shown to be linear in the concentrations found in the brain tissue (Löfström et al. 1976; Andersson et al. 1984). In the H44/68 experiments the rate constants of catecholamines were determined using regression analysis of the natural logarithmic values of the catecholamine determinations obtained in each rat at various time intervals following the injection of the α-methyl-dl-p-tyrosine methyl ester. As deduced from the F value there was no significant deviation of the logarithmic values from a linear decay. The increases obtained following Catron treatment at the various time intervals following monoamine oxidase inhibition using histochemical and biochemical analysis were compared using the Mann-Whitney U-test (Snedecor & Cochran 1980).

Abbreviations used: SEL= subependymal layer of the median eminence; MPZ= medial palisade zone of the median eminence; LPZ= lateral palisade zone of the median eminence; PV II= posterior periventricular hypothalamic region; DM= dorsomedial hypothalamic nucleus; BZ= border zone, an area located at the border of the medial and lateral hypothalamus located between the fornix and the ventral surface of the brain; PV I= anterior periventricular hypothalamic region; PA FP= parvocellular part of the paraventricular hypothalamic nucleus; PA FM= magnocellular part of the paraventricular hypothalamic nucleus; SO= nucleus supraopticus; POP= periventricular preoptic region; POSC= nucleus preopticus suprachiasmaticus; POM= nucleus preopticus medialis; Nist vent.= nucleus interstitialis striae terminalis, pars ventralis, CAUD marg.= marginal zone of the caudate nucleus; CAUD med.= medial part of the caudate nucleus; CAUD cent.= central part of the caudate nucleus; ACC dif.= diffuse type of dopamine fluorescence in anterior nucleus accumbens; ACC dot.= dotted type of dopamine fluorescence in the dorsomedial posterior nucleus accumbens; TUB dot.= dotted type of dopamine fluorescence in medial-posterior tuberculum olfactorium; TUB dif.= diffuse type of dopamine fluorescence in lateral-posterior tuberculum olfactorium.

Experiments on semiquantitative determination of immunoreactivity

In the present immunocytochemical analysis of the substantia nigra a monoclonal antibody against substance P (SP) was used (Cuello et al. 1979). The dilution of the antiserum against SP varied from 1 : 200 to 1 : 4000 in order to perform a saturation analysis (see Result section). As control antiserum was used various types of diluted antisera which had been dissolved with SP in a concentration of 50 µg/ml of the respective diluted antiserum. This procedure completely prevented demonstration of SP immunoreactive nerve terminals within the substantia nigra. The unlabelled immunoperoxidase procedure method was used. The peroxidase-antiperoxidase complex was used to demonstrate SP immunoreactivity within the substantia nigra. The procedure was performed in principle according to Sternberger (1979). For details, see Fuxe et al., this symposium. In the present analysis anti-rat goat immunoglobulin was used as well as rat PAP. After reaction with diaminobenzidine and H_2O_2 the sections were defatted and coverslipped in entellan. In most of the experiments reported here thin freeze-microtome section (14 μm thick) were used and the fixation procedure involved the use of a 4% formaldehyde buffer solution (see Fuxe et al. this symposium). The animals with unilateral partial hemitransections were taken for immunocyto-chemistry 7 - 10 days following the operation.

For radioimmunoassay (RIA) determination of SP in substantia nigra frozen brain tissue was punched out from the medial and lateral part of the substantia nigra using a circular punch (1 mm in diameter). Extraction of tissue pieces was made with 0.01 M HCl, and the resulting solution was added to cold aceton (0.5 ml extract per 1 ml aceton). The clarified extract was taken to dryness under a stream of nitrogen. The extract was reconstituted in acid buffer (BSA-peptine buffer) and 200 μl samples were analized by radioimmunoassay using reagents from Immunonuclear Corporation (Stillwater, Minn., USA). The assay system displayed no significant crossreactivities with other peptides such as β-endorphin, leu-enkephalin and met-enkephalin (< 0.05%). Samples were run in duplicates. The variation between these duplicates were below 4%.

RESULTS

Quantitative microfluorimetrical studies of catecholamine nerve terminal systems: comparisons with biochemistry

Analysis of catecholamine pool sizes, rate constants and turnover rates. The results obtained in the peri- and paraventricular regions and within the Nist vent. are summarized in Table 1 and in figs. 1-2. It is seen that an apparent monophasic disappearance of the catecholamine stores takes place in the noradrenaline nerve terminal networks of the PV II, PV I, PA FP, PA FM and Nist vent. The half-lives in the peri- and paraventricular region range from 188 - 233 minutes (PV II and PA FM respectively). The biochemical analysis of noradrenaline pool sizes, rate constants and turnover rate in the peri- and paraventricular region shows a similar noradrenaline half-life (206 min) as seen in the quantitative microfluorimetrical studies but the pool size is 20 - 30 times lower. This finding is expected, since the "dilution" of catecholamine rich areas with catecholamine poor areas in the biochemical sampling cannot be avoided to the same extent as it can be in the histochemical studies. Consequently also the catecholamine turnover rates become 20 - 30 times higher than determined by quantitative microfluorimetrical analysis. Thus, when discussing the absolute pool size and turnover rate of catecholamine

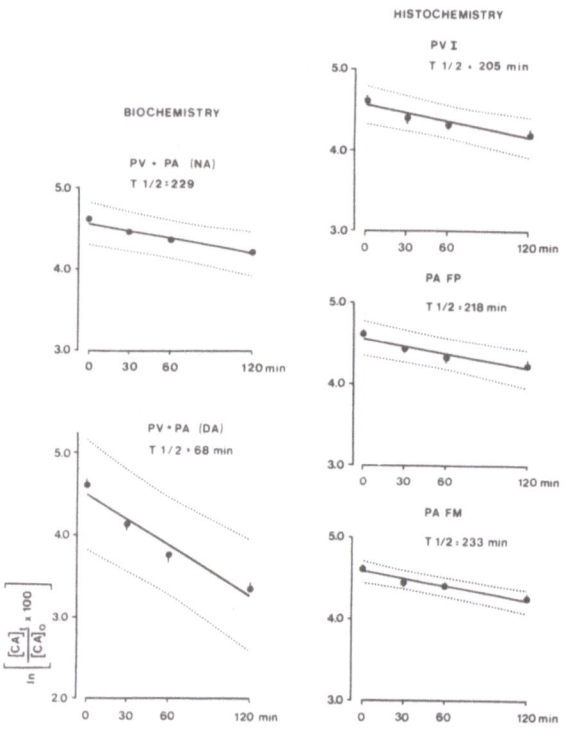

Fig. 1 The disappearance of the catecholamine levels after tyrosine hydroxylase inhibition is shown in PV I, PA FP and PA FM
using histochemical analysis and in the PV + PA using biochemical
analysis. H44/68 (250 mg/kg, i.p.) was given 30, 60 or 120 min
before decapitation. A semilogarithmic plot is shown. Means +
s.e.m. are given. The values have been expressed in per cent of
the control group mean value after which the transformation into
the natural logarithm was performed. n= 6-8 in all groups. The
dotted line indicate the 95% confidence interval around the best
fit line. For details on pool sizes, half-lives, rate constants,
turnover rates and linear fitting as well as for abbreviations,
see Table 1.

obtained using quantitative microfluorimetry it must be realized
that these numbers only reflect the situation in the sampled
regions within the nucleus. Usually within each nucleus there is
an unevenness of distribution of the catecholamine nerve terminal
networks (see Fuxe 1965). Very similar results have been obtained
within the noradrenaline nerve terminal networks in the Nist vent.
as seen in the Table 1 and in fig. 2. It must also be pointed out
as seen in the biochemical analysis (see Table 1 and fig. 1) that

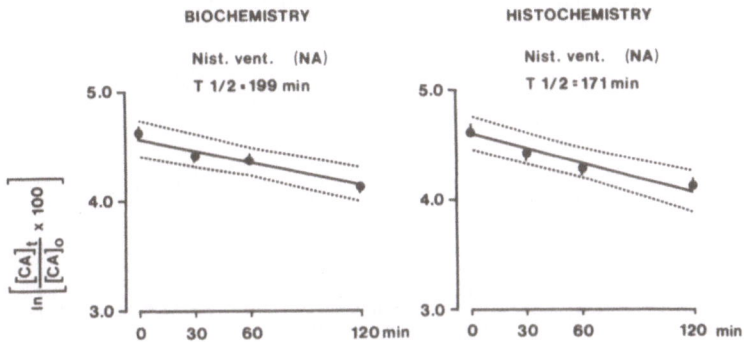

Fig. 2. Disappearance of the catecholamine levels after tyrosine hydroxylase inhibition in the ventral part of the nucleus inter- stitialis striae terminalis as determined by biochemical or histochemical analysis. For further details, see text to fig. 1. For abbreviations see text to Table 1.

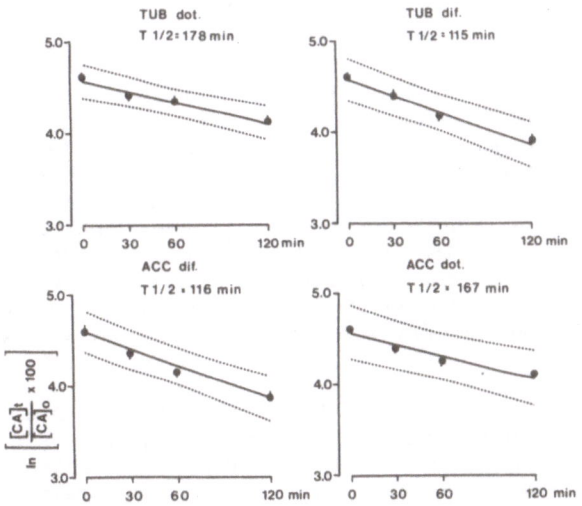

Fig. 3. Disappearance of the catecholamine levels after tyrosine hydroxylase inhibition in TUB dot. and dif. and ACC dot. and dif. of the tuberculum olfactorium and nuc. accumbens as determined by quantitative microfluorimetry. For further details, see text to fig. 1. For abbreviations see text to Table 1.

K. Fuxe, et al.

Table 1.

POOL SIZE, HALF-LIFE, RATE CONSTANT, TURNOVER RATE AND LINEAR FIT INDEX
OBTAINED IN DISCRETE HYPOTHALAMIC AND PREOPTIC CA NERVE TERMINAL NETWORKS
OF THE NORMAL MALE RAT

	Pool size nmol/g means ± s.e.m.	T½ min	Rate constant min $^{-1}$ ± 95 % confidence interval	Turnover rate nmol/g/min	F-value
PV II	133± 5	188	-0.0037±0.00117	0.491	$F_{2/28}$ 2.98
PV I	157± 5	233	-0.0030±0.00081	0.471	$F_{2/28}$ 2.42
PA FP	219± 7	272	-0.0026±0.00081	0.569	$F_{2/28}$ 2.10
PA FM	291± 9	233	-0.0030±0.00083	0.873	$F_{2/28}$ 0.51
Nist vent	613±44	171	-0.0041±0.00067	2.513	$F_{2/22}$ 1.32

Biochemistry

PV + PA NA	9.16±0.31	206	-0.0034±0.00095	0.311	$F_{2/53}$ 2.23
PV + PA DA	0.656±0.022	63	-0.0110±0.00114	0.00722	$F_{2/53}$ 9.13
Nist vent NA	27.4±1.50	195	-0.0036±0.00118	0.0986	$F_{2/43}$ 0.44

Table 2

POOL SIZE, HALF-LIFE, RATE CONSTANT, TURNOVER RATE AND LINEAR FIT INDEX OBTAINED IN DISCRETE FOREBRAIN CA
NERVE TERMINAL NETWORKS OF THE NORMAL MALE RAT

Histochemistry	Pool size nmol/g means ± s.e.m.	$T_{1/2}$ min	Rate constant min $^{-1}$ ± 95 % confidence interval	Turnover rate nmol/g/min	F-value
ACC dif.	155 ± 5	116	-0.0060 ± 0.00100	0.930	$F_{2/27}$ 1.10
ACC dot.	330 ± 9	167	-0.0042 ± 0.00079	1.386	$F_{2/27}$ 2.80
TUB dot.	280 ± 9	178	-0.0039 ± 0.00077	0.092	$F_{2/27}$ 1.09
TUB dif.	174 ± 7	115	-0.0060 ± 0.00100	1.044	$F_{2/27}$ 1.10

there exists a small pool of dopamine within the peri- and paraventricular region having high turnover (half-life of 68 min).
This store of dopamine is probably present within the dopamine
nerve terminals in the peri- and paraventricular region, but they
cannot be visualized using the present fluorescence histochemical
procedure as also confirmed in the present analysis. The rate
constants using quantitative microfluorimetry is very similar to
the rate constant of noradrenaline as determined by biochemistry.
A multiphasic disappearance of the dopamine store in the peri- and
paraventricular region may also take place in view of the high F
value (Table 1).

The pool sizes and half-lives of catecholamines have also
been determined within the dopamine nerve terminal systems of the
nucleus accumbens and tuberculum olfactorium (ventral striatum).
The dotted and diffuse types of dopamine nerve terminal networks
were analyzed. As seen in fig. 3 and Table 2 the dotted cholecystokinin immunoreactive type of dopamine nerve terminal have a
higher pool size and a longer half-life of dopamine than the
diffuse cholecystokinin negative types of dopamine nerve terminal
networks within the nucleus accumbens and tuberculum olfactorium.
An apparent monophasic disappearance of the CA stores is obtained
in the nuc. accumbens and tuberculum olfactorium following the
tyrosine hydroxylase inhibition.

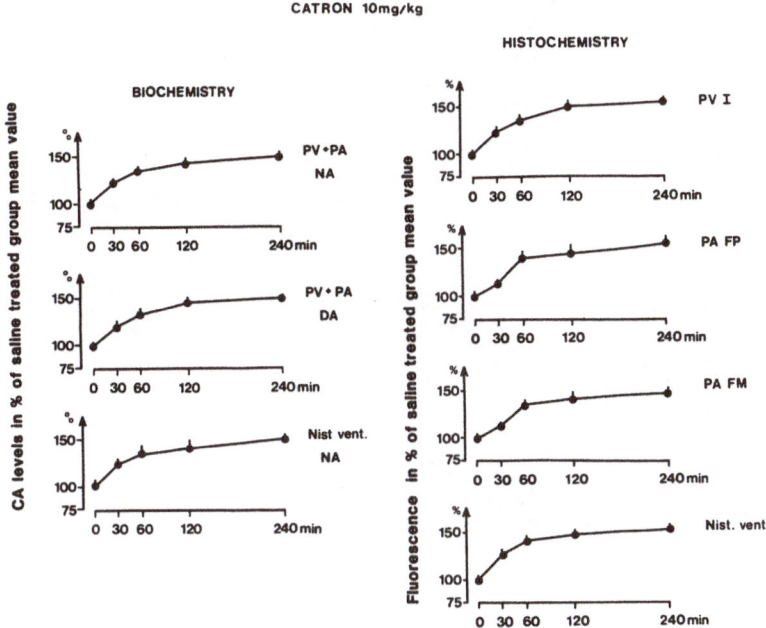

Fig. 4. Effects of the monoamine oxidase inhibitor catron on catecholamine levels in PV I, PA FP, PA FM and Nist vent. using histochemical analysis based on quantitative microfluorimetry. In the PV + PA and Nist vent. using biochemical analysis was employed. Catron (10 mg/kg, i.p.) was given 30, 60, 120 and 240 min before decapitation. The values are expressed in per cent of control group mean values. Means + s.e.m. are shown. n= 6-8 in all groups. For abbreviations, see Table 1.

Experiments with the monoamine oxidase inhibitor Catron; comparison between biochemistry and quantitative microfluorimetry. As seen in fig. 4 very similar results are obtained when measuring the catecholamine levels within the peri- and paraventricular region and within the Nist vent. by biochemistry or by quantitative microfluorimetry following treatment with Catron 10 mg/kg, i.p. (see fig. 5). Increases in noradrenaline and dopamine levels both follow a hyperbolic function reaching maximal increase after 2 to 4 hours following the injection of Catron. As shown by both biochemistry and quantitative microfluorimetry the maximal increases were in the order of 50 - 60%. Furthermore, at no time interval studied following Catron was there a significant difference between the percentage increase as determined by biochemistry or by quantitative microfluorimetry.

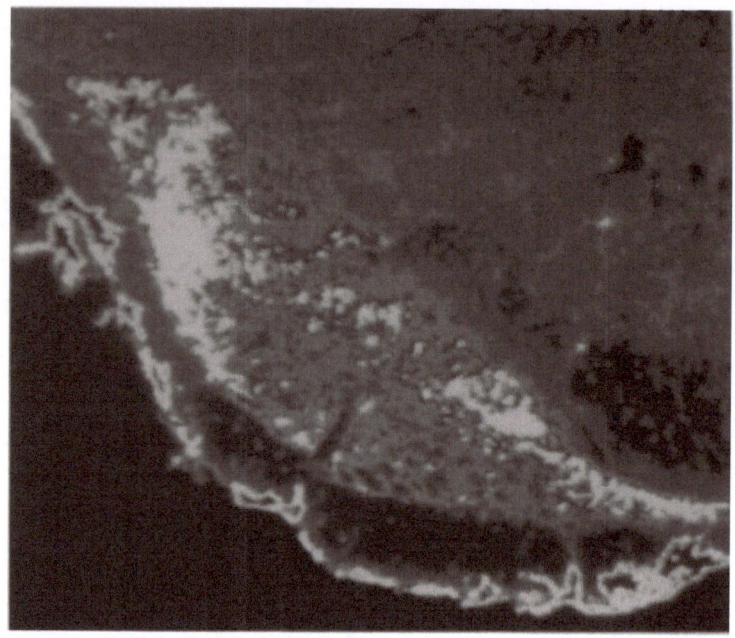

Fig. 5. A colour coded image of Substance P immunoreactivity in the substantia nigra of the normal male rat. High tones are shown in yellow-white colours. Areas with high tones are substantially more frequent in the lateral than in the medial part of the substantia nigra and the zona reticulata. Magnification 20 x.

The present method for the semiquantitative evaluation of immunoreactivity in neuronal structures; studies on substance P immunoreactivity in nerve terminal networks in the substantia nigra

In view of the fact that the monoclonal antibody antigen reaction should follow the law of mass action it should be possible to calculate the "B_{max} value" of the antigen contents by means of a saturation study followed by a Scatchard analysis or a nonlinear fitting procedure. In this way it should be possible to obtain a reliable relative quantitative value of the contents of antigen in the analyzed structures. Provided no cooperativity and/or heterogeneity of sites exists in the interaction between the monoclonal antibody and antigen a straight line will be observed in the Scatchard plot. Based on these theoretical considerations we have analyzed SP immunoreactivity in the substantia nigra of intact and partially hemitransected rats. Thus, we have used different dilutions of the SP antibody to study the SP innervation of the substantia nigra and by means of the densitometric programs in the IBAS Image Analyzer (Zeiss Kontron) we can measure absorption, transmittance and optical densities in selected regions of the SP immunoreactive areas of the substantia nigra. Also differences in absorption can be shown in false colors (see fig. 5). It is seen that higher tones (white-yellow color) are much more frequent within the lateral part of the substantia nigra than in the medial part. In fig. 6 in which the coding is obtained by means of gray tones it is observed that the SP immunoreactive regions with high tones (black areas) are markedly diminished as the dilution of the SP antibody is increased. In the measurements of absorption within the substantia nigra we have in addition also used an automatic procedure in the IBAS similar to the one introduced by Agnati et al. (1978). In this case we have determined absorption in selected regions of the substantia nigra by means of a tone discrimination through five different grey tone levels. After each different discrimination the immunoreactive region has been determined and expressed as percentage of the total immunoreactive area (no discrimination). A relative value of absorption is then obtained via calculation of the grey tone level causing a 50% reduction of the SP immunoreactive area.

In order to study the validity of these procedures to measure in a relative way the SP immunoreactivity we have compared the results with those obtained by radioimmunoassay determinations of SP immunoreactivity in the medial and lateral part of the substantia nigra. The results are shown in fig. 7. It is seen that similar results are obtained by means of the immunocytochemical procedure and the radioimmunoassay procedure. Thus, taking the lateral part in per cent of the medial part 200 - 250 % higher amounts of the SP immunoreactivity are found within the lateral substantia nigra as revealed both histo and biochemically (fig. 7). These immunohistochemical procedures have now been used to study SP immunoreactivity within the substantia nigra at different dilutions of the SP antiserum in intact animals and in animals having partial unilateral hemitransections. As seen in fig. 8, the results obtained by the saturation analysis reveals the existence only of one binding site and also shows a marked reduction in the B_{max} values in the order of 75% for SP immunoreactivity on the lesioned side compared with the intact side which also is in agreement with radioimmunoassay measurements. However, the saturation analysis also reveals that the affinity of the antibody for the antigen is not clearly modulated by the lesion, since the dilution of the antibody producing a 50% reduction in the immunoreactivity compared with the B_{max} value appears similar; 1/1700 of the intact side and 1/2100 on the lesioned side. A lack of changes in the affinity is also further illustrated in fig. 9, where the antibody dilution producing a 50% reduction in comparison with the B_{max} value is shown for the medial and lateral part of the substantia nigra on the intact and lesioned side. No significant differences have been observed when comparing these values.

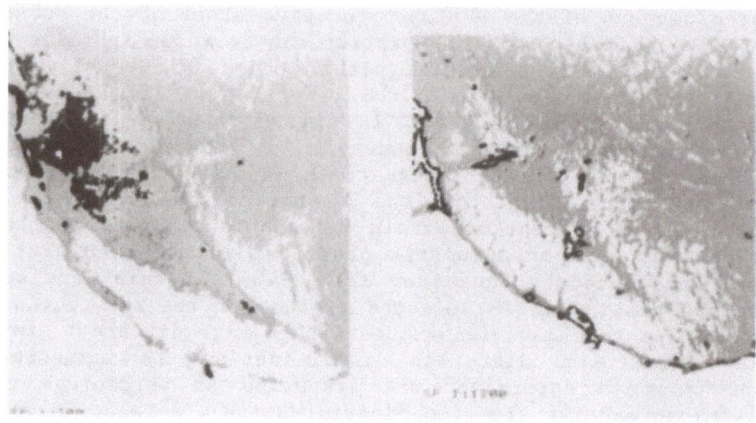

Fig. 6. Optical density map of Substance P-like immunoreactivity
in the substantia nigra in the normal male rat using two dilutions
of the Substance P antiserum (1:300, 1:1200). Eight different
tones can be distinguished. The darkest tone shown represent the
highest tones and exist predominantly within the lateral part of
the substantia nigra. It is seen that in the high dilution of the
antiserum the highest tones disappears. A frontal section of the
ventral half of one substantia nigra is shown. Magnification 20 x.

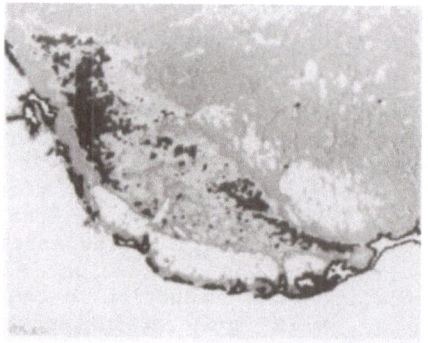
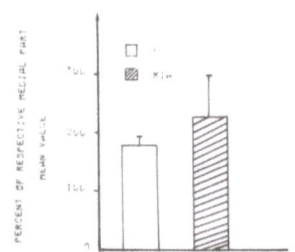

Fig. 7. An optical density map of Substance P-like immunoreactivity is shown in the left part of the picture. In the right part
of the picture the ratio of Substance P-like immunoreactivity
found in the lateral and the medial part of the substantia nigra
is shown as revealed by immunocytochemistry and by radioimmunoassay procedures (ICC versus RIA). Means ± s.e.m. are shown out of
five rats. No significant differences are observed.

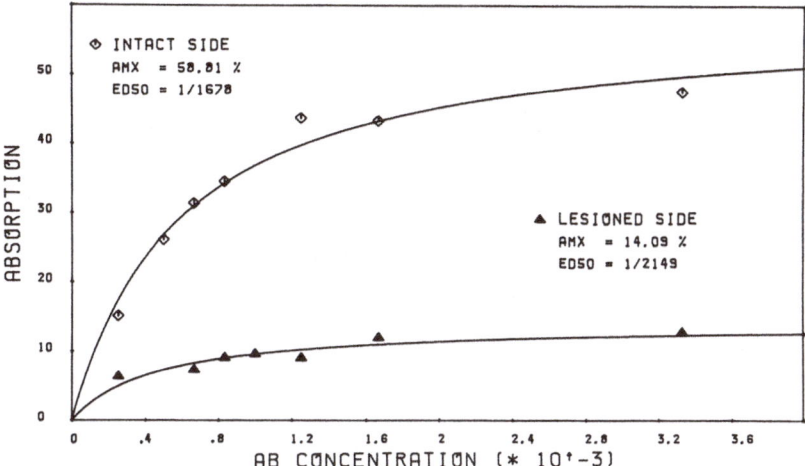

Fig. 8. Semiquantitative evaluation of Substance P-like immunoreactivity within the substantia nigra of the male rat following
partial hemitransection in front of the substantia nigra. For
further details, see text. A saturation analysis using different
concentrations of the Substance P antiserum has been performed on
the intact and lesioned side. A nonlinear fitting procedure has
been used. It is shown that the B_{max} values on the lesioned side
are markedly reduced compared with the corresponding values in the
intact side. Instead the ED 50 values are of a similar magnitude
indicating no change in the affinity of the antibody for the
antigen Substance P.

QUANTITATIVE IMMUNOCYTOCHEMISTRY OF SP-POSITIVE TERMINALS IN SUBSTANTIA NIGRA AFTER HEMITRANSECTION

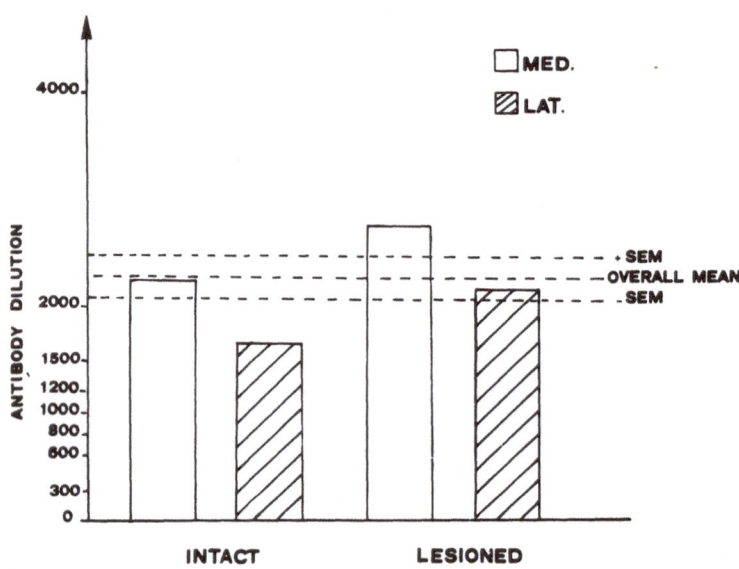

Fig. 9. Semiquantitative immunocytochemistry using saturation analysis to analyse Substance P immunoreactive nerve terminal networks in the substantia nigra of the intact and lesioned side following a partial unilateral hemitransection. Similar ED 50 values are obtained on both the medial and lateral part of the substantia nigra on the intact and lesioned side.

DISCUSSION

The present quantitative microfluorimetrical results give further evidence that it is possible to determine the pool sizes and the half-lives in discrete dopamine and noradrenaline nerve terminal systems all over the hypothalamus, preoptic area and the telencephalon. Also it is shown that this is possible within regions, such as the ventral part of the nucleus interstitialis striae terminalis, which has a very large pool size. The absence of catecholamine concentration dependent quenching of catechol-amine fluorescence is also further amplified in the experiments with the monoamine oxidase inhibitor, since similar increases of catecholamine levels are observed in the biochemical and the quantitative microfluorimetrical analysis. Furthermore, evidence is presented for the existence of only one pool of catecholamines in view of the apparent monophasic disappearance of the catechol-amine stores in all the various catecholamine nerve terminal systems analyzed following tyrosine hydroxylase inhibition. These results are in agreement with those of Paden (1979) showing that the ^{3}H-dopamine stores and the total dopamine stores disappear with the same rate constant following tyrosine hydroxylase inhibi-tion. Finally, the present results have given further evidence that the dotted and the diffuse types of dopamine nerve terminal networks within the nucleus accumbens and tuberculum olfactorium represent functionally distinct types of dopamine nerve terminal networks (Fuxe et al. 1979). These types of dopamine nerve termi-nals also show a difference with regard to their biochemistry, since only the dotted types of terminals appear to contain chole-cystokinin immunoreactivity (Hökfelt et al. 1980).

The present semiquantitative method to study immunoreactivity in biological structures offers a new way to reliably determine in a semiquantitative way differences in antigen content. Thus, it becomes possible to obtain relative measurements of antigen contents, which can be matched with the binding characteristics obtained in the same region for the receptor of the antigen, in this case the substance P receptor. These values are obtained by quantitative receptor autoradiography (see e.g. Benfenati et al. this symposium). The present semiquantitative method for the determination of antigen contents seems also to be highly comple-mentary to the quantitative immunocytochemical procedures intro-duced by Schipper et al. (1984) and Benno et al. (1982).

ACKNOWLEDGEMENTS

This work has been supported by a grant (MH25504) from the NIH, by grants (04X-715) from the Swedish and British Medical Research Council, by a CNR international grant, by a grant from Knut & Alice Wallenberg's Foundation and by a grant from the Wellcome Trust Foundation. For excellent technical assistance we

are grateful to Mrs Beth Andbjer, Mrs Ulla-Britt Finnman, Miss
Katarina Nilsson, Mrs Siv Nilsson, Mrs Birgitta Nyberg, Miss
Barbro Tinner, Mr Mauro Ferri and Carlo Brusiani. For excellent
secreterial assistance we are grateful to Mrs Anne Edgren.

REFERENCES

Agnati, L.F., Benfenati, F., Cortelli, P. and D'Alessandro, R.
(1978). A new method to quantify catecholamine stores visualized
by means of the Falck-Hillarp technique. Neurosci. Lett., 10,
11-17.

Agnati, L.F., Andersson, K., Wiesel, F. and Fuxe, K. (1979). A
method to determine dopamine levels and turnover rate in discrete
dopamine nerve terminal systems by quantitative use of dopamine
fluorescence obtained by Falck-Hillarp methodology. J Neurosci
Meth., 1, 365-373.

Agnati, L.F., Fuxe, K., Calza, L., Benfenati, F., Cavicchioli, L.,
Toffano, G. and Goldstein, M. (1983). Gangliosides increase the
survival of lesioned nigral dopamine neurons and favour the
recovery of dopaminergic synaptic function in striatum of rats by
collateral sprouting. Acta Physiol Scand., 119, 347-363.

Andersson, K., Fuxe, K. and Agnati, L.F. (1984). Determinations of
catecholamine half-lives and turnover rates in discrete catecholamine nerve terminal systems of the hypothalamus, the preoptic
region and the forebrain by quantitative microfluorimetry. Acta
Physiol Scand., in press.

Benfenati, F., Agnati, L.F., Fuxe, K., Cimino, M., Battistini, N.,
Merlo Pich, E., Farabegoli, C. and Zini, I. (1984). Quantitative
autoradiography as a tool to study receptors in neural tissue.
Studies on 3H-ouabain binding sites and correlation with synaptic
protein phosphorylation in different brain areas. Mac Millan
Press, Hampshire, England, in press.

Benno, R.H., Tricker, L.W., Joh, T.H. and Reis, D.J. (1982).
Quantitative immunocytochemistry of tyrosine hydrocylase in rat
brain. Brain Res., 246, 225-236.

Cuello, A.C., Galfre, G. and Milstein, C. (1979). Detection of
substance P in the central nervous system by a monoclonal antibody. Proc. Natl. Acad. Sci., USA, 76, 3532-3536.

Fuxe, K. (1965). Evidence for the existence of monoamine neurons
in the central nervous system: IV. The distribution of monoamine
nerve terminals in the central nervous system. Acta Physiol
Scand., 64, Suppl. 247, 39-85.

Fuxe, K., Agnati, L.F., Andersson, K., Calza, L., Benfenati, F., Zini, I., Battistini, N., Köhler, C., Ögren, S.-O. and Hökfelt, T. (1983). Analysis of transmitter-identified neurons by morphometry and quantitative microfluorimetry. Evaluation of the actions of psychoactive drugs, especially sulpiride. In: Special Aspects of Psychopharmacology. (M. Ackenheil & N. Matussek, eds.). Espansion Scientifique Francaise, Paris, 13-32.

Fuxe, K., Agnati, L.F., Zoli, M., Härfstrand, A., Grimaldi, R., Tucci, F. and Goldstein, M. (1984). Development of quantitative methods for the evaluation of the entity of coexistence of neuro-active substances in nerve terminal populations in discrete areas of the central nervous system: Evidence for hormonal regulation of cotransmission. Mac Millan Press, England, in press.

Fuxe, K., Andersson, K., Schwarcz, R., Agnati, L.F., Pérez de la Mora, M., Hökflet, T., Goldstein, M., Ferland, L., Possani, L. and Tapia, R. (1979). Studies on different types of dopamine nerve terminals in the forebrain and their possible interactions with hormones and with neurons containing GABA, glutamate and opioid peptides. In: Advances in Neurology, (L.J. Piorier, T.L. Sourkes & P.J. Bédard, eds.), Raven Press, New York, Vol. 24, 199-214.

Fuxe, K., Hökfelt, T., Jonsson, G. and Ungerstedt, U. (1970). Fluorescence microscopy in neuroanatomy. In: Contemporary Research Methods in Neuroanatomy (W.J.H. Nauta & S.O.E. Ebbesson, eds.), Springer-Verlag, Berlin, Heidelberg & New York, pp. 275-314.

Fuxe, K. and Jonsson, G. (1973). The histochemical fluorescence method for the demonstration of catecholamines. Theory, practice and application. J Histochem Cytochem., Vol 21, No 4, 293-311.
Hökfelt, T., Skirboll, L., Rehfeld, J.F., Goldstein, M., Markey, K. and Dann, O. (1980). A subpopulation of mesencephalic dopamine neurons projecting to limbic areas contains a cholecystokinin-like peptide: Evidence from immunohistochemistry combined with retro-grade tracing. Neurosci 5, 2093-2124.

Jonsson, G., Hallman, H., Mefford, I. and Adams, N. (1980). The use of liguid chromatography with electrochemical detection for the determination of adrenaline and other biogenic monoamines in the CNS. In: Central Adrenaline Neurons: Basic Aspects and Their Role in Cardiovascular Disease (K. Fuxe, M. Goldstein, Hökfelt, B. & Hökfelt, T., eds), Pergamon Press, New York, pp. 59-71.

Kanazawa, I., Emson, P.C. and Cuello, A.C. (1977). Evidence for the existence of substance P-containing fibres in striato-nigral and pallido-nigral pathways in rat brain. Brain Res., 119, 447-453.

Keller, R., Oke, A., Mefford, I. and Adams, R.N. (1976). Liquid chromatographic analysis of catecholamines. Routine assay for regional brain mapping. Life Sci 19: 995-1004.

Löfström, A., Jonsson, G. and Fuxe, K. (1976a). Microfluorimetric quantitation of catecholamine fluorescence in rat median eminence. I. Aspects on the distribution of dopamine and noradrenaline nerve terminals. J Histochem Cytochem 24: 415-429.
Löfström, A., Jonsson, G., Wiesel, F.A. and Fuxe, K. (1976b). Microfluorimetric quantitation of catecholamine fluorescence in rat median eminence. II. Turnover changes in hormonal states. J Histochem Cytochem 24: 430-442.

Paden, C.M. (1979). Disappearance of newly synthesized and total dopamine from the striatum of the rat after inhibition of synthesis: Evidence for a homogeneous kinetic compartment. J Neurochem 33: 471-479.

Schipper, J., Werkman, T.R. and Tilders, F.J.H. (1984). Quantitative immunocytochemistry of corticotropin-releasing factor (CRF). Studies on nonbiological models and on hypothalamic tissues of rats after hypophysectomy, adrenalectomy and dexamethasone treatment. Brain Res., 293, 111-118.

Snedecor, G.W. and Cochran, W.G. (1980). Statistical Methods, the Iowa state university press, Ames, Iowa, 7th edition.

Sternberger, L.A. (1979). Immunocytochemistry. 2nd ed., Wiley, New York.

QUANTITATIVE RECEPTOR AND 2–DEOXYGLUCOSE AUTORADIOGRAPHY

QUANTITATIVE AUTORADIOGRAPHIC LOCALIZATION OF DOPAMINE RECEPTORS AND UPTAKE SITES IN THE RAT AND HUMAN CNS

B. SCATTON, A. DUBOIS, A. CAMUS, F. JAVOY–AGID, N.R. ZAHNISER, M.L. DUBOCOVICH and A. CUDENNEC

Biochemical Pharmacology Group, Synthélabo—L.E.R.S., 31 Avenue Paul Vaillant Couturier, F–92220 Bagneux, France

INTRODUCTION

Dopamine (DA) receptors have been identified in various areas of the mammalian central nervous system (CNS) by conventional in vitro binding techniques using radiolabelled DA agonists and antagonists (see Seeman, 1980 for a review). However, in order to better understand the organization of DA-containing systems within the CNS, DA receptors need to be localized more precisely within the various DA-rich areas. The recent development of light microscopic receptor autoradiography (Young and Kuhar, 1979 ; Murrin, 1981 ; Palacios, 1981) has provided a means to analyze the localization of specific sites quantitatively and with a high degree of anatomical resolution in the CNS. In the present study we have used the technique of autoradiography to investigate in detail the anatomical distribution of DA receptors in the rat brain and spinal cord subregions and in selected regions of the human brain using ^{3}H-N-propyl-norapomorphine, a potent DA receptor agonist, as a ligand.

High affinity binding sites for tricyclic antidepressants such as imipramine and desipramine have been identified in membrane fractions and in sections of the mammalian brain (Langer et al, 1980 ; Raisman et al, 1980 ; Lee and Snyder, 1981 ; Biegon and Rainbow, 1983 ; Fuxe et al, 1983). Evidence has been provided that the binding sites for ^{3}H-imipramine and ^{3}H-desipramine may be associated with the presynaptic uptake site for serotonin and norepinephrine, respectively. Since a close parallelism has been observed between the distribution of tricyclic binding sites in brain regions and the distribution of monoaminergic terminals, the ability to label monoamine uptake sites with potent labelled antidepressants offers a new tool to study the organization and topographical distribution of monoaminergic nerve terminals in the CNS. Recently, the potent DA uptake inhibitors, ^{3}H-cocaine, ^{3}H-mazindol and ^{3}H-nomifensine, have been used to label a site associated with the neuronal uptake

of DA in striatal membranes (Dubocovich and Zahniser, 1983 ; Javitch
et al, 1983 ; Kennedy and Hanbauer, 1983 ; Pimoule et al, 1983 ;
Schoemaker et al, 1984). In an effort to evaluate the validity of
using DA uptake inhibitor radioligands to label DA nerve terminals
in the CNS we have analyzed, by quantitative autoradiography, the
regional distribution of ^{3}H-nomifensine binding sites and their location, especially in relation to the different ascending and des-
cending dopaminergic pathways, in the rat brain and spinal cord.
Preliminary data concerning the distribution of ^{3}H-nomifensine binding sites in the normal and pathological human caudate nucleus will
also be presented.

METHODS

<u>^{3}H-NPA binding in tissue sections</u>

      Male Sprague Dawley rats (COBS CD strain, Charles River, Fran-
ce) weighing 200 g were anaesthetized with 2% halothane, heparinized
and perfused transcardially with cold 0.05 M phosphate-buffered saline (pH 7.4) containing 0.01% formaldehyde for at least 15 min. The
brain and spinal cord were rapidly removed, mounted onto brass microtome chucks and frozen. Coronal tissue sections (16 µm in thickness) were prepared on a cryostat (Bright) at -20°C and thaw-mounted
onto subbed (chrome alum and gelatin) microscope slides. They were
then dried rapidly under vacuum and kept frozen until the binding
experiment. For binding assay, the tissue sections were brought to
room temperature and incubated for 1 hour in a previously deoxygena-
ted buffer solution (50 mM Tris HCl, 1 mM EDTA, 0.01% ascorbic acid,
pH 7.4) containing various concentrations (0.06 - 2 nM) of ^{3}H-N-pro-
pylnorapomorphine (^{3}H-NPA) (58.8 Ci/mmol, New England Nuclear, Boston MA). Non-specific binding was defined as the binding in the presence of DA (10 µM) or (+)butaclamol (1 µM). Under our experimental
conditions, specifically bound ^{3}H-NPA accounted for 70% of the total binding at a ligand concentration of 0.25 nM. The incubation was
terminated by rinsing the sections for twice 10 min in cold phosphate-buffered saline, then for 10 sec in cold distilled water. Each
section was dried rapidly under a stream of cold air. The microscope
glasses with the sections were fixed to cardboard with double sided
Scotch tape and placed in a light-tight X-Ray cassette. Autoradiographs were prepared by apposing the slide-mounted tissue sections
to tritium-sensitive film (LKB Ultrofilm) for 7-8 weeks at 4°C. The
film was developed for 1.5 min (Radio Kodak AL-4), rinsed, immersed
in a fixative (Radio Kodak AL-4) for 5 min, rinsed again in running
water for 30 min and dried in a fresh air stream. The optical density of the autoradiograms was quantified by using an automatic image analyzer and densitometer (Quantimet 720, Cambridge, Instruments)
with a variable frame (0.006 to 0.02 mm^2) connected to a PDP 11/03
computer. The determination of optical density was made in 12 sites
in the regions of interest on each section. The optical density of
the different CNS areas was converted into fmoles ^{3}H-NPA specifically bound per mg of protein by using ^{3}H-ornithine standards made
from rat brain mash (Unnerstall et al, 1982).

Human brains were obtained from 5 control patients (without a history of neurological or psychiatric illness), 2 patients who suffered from Parkinson's disease and one patient who died from supranuclear palsy. The postmortem delay varied from 3-10 hr. The brains were removed less than 2 h after autopsy and stored frozen at -70°C until dissection. Slide-mounted brain coronal sections were prepared and labelled with ³H-NPA as described above for the rat. However, prior to labelling, human brain sections were preincubated for 30 min at 22°C in 50 mM Tris/HCl pH 7.4 containing 1 mM EDTA and 0.01% ascorbic acid. This preincubation step increased the ratio of specific to non-specific binding of ³H-NPA presumably by removing the endogenous transmitter from the tissue sections. After labelling with ³H-NPA, the sections were continuously rinsed for 30 min in fresh buffer at 4°C in order to reduce the levels of non-specific binding. Under our standard assay conditions, the specific binding of ³H-NPA (0.25 nM) to human caudate nucleus sections represented 80% of the total binding. Autoradiographs were generated and analyzed as described above.

³H-nomifensine binding in tissue sections

The procedure used for the preparation of slide-mounted coronal sections (16 μm thick) from rat brain and spinal cord and human brain was similar to that described above. Incubation was performed for 1 hr at 4°C directly on slices by covering the sections with a drop of Tris/HCl buffer pH 7.4 containing 120 mM NaCl, 5 mM KCl and 2.5 - 100 nM ³H-nomifensine (S.A. 44 Ci/mmol, New England Nuclear, Boston MA). Non-specific binding was defined by the use of 100 μM benztropine. In sections from the rat caudate nucleus, specific binding accounted for 60% of the total binding in the presence of 10 nM ³H-nomifensine. Following incubation, the sections were rapidly rinsed for 1 min and then for 2.5 min in fresh buffer, and finally dipped in cold distilled water for 10 sec. After drying, sections were juxtaposed to tritium-sensitive film for 4 weeks at 4°C. Autoradiographs were then measured by quantitative densitometric analysis as described above.

AUTORADIOGRAPHIC LOCALIZATION OF ³H-NPA BINDING SITES IN THE RAT AND HUMAN CNS.

Rat brain

Kinetic analysis and displacement studies of ³H-NPA binding performed in caudate-putamen sections revealed that ³H-NPA binding in this area was saturable and of high affinity. Scatchard analysis indicated that, in the ventrolateral part of the striatum, ³H-NPA labelled a single population of non interacting sites (Hill coefficient 0.99) with an apparent dissociation constant of 1.1 nM and a maximal capacity of 2195 fmol/mg protein. These results agree well with previous data obtained on rat striatal membrane preparations (Hall et al, 1983). Displacement experiments with different neuroleptic agents and DA agonists showed that the binding sites labelled with ^{3}H-NPA in the striatum have the pharmacological characteristics of DA receptors of the D_2 type (Scatton et al, in preparation).

The autoradiographs prepared from serial coronal brain sections (incubated in the presence of 0.25 nM ^{3}H-NPA) showed that high densities of autoradiographic grains were present throughout the forebrain. As expected, the highest levels of specific binding were observed in the caudate-putamen, nucleus accumbens and the olfactory tubercle (Fig. 1). In these areas, the grains were distributed heterogeneously. In the rostral part of the caudate nucleus, the highest density of ^{3}H-NPA binding was found laterally while lower, although still substantial, densities were measured in the dorso-medial and medial parts. Areas with intermediate levels of binding included the lateral septum, the central nucleus of the amygdala, the substantia nigra and the hippocampus. No specific ^{3}H-NPA binding sites were found in the globus pallidus and cerebellum. The cerebral cortex also showed elevated grain densities. At the level of the forceps minor, there were substantial densities of (+)butaclamol-displaceable ^{3}H-NPA binding in the superficial layers of the cingulate cortex. In more posterior areas of the cortex, there was a slight enrichment in grain densities in lamina IV but these cortical binding sites were not displaceable by (+)butaclamol.

Figure 1. Autoradiographic distribution of ^{3}H-NPA binding
            sites in a coronal section of rat brain.

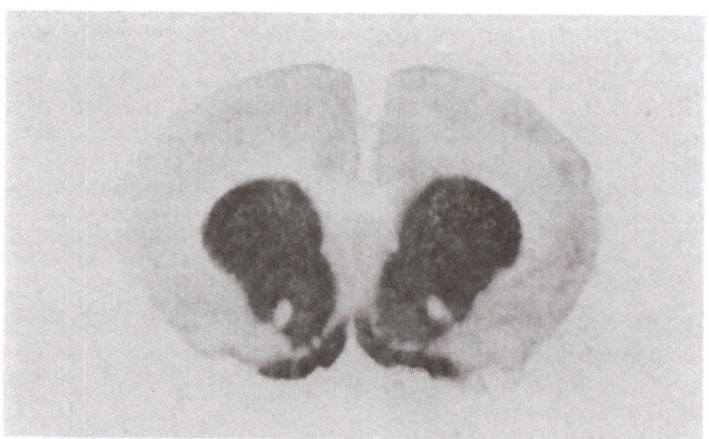

Sections were incubated with 0.25 nM of ^{3}H-NPA and exposed
for 4 weeks to tritium-sensitive film. The degree of greyness is directly proportional to the receptor density.

From these data it appears that there is a good agreement between the distribution of ³H-NPA binding sites and the distribution
of DA-containing fibers and terminals found by histochemistry and
biochemistry (Versteeg et al, 1976 ; Lindvall and Björklund, 1978).
Moreover, the autoradiographic distribution of ³H-NPA binding sites
correlates quite well with that previously reported with conventional in vitro binding techniques or autoradiography using ³H-neuro-
leptics as ligands (Seeman, 1980).

The resolution of the autoradiographic method used here is not
sufficient to allow conclusions to be drawn as to whether or not the
receptor sites are associated with pre- or postsynaptic elements.
In an attempt to establish more precisely the localization of ³H-NPA
binding sites in the striatum, we have examined the effects of a
6-hydroxydopamine-induced lesion of the nigro-striatal dopaminergic
pathway on the density of these binding sites. This lesion resulted
in a significant increase (+20%) in the density of ³H-NPA binding
in the ventrolateral and medial but not in the dorsomedial part of
the striatum (in preparation). This increase is consistent with
other studies showing a denervation supersensitivity following lesion of the dopaminergic input to the caudate-putamen (see Seeman,
1980). From these data, it appears that the bulk of ³H-NPA binding
sites is located postsynaptically on DA target cells. However, a
minor population of ³H-NPA binding sites may also be present on dopaminergic terminals. If presynaptic ³H-NPA binding sites exist and
were decreased this may have been masked by the proliferation of
postsynaptic binding sites subsequent to the development of DA target cell disuse supersensitivity. The present data also reveal that
the increase in striatal DA receptors occurs only in certain subregions of the caudate-putamen. A similar regional heterogeneity of
the increase of ³H-NPA binding sites in the striatum has been observed after chronic neuroleptic treatment (in preparation). This em-
phasizes the usefulness of quantitative receptor autoradiography in
revealing localized changes in DA receptor number within a given
brain area in response to psychoactive agents.

Rat spinal cord

Anatomical, biochemical and electrophysiological studies have
provided evidence for the existence of a dopaminergic innervation
of the mammalian spinal cord (see Commissiong and Neff, 1979 for a
review). The substantia nigra as well as the diencephalic A11 cell
group have been proposed as the possible sources of this innervation (Björklund and Skagerberg, 1979 ; Commissiong et al, 1979).
Conventional in vitro binding studies using ³H-haloperidol as a ligand have suggested the presence of DA receptors in the rat spinal
cord (Demenge et al, 1981). However, the precise anatomical localization of these receptors is still unknown. Since ³H-NPA appears to
be a suitable ligand to label DA receptors in the brain (see above)
we have used it to localize autoradiographically DA receptors in the
rat spinal cord.

The autoradiographs prepared from sections of the cervical spinal cord (in the presence of 0.25 nM of ^{3}H-NPA) showed that ^{3}H-NPA
binding sites were contained exclusively within the gray matter although with a heterogeneous distribution (Fig. 2)(Scatton et al,
1984). Particularly high densities were found in the substantia gelatinosa (Lamina II of the Rexed classification) and in a zone just
dorsal to the central canal. Substantial but lower densities of [3]H-
NPA binding sites were seen in the other gray matter laminae. A similar distribution pattern of ^{3}H-NPA was observed in the thoracic
and lumbar portions of the spinal cord (unpublished data). Scatchard
analysis of the binding to different laminae of the cervical spinal
cord indicated that ^{3}H-NPA labelled a population of non-interacting
sites with a Kd (0.8 nM) similar to that found in the caudate-putamen. However, the maximal capacity found in the substantia gelatinosa (584 fmol/mg prot) represented about one fourth of that measured
in the ventrolateral caudate-putamen. Competition experiments also
showed that the binding sites labelled with ^{3}H-NPA in the spinal
cord have the pharmacological characteristics of DA receptors
(Scatton et al, 1984).

Thus, from autoradiographic data, DA receptors appear to be distributed heterogeneously in the spinal cord. The localization of
[3]H-NPA binding sites correlates well with the pattern of distribution of DA and its metabolites which are concentrated in the dorsal
horn of the spinal cord (Commissiong and Neff, 1979). The presence
of DA receptors in the spinal cord, supports the existence of a dopaminergic innervation of this structure. The light microscopic level of anatomical resolution of the receptor autoradiography technique does not allow the determination of the precise neuronal localization of the ^{3}H-NPA binding sites in the cord. The binding sites
found in lamina II may be localized on intrinsic neurons or on primary afferent terminals. Experiments of dorsal rhizotomy and of spinal cord transection are in progress to establish the exact neuronal
localization of DA receptors in the spinal cord. Although, the exact
functional role of DA in the spinal cord has to be further investigated, the preferential localization of ^{3}H-NPA binding sites in the
substantia gelatinosa is compatible with a possible role for spinal
cord DA in the processing of sensory information.

Figure 2. Distribution of ^{3}H-NPA binding sites in the cervical spinal cord of the rat.

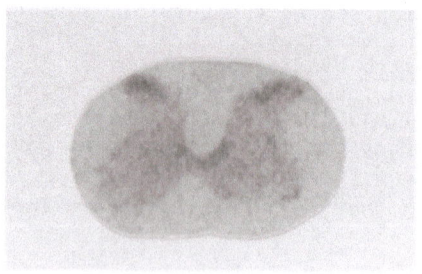

Human brain

    DA receptors have been studied in the normal and pathological human brain by conventional binding techniques using membrane preparations (see Seeman, 1980). However, these biochemical techniques have a limited anatomical resolution. Quantitative receptor autoradiography can provide more precise information about the localization and properties of receptors in numerous regions and subregions of normal human brain and about the regional localization of receptor changes in pathological states. We have therefore used quantitative autoradiography to study the precise distribution of DA receptors in human brain regions using ^{3}H-NPA as a ligand (Camus et al, in preparation). The use of ^{3}H-NPA was of particular interest since previous biochemical studies on human brain have almost exclusively been performed with radiolabelled DA antagonists as ligands (Seeman, 1980).

    Initial experiments performed in caudate-putamen sections from normal patients indicated that, under the experimental conditions used, ^{3}H-NPA labelled a single population of non-interacting sites with a Kd of 0.2 nM and a maximal capacity ranging from 180-556 fmol/mg protein. That the affinity of ^{3}H-NPA for its binding sites is four times greater in the human than in the rat caudate nucleus may be related to the fact that, unlike rat brain sections, human brain sections were preincubated prior to labelling with ^{3}H-NPA to remove as much as possible of the endogenous ligand. Displacement experiments with a variety of DA agonists and antagonists showed that the binding sites labelled by ^{3}H-NPA in the human caudate nucleus possess the pharmacological features of DA receptors of the D_2 type. Thus, ^{3}H-NPA binding was displaced by (+)butaclamol (IC_{50} 0.05 μM), sulpiride, a specific D_2 antagonist (IC_{50} 1 μM) and LY 141865, a specific D_2 agonist (IC_{50} 0.06 μM) but not by SKF 38393, a specific D_1 antagonist (IC_{50} > 10 μM). The binding of ^{3}H-NPA was also found to be stereospecific: (+)- but not (-)butaclamol displaced the ^{3}H-ligand.

    The autoradiographs prepared from various coronal sections of human brain (in the presence of 0.25 nM ^{3}H-NPA) showed the presence of ^{3}H-NPA binding sites in several DA-rich areas. The highest levels of labelling were found in the caudate nucleus followed by the putamen and pallidum (Fig. 3). In the caudate nucleus, the density of ^{3}H-NPA binding sites was higher in the anterior (Bmax 569 fmol/mg protein) than in the posterior (Bmax 188 fmol/mg protein) part. In these basal ganglion areas, particularly in the caudate nucleus, the distribution of autoradiographic grains was heterogenous and punctate (areas of both high and low density were present, see Fig. 3). Particularly high densities of autoradiographic grains were also present in the nucleus accumbens and in the substantia nigra. A slight labelling was also seen in areas e.g. the hippocampus and cerebral cortex that have been suggested to receive a discrete dopaminergic innervation in animals (Lindvall and Björklund, 1978 ; Scatton et al, 1980).

B. Scatton, et al.

Figure 3. Autoradiographic localization of ³H—NPA binding
          sites in a caudate-putamen section from a control
          human brain.

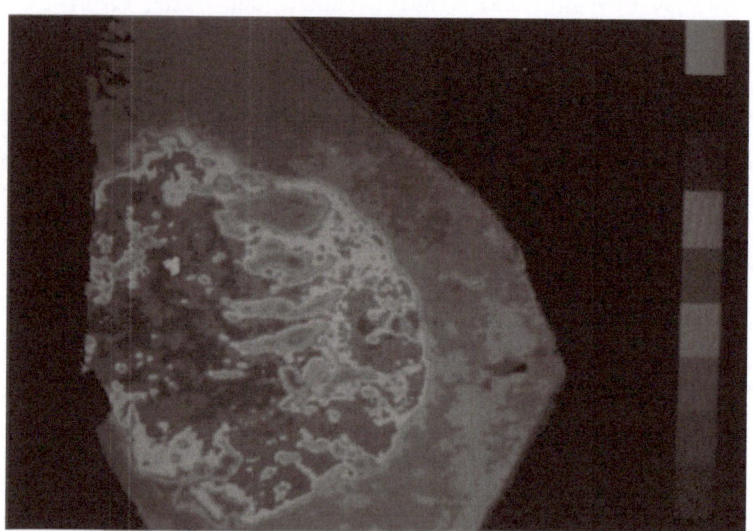

Color coding: white, highest densities ; violet, background.

    In preliminary experiments, performed on a limited number of
patients suffering from Parkinsonism, ³H—NPA binding densities in
the caudate nucleus, putamen and nucleus accumbens were found not to
differ from control values. In contrast, the density of ³H—NPA binding sites in these areas was subnormal in one patient suffering
from supranuclear palsy, a disorder that resembles Parkinson's disease and also characterized by subcortical lesions that include the
nigrostriatal system (Steele et al, 1964) but which responds poorly
to levodopa therapy. These results obtained in tissue sections con-
firm the findings of studies on homogenate of basal ganglia in Parkinson's disease (Bokobza et al, 1984). The loss of ³H—NPA binding
sites observed in the caudate nucleus in supranuclear palsy is also
in agreement with the fall of ³H-spiperone binding that we have recently observed in these subjects in membrane preparations from the
caudate nucleus and putamen (Bokobza et al, 1984). The fact that patients suffering from supranuclear palsy do not respond to levodopa
therapy can probably be attributed to the decrease of DA receptors
in basal ganglion regions.

    In conclusion, the present study indicates that it is possible
to localize and quantify DA receptors in rat and human postmortem
brain sections with a light microscopic degree of resolution using
a radio-labelled DA agonist as a ligand. The autoradiographic study
of the density and distribution of DA receptors in human brain can
be used to examine the discrete changes in these receptors in various pathological states. Our preliminary data suggest that DA

receptors in the basal ganglia may not be altered in Parkinson's
disease but may be diminished in supranuclear palsy.

AUTORADIOGRAPHIC LOCALIZATION OF ^{3}H-NOMIFENSINE BINDING SITES IN
THE RAT AND HUMAN CNS.

The successful use of ^{3}H-nomifensine, a potent DA uptake inhibitor, to label a site associated with the neuronal uptake of DA in
rat striatal membranes (Dubocovich and Zahniser, 1983) has prompted
us to localize and quantify autoradiographically ^{3}H-nomifensine binding sites in the rat and human (normal and pathological) brain.

The characteristics of ^{3}H-nomifensine binding to rat striatal
sections were determined prior to the autoradiographic studies. Saturation studies in rat caudate nucleus sections revealed the existence of a single population of non-interacting (nH= 0.98) ^{3}H-nomi-
fensine binding sites with a Kd of 56 nM and a Bmax of 2800 fmoles/
mg protein. The binding of ^{3}H-nomifensine was found to be fully so-
dium-dependent like the carrier-mediated transport of DA. Displacement of sodium dependent specific ^{3}H-nomifensine binding in rat
striatal sections by different monoamine uptake inhibitors correlated well with their capacity to inhibit DA uptake in striatal synaptosomes. Thus, under our conditions, ^{3}H-nomifensine apparently labelled the recognition site for the neuronal uptake of DA in striatal
sections with a high degree of selectivity and potency.

The study of the regional localization of ^{3}H-nomifensine in rat
brain coronal sections revealed that the distribution of the ^{3}H-li-
gand coincides to a large extent with the distribution of dopaminergic nerve terminals (as determined by histofluorescence and biochemical studies) and with the distribution of ^{3}H-NPA binding sites
(see above). Thus, the highest binding was found in the striatum,
nucleus accumbens and olfactory tubercle (Fig. 4 ; Table 1). In these areas, especially in the striatum, the grains were distributed in
clusters. Intermediate binding was also found in hypothalamus (ar-
cuate, ventromedial and dorso-medial nuclei), frontal cortex, anterior and posterior cingulate cortex, thalamus (especially in the la-
tero-dorsal, mediodorsal and anteroventral nuclei) and amygdaloid
complex (Table 1). The hippocampus was also labelled substantially
by the ^{3}H-ligand in all the CA fields and in the dentate gyrus. Within the neocortex and cingulate cortex, ^{3}H-nomifensine binding was
generally higher in the superficial (laminae I-IV) as opposed to
the deep laminae. Substantial grain densities were also observed in
the superficial gray of the superior colliculus, in the bed nucleus
of the stria terminalis, the lateral septum, the zona incerta and
parietal cortex (Table 1). ^{3}H-nomifensine binding sites were also
found in substantial concentrations in the spinal cord (lamina II)
(not shown).

In those brain areas which receive a dopaminergic innervation,
desipramine (10 µM) failed to displace specific ^{3}H-nomifensine binding suggesting that ^{3}H-nomifensine specifically labels the DA upta-

Table 1. REGIONAL DISTRIBUTION OF ^{3}H-NOMIFENSINE BINDING IN
THE RAT BRAIN

Structure	^{3}H-nomifensine specifically bound (fmoles/mg prot.)
Caudate-putamen	274
Nucleus accumbens	230
Olfactory tubercle	257
Hypothalamus – arcuate, ventromedial nucleus	148
– dorso-medial nucleus	143
Amygdala – nucleus amygdaloid centralis	104
– medial amygdaloid nucleus	130
Cingulate cortex	152
Hippocampus CA$_1$	121
CA$_2$	90
CA$_3$	102
Dentate gyrus	130
Thalamus – parafascicular nucleus	90
– paraventricular nucleus	133
– laterodorsal nucleus	273
– mediodorsal nucleus	223
– anteroventral nucleus	223
Superior colliculus	116
Bed. nucleus stria terminalis	189
Septum	160
Zona incerta	138
Parietal cortex	69

Brain sections were labelled with 10 nM of ^{3}H-nomifensine
and exposed for 4 weeks to tritium-sensitive film. The non
specific binding was defined by benztropine (100 μM).

-ke site. This view is corroborated by the observation of a reduction (about 80%) of the ^{3}H-ligand binding in the striatum, nucleus
accumbens, olfactory tubercle and anterior cingulate cortex 3 weeks
after 6-hydroxydopamine-induced lesion of the ascending (nigro-
striatal and meso-limbic) dopaminergic pathways (Fig. 4). However,
in areas which receive a noradrenergic but not a dopaminergic innervation e.g. the thalamic nuclei and parietal cortex (Versteeg et al,
1976 ; Jones and Moore, 1977), desipramine (10 μM) inhibited totally the binding of the ligand. This suggests that in these areas ^{3}H-
nomifensine labelled norepinephrine rather than DA uptake sites.
This is consistent with the fact that nomifensine is also a potent
inhibitor of norepinephrine uptake in brain synaptosomes (Hunt et
al, 1974). Incubation of brain sections with desipramine provides a
means to eliminate the binding of ^{3}H-nomifensine to the norepinephrine uptake recognition site.

Preliminary experiments suggest that ^{3}H-nomifensine may label
the DA uptake recognition site in human brain as well. Thus, heavy
^{3}H-nomifensine binding was found in the caudate nucleus and putamen

Figure 4. Autoradiographic localization of ³H-nomifensine binding sites in a coronal section (at the level of the striatum) after a unilateral chemical destruction of the ascending dopaminergic pathways in the rat.

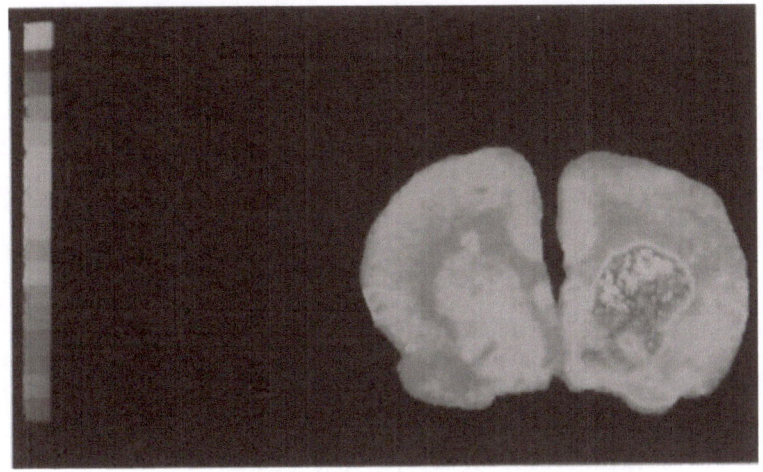

Left: lesioned side, right: unlesioned side. The color scale corresponds to the optical densities: white is the highest detected, violet is background level.

from normal humans. The autoradiographic grains were distributed heterogeneously with areas of high and low densities. Moreover, in coronal caudate nucleus sections from a patient with Parkinson's disease, the ³H-ligand binding was found to be markedly lower than control values.

From the above data it appears that ³H-nomifensine may be a suitable ligand to label sites associated with the DA uptake pump and consequently dopaminergic terminals (provided that desipramine is included into the incubation medium). Thus, most of the brain regions exhibiting high density of ³H-nomifensine binding sites (non displaceable by desipramine) contain or are suspected to contain dopaminergic nerve terminals. In this respect, the presence of ³H-nomifensine binding sites in the hippocampus, an area which has been postulated but not proved definitely to receive dopaminergic afferents (Scatton et al, 1980) is of particular interest.

The hypothesis that the high affinity ³H-nomifensine binding site is associated with the neuronal DA transporter is further supported by: firstly, the marked decrease in ³H-nomifensine binding in DA-rich brain areas following chemical lesion of the dopaminergic neurons in the rat ; and secondly, the pronounced decrease of the ³H-ligand binding (that parallels the severe degeneration of the dopaminergic innervation) which was also found in the caudate nucleus

in Parkinson's disease. However, ^{3}H-nomifensine binding sites may
also be located on the DA nerve cell membrane of the axons since the
DA uptake mechanism is also present on this part of the neuron.

The relationship between DA terminal networks and the ^{3}H-nomi-
fensine binding sites in some brain areas (e.g. the nuclei of the
thalamus or the parietal cortex) is unclear in view of the lack of
dopaminergic innervation of these areas (Versteeg et al, 1976). However, the facts that these areas receive a dense noradrenergic innervation (Versteeg et al, 1976 ; Jones and Moore, 1977) and that
the potent norepinephrine uptake blocker desipramine displaces ^{3}H-
nomifensine binding in these regions strongly suggest that the ^{3}H-
ligand binding may be linked to the norepinephrine uptake carrier.

In conclusion, the present results demonstrate that the distribution of ^{3}H-nomifensine binding sites in the rat and human brain
is in good agreement with the distribution of dopaminergic terminals
providing additional evidence that ^{3}H-nomifensine labels some component of the presynaptic DA uptake site (Dubocovich and Zahniser,
1983). This offers the possibility of applying in vitro autoradiography and computerized image analysis to provide detailed maps of
dopaminergic nerve densities in the CNS and to study their alterations in neurological or psychiatric disorders.

CONCLUDING REMARKS

The present results show that with the technique of receptor
autoradiography, it is possible to perform functional studies at
both pre- and postsynaptic levels of dopaminergic neurons in the rat
and human CNS using ^{3}H-nomifensine and ^{3}H-NPA as ligands. This technique allows the quantification of the distribution of binding sites
in tissue sections by means of computerized densitometry and morphometrical evaluation of the size and distribution of the radioligand
positive areas. We have observed that there exists an heterogeneity
of the distribution of ^{3}H-NPA and ^{3}H-nomifensine binding sites in
rat and human DA-rich brain areas. By means of quantitative receptor autoradiography it is possible to study the localized alterations in the density of DA nerve terminals and receptors in discrete brain regions and subregions in neuropsychiatric disorders and
to discover possible heterogeneities in the responses to psychoactive agents when administered acutely or chronically.

REFERENCES

Biegon, A. and Rainbow, T.C. (1983). Localization and characterization of ^{3}H-desmethylimipramine binding sites in rat brain by quantitative autoradiography. J. Neurosci. 3, 1069-1076.
Björklund, A. and Skagerberg, G. (1979). Evidence for a major spinal
cord projection from the diencephalic A11 dopamine cell group in the
rat using transmitter specific fluorescent retrograde tracing.
Brain Res. 177, 170-175.
Bokobza, B., Ruberg, M., Scatton, B., Javoy-Agid, F. and Agid, Y.

(1984). ³H-spiperone binding, dopamine and HVA concentrations in Parkinson's disease and supranuclear palsy. Europ. J. Pharmacol. 99, 167-175.

Commissiong, J.W., Gentleman, S. and Neff, N.H. (1979). Spinal cord dopaminergic neurons: evidence for an uncrossed nigrospinal pathway Neuropharmacology 18, 565-568.

Commissiong, J.W. and Neff, N.H. (1979). Current status of dopamine in the mammalian spinal cord. Biochem. Pharmacol. 28, 1569-1573.

Demenge, P., Mouchet, P., Guérin, B. and Feuerstein, C. (1981). Identification and distribution of neuroleptic binding sites in the rat spinal cord. J. Neurochem. 37, 53-59.

Dubocovich, M.L. and Zahniser, N.R. (1983). Binding of the dopamine uptake inhibitor ³H-nomifensine to striatal membranes. Soc. Neurosci. Abstr. 9, 564.

Fuxe, K., Calza, L., Benfenati, F., Zini, I. and Agnati, L.F.(1983). Quantitative autoradiographic localization of ³H-imipramine binding sites in the brain of the rat: relationship to ascending 5-hydroxytryptamine neuron systems. Proc. Natl. Acad. Sci. USA, 80, 3836-3840.

Hall, M.D., Jenner, P., Kelly, E. and Marsden, C.D. (1983). Differential anatomical location of ³H-N,n-propylnorapomorphine and ³H-spiperone binding sites in the striatum and substantia nigra of the rat. Brit. J. Pharmacol. 79, 599-610.

Hunt, P., Kannengiesser, M.H. and Raynaud, J.P. (1974). Nomifensine: a new potent inhibitor of dopamine uptake into synaptosomes from rat brain corpus striatum. J. Pharm. Pharmac. 26, 370-371.

Javitch, J.A., Blaustein, R.O. and Snyder, S.H. (1983). ³H-mazindol binding associated with neuronal dopamine uptake sites in corpus striatum membranes. Europ. J. Pharmacol. 90, 461-462.

Jones, B.E. and Moore, R.Y. (1977). Ascending projections of the locus coeruleus in the rat. II. Autoradiographic studies. Brain Res. 127, 23-53.

Kennedy, L.T. and Hanbauer, I. (1983). Sodium-sensitive cocaine binding to rat striatal membrane: possible relationship to dopamine uptake sites. J. Neurochem. 41, 172-178.

Langer, S.Z., Moret, C., Raisman, R., Dubocovich, M.L. and Briley, M. (1980). High affinity ³H-imipramine binding in rat hypothalamus is associated with the uptake of serotonin but not norepinephrine, Science 210, 1133-1135.

Lee, C.M. and Snyder, S.H. (1981). Norepinephrine neuronal uptake binding sites in rat brain membranes labelled with ³H-desipramine. Proc. Natl. Acad. Sci. USA 78, 5250-5254.

Lindvall, O. and Björklund, A. (1978). Anatomy of the dopaminergic neuron system in the rat brain. In Dopamine Advances in Biochemical Psychopharmacology, Vol. 19 (eds P.J. Roberts, G.N. Woodruff and L. L. Iversen). Raven Press, New York, pp 1-23.

Murrin, C.L. (1981). Neurotransmitter receptors: neuroanatomical localization through autoradiography. Int. Review of Neurobiol. 22, 111-171.

Palacios, J., Niehoff, D., Kuhar, M. (1981). Receptor autoradiography with tritium-sensitive film. Potential for computerized densitometry. Neurosci. Lett. 25, 101-105.

Pimoule, C., Schoemaker, H., Javoy-Agid, F., Scatton, B., Agid, Y.

and Langer, S.Z. (1983). Decrease in ³H-cocaine binding to the dopamine transporter in Parkinson's disease. Europ. J. Pharmacol. 95,
145-146.
Raisman, R., Briley, M.S. and Langer, S.Z. (1980). Specific tricyclic antidepressant binding sites in rat brain characterized by high
affinity ³H-imipramine binding. Europ. J. Pharmacol. 61, 373-380.
Scatton, B., Dubois, A. and Cudennec, A. (1984). Autoradiographic
localization of dopamine receptors in the spinal cord of the rat
using ³H-N-propylnorapomorphine. J. Neural Transm. 59, in press.
Scatton, B., Simon, H., Le Moal, M. and Bischoff, S. (1980). Origin
of dopaminergic innervation of the rat hippocampal formation. Neurosci. Lett. 18, 125-131.
Schoemaker, H., Pimoule, C., Arbilla, S., Scatton, B., Javoy-Agid,
F., Agid, Y. and Langer, S.Z. (1984). Sodium dependent ³H-cocaine
binding associated with dopamine uptake sites parallels dopaminergic denervation in Parkinson's disease. Submitted.
Seeman, P. (1980). Brain dopamine receptors. Pharmacol. Reviews 32,
229-313.
Steele, J.C., Richardson, J.C. and Olzewski, J. (1964). Progressive
supranuclear palsy. A heterogeneous degeneration involving the brain
stem, basal ganglia and cerebellum with vertical gaze and pseudo-
bulbar palsy, nuchal dystonia and dementia. Arch. Neurol. 10, 333-
343.
Unnerstall, J.R., Niehoff, D.L., Kuhar, M.J. and Palacios, J.M.
(1982). Quantitative receptor autoradiography using ³H-ultrofilm:
application to multiple benzodiazepine receptors. J. Neurosci. Meth.
6, 59-73.
Versteeg, D.H., Van der Gugten, J., De Jong, Z. and Palkovits, M.
(1976). Regional concentrations of noradrenaline and dopamine in rat
brain. Brain Res. 113, 563-574.
Young, W.S. and Kuhar, M.J. (1979). A new method for receptor autoradiography: ³H-opioid receptor labelling in mounted tissue sections. Brain Res. 179, 255-270.

THE USE OF QUANTITATIVE AUTORADIOGRAPHIC TECHNIQUES ON THE STUDY OF DRUG ACTION IN THE BRAIN: RECEPTOR AUTORADIOGRAPHY AND 2-DEOXYGLUCOSE TECHNIQUE

JOSÉ M. PALACIOS and KARL-HEINZ WIEDERHOLD

Preclinical Research, Sandoz Ltd, Basle, CH-4002, Switzerland

INTRODUCTION

The main goal of pharmaceutical research is the discovery and development of new, efficacious and safe drugs for the treatment of disease. Rational drug-development is based on the understanding of the sites where drugs act and of the mechanisms involved in drug action. Sites of action for drugs are very diverse and include membrane receptors, enzymes, transport systems and others. Many physiological mechanisms can be modified as a consequence of drug action and drugs acting at different levels can lead to similar end-effects acting upon different pathways.

In many instances a knowledge of the anatomical structures and cell types involved in the action of a drug are required. This is particularly important in Neuropsychopharmacology, because the target organ, namely the brain, is particularly complex both from an anatomical and cellular point of view. Because many drugs are available labeled with radioisotopes the technique of autoradiography becomes a natural choice. Because quantitative comparisons are important, quantitative autoradiographic techniques are also necessary.

In recent years a number of such quantitative autoradiographic techniques for the study of brain function have been described. In some of them the label is administered in vivo while in others in vitro techniques are used. In vivo techniques are available today for the study of regional blood flow (Reivich et al., 1976), glucose utilization (Sokoloff et al., 1977) protein synthesis (Smith et al., 1980) receptor binding for hormones (Stumpf, 1971) and neurotransmitter receptors (Kuhar and Yamamura, 1975) lipid biosynthesis (Kimes et al., 1983) and phosphatidylinositol turnover (Anderson & Hollyfield,

1981). In vitro techniques (Young and Kuhar, 1979) have been
used for receptors (Kuhar, 1982; Wamsley and Palacios, 1983)
uptake sites (Rainbow et al., 1982), ionic channels (Murphy et
al., 1982), and neurotransmitters by immunohistochemical
techniques using ^{3}H-biotin (Hunt & Mantyh, 1984) or internally
labeled monoclonal antibodies (Cuello, Milstein and Palacios,
1984).

In this paper the use in drug research of two of such
techniques the in vivo 2-deoxyglucose technique of Sokoloff et
al., 1977 and the in vitro receptor autoradiography of (Young
& Kuhar, 1979) will be illustrated.

METHODOLOGY

In the experiment described in this paper we have used
the techniques of in vitro receptor autoradiography (Young and
Kuhar, 1979) and 2-^{14}-C-deoxyglucose (Sokoloff et al., 1977) as
modified in our laboratory or by others (Unnerstall et al., 1982
and Wamsley and Palacios, 1983, and Meibach et al., 1980, and
Palacios and Wiederhold, 1984), for the receptors and 2-deoxy-
glucose techniques, respectively.

Autoradiograms generated using ^{3}H-Ultrofilm (LKB, Sweden)
and Kodak SB5 films (Kodak, Lausanne, Switzerland) were
developed using manual and/or automatic procedures following
standardized conditions which result in reproducible optical
density levels. The latter are controlled using standards.

For the analysis and quantification of optical densities in
the autoradiograms a computer-assisted image analysis system
(similar to the ASBA system commercialized by Wild-Leitz,
Zürich, Switzerland, and developed in our laboratories by A.
Schweitzer and collaborators) was used (Palacios, 1983). This
is a system using a TV camera for digitalizing the autoradiographic picture into 256x256 picture units. The image is
displayed at the same time on the screen of a TV monitor,
where the areas or nuclei of interest are selected using a light
sensitive pen or a graphic tablet. After adjustment of the
working range of the densitometer with a series of grey filters
a calibration curve is generated using standards containing
known amounts of radioactivity. Grey or white matter standards, prepared as described by Unnerstall et al., 1980,
containing a non-volatile ^{3}H-labeled substance are used in
receptor autoradiography. The optical densities of these areas
were computed and transformed into receptor densities by
comparison with the optical densities of the standards, which
provide a factor transformation relating optical densities to the
receptor densities expressed in fmol/mg of protein.

Data from competition experiments are stored on a diskette and transferred to another computer where they are analyzed using programs for the analysis of binding data such as the one described by Tobler and Engel (1983). Fig.1 is an example of such an analysis.

In the 2-DG experiments the operation begins by reading the calibrated standards which allows transformation of optical densities into glucose equivalents, as described by Sokoloff et al. (1977). Because no blood values were determined for ^{14}C-2-DG and glucose, we measured only indices of 2-DG uptake (GU). As a reference value, to correct for inter-individual variations as well as non-specific drug effects, we used the "mean brain GU" rather than white matter values. The "mean brain GU" is calculated by determining the mean GU from autoradiograms of the 6 coronal sections, at the levels used in reading the areas and nuclei of interest. Data from each experiment are stored on floppy disks until the end of the experiment. The computer is programmed to calculate the individual relative values, mean and standard deviation from each experimental group as well as to perform statistical comparisons with control values. Results are printed and a graphic "profile" is generated by the computer along with a list of the structures measured, mean control values, statistical analyis of the comparison of the two groups and as bars, corresponding to each region. The length of the bars is proportional to the per cent change with respect to control values and the width indicates the statistical significance of these comparisons (see Fig.3 for an example).

RECEPTOR AUTORADIOGRAPHY: LOCALIZING MULTIPLE RECEPTORS FOR DRUGS AND NEUROTRANSMITTERS

It has been known for some time that multiple receptors for some neurotransmitters exist. Due to the higher resolution of modern techniques and to the synthesis of new molecules with higher selectivity the number of receptor subtypes has multiplied in the recent past (Snyder and Goodman, 1980). This fact, although contested by some (Laduron, 1983), could represent an important starting point for the development of more selective drugs. Receptor subtype charaterization is however, hampered by i) the lack of completely selective drugs and ii) the unavailability of a tissue or organ enriched in only one given receptor subtype. We have tried to overcome some of these barriers by using quantitative receptor auto-radiography. Examples of the application of this technique to the localization of multiple receptor subtypes exist in the literature (Unnerstall et al., 1982; Goodman et al., 1980). Here we will illustrate this for the case of multiple muscarinic cholinergic receptors.

At least two different subpopulations of muscarinic cholinergic receptors appear to be present in the mammalian brain. Both agonists and antagonists can be used to differentiate these two subpopulations. The newly developed antagonist pirenzepine (Hammer et al., 1980) appears to label preferentially the M_1 subtype, whereas agonists such as oxotremorine or carbachol recognize M_2 receptors with a high affinity. Fig.1 illustrates the distribution of these subtypes on sections of the rat brain. Sections were incubated with ^{3}H–N–methylscopolamine (^{3}H–NMS) or ^{3}H–pirenzepine as described by Wamsley et al., (1981) and Yamamura et al., (1983) respectively.

In order to obtain better characterization of the sites labeled by ^{3}H–NMS complete competition curves were generated by incubating consecutive sagittal sections with 0.1nM ^{3}H–NMS and increasing concentrations of oxotremorine, carbachol, atropine and pirenzepine. The blockade of the binding of ^{3}H–NMS was measured microdensitometrically and the data analyzed using the program of Tobler and Engel (1983). Examples of such competition curves are presented on Fig.1. This figure illustrates the possibility of characterizing receptors pharmacologically in very small brain areas or nuclei. For example it can be seen how pirenzepine has a higher affinity for the sites labeled by ^{3}H–NMS in the CA1 field of the hippocampal formation as compared to the receptors in the superior colliculus in the midbrain. In contrast with pirenzepine, oxotremorine and carbachol have higher affinities for the midbrain area compared to the hippocampal formation.

These, and other studies (for example Goodman et al., 1980, for multiple opiate receptors, Unnerstall et al., 1982, for multiple benzodiazepine and Cortés et al., 1984, for multiple serotonin receptors) have clearly demonstrated the usefulness of the receptor autoradiographic technique in the investigation of new drugs and neurotransmitter recognition sites. In addition, these studies have suggested that certain brain areas are enriched in one or the other subtype. Further biochemical and physiological experiments could be performed which may eventually lead to the development of more specific pharmacological agents.

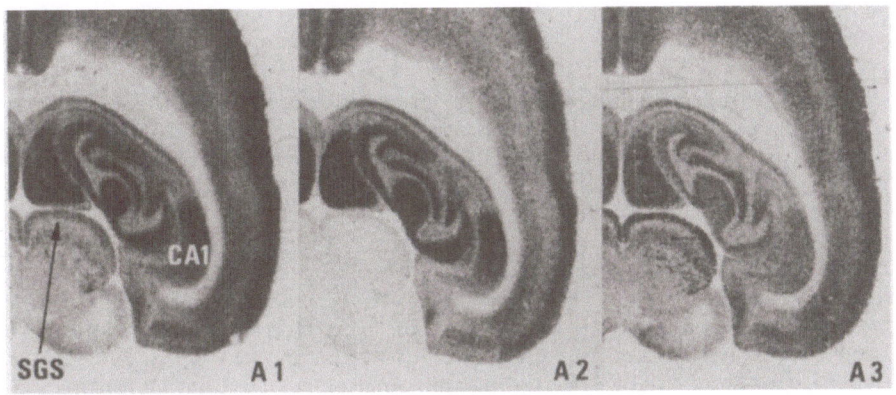

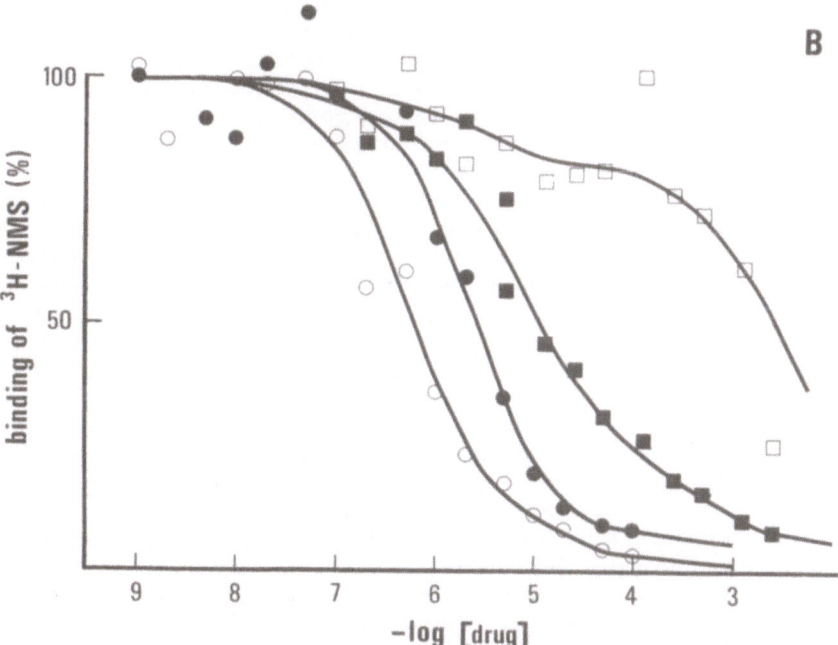

Figure 1 Autoradiographic localization and characterization of multiple muscarinic receptors in rat brain. A1-3 are photomicrographs from autoradiograms from rat brain horizontal sections incubated with ^{3}H-NMS alone (A1) or in the presence of 100µM carbachol (A2) or 300µM pirenzepine (A3). In B competition for ^{3}H-NMS binding by carbachol (□,CA1, ■,SGS) and pirenzepine (○,CA1, ●,SGS) obtained by autoradiography and analyzed as described by Tobler and Engel (1983). CA1 is for CA1 field of the hippocampus; SGS: striatum griseum superficialis of the superior colliculus.

"FINGER-PRINTING" DRUG ACTION IN BRAIN WITH THE 2-DEOXY-
GLUCOSE TECHNIQUE

The autoradiographic $2-^{14}C$-deoxyglucose (2DG) technique of
Sokoloff et al. (1977) is a powerful tool for the investigation of
brain function (Sokoloff, 1980). This technique is based on the
direct relationship between the functional activity and energy
consumption of any CNS region and the fact that the brain uses
glucose as its main energy source. Because centrally acting
drugs modify neuronal activity one could assume that drug
administration will be followed by modification of brain
metabolism and that the pattern of modification would indicate
the brain regions involved in the action of a given drug.
Descriptions of the effects of a number of drugs belonging to
different pharmacological classes are present in the literature
(for a recent review see McCulloch, 1982). These results have
clearly indicated the feasibility of using the 2DG technique in
drug research. However, only few drugs of each class have
been investigated which makes it difficult to establish
comparisons. When we started the experiments described below
we asked the following questions: 1) Does the administration of
different drugs belonging to a given pharmacological class
result in a similar pattern of alteration of regional brain
glucose uptake (GU)? 2) Is this pattern of alteration (what we
call "2DG-fingerprint") related to the pharmacological activity
of the drug, for example to its interaction with specific
receptors? and 3) Is the "2DG-fingerprint" specific for a class
of drugs or do the administration of different drugs acting on
different systems result in the same "fingerprint"?

Initially we selected the dopaminergic system because of
the availability of a large number of drugs acting at different
levels of the dopaminergic neuron and of a large body of
information on behavioral, biochemical and electrophysiological
effects of these drugs and because the 2DG technique has been
extensively used in the study of several parameters of the
dopaminergic system including lesion and drug effects (McCulloch, 1982). We looked at the effects of about 50 different
dopaminergic drugs. This included the precursor L-DOPA,
inhibitors of the biosynthesis of dopamine (DA), depletors of
endogenous DA stores, uptake inhibitors, pre- and postsynaptic
agonists, both D_1 and D_2, and postsynaptic antagonists,
typical and atypical neuroleptics. The dose dependency of the
effects was also investigated and in some cases the effects of
other pretreatments were also looked at.

Two general pattern of GU modification after administration
of dopaminergic drugs emerged from our results. The "neuroleptic 2DG fingerprint" (Fig.3 and 4) was generated not only by

the administration of dopamine receptor blockers (neuroleptics) but, also by other treatments leading to the interruption or attenuation of dopaminergic neurotransmission for example, by inhibition of DA biosynthesis by treatment with α-methyl-p-tyrosine, depletion of endogenous DA with reserpine and interestingly by the administration of postulated presynaptic agonists such as (-) 3PPP or the ergot derivative CF-25 397.

The main feature of the "neuroleptic 2DG finger-print" was a marked increase on the GU of the nucleus of the lateral habenula (Figs.2 and 3) a component of the epithalamus. Other characteristics were decreases in areas of the thalamus the subthalamic nucleus and neocortex and increases on GU in the nucleus accumbens and substantia nigra compacta.

This pattern of modification of GU utilization was seen after treatment with "typical" and "atypical" neuroleptics. It was dose dependent, presenting a maximal effect, which was lower for some atypical neuroleptics (clozapine, thioridazine and sulpiride). The rank-order of activity of the neuroleptics corresponded with that of these drugs acting on D_2-DA receptors (Seeman, 1981). The apparent involvement of D_2-DA on the modification of the pattern of brain GU is further supported by the lack of effects of administration of bulbocapnine, a relatively specific D_1-antagonist. The peripheral "neuroleptic" domperidone did not produce this finger-print indicating the involvement of central DA-receptors.

Interestingly other pharmacological manipulations which result in the blockade of dopaminergic neurotransmission also result in a "neuroleptic 2DG finger-print". Such is the case after inhibition of DA synthesis with α-methyl-p-tyrosine or DA-depletion with reserpine. More interestingly the administration of putative presynaptic agonists such as (-) 3PPP or the ergot derivative CF-25 397 also resulted in a "neuroleptic 2DG finger-print".

In contrast to the neuroleptics which tended to decrease the metabolic activity of many brain areas, dopaminergic agonists induced an increase of GU in a large number of brain structures. The main features of the "DA-agonist 2DG finger-print" are illustrated in Fig.4 and 5. Marked increases of GU are seen in: areas of the neocortex, some thalamic nuclei, the nucleus subthalamicus, a region in the lateral hypothalamus, probably a part of the medial forebrain bundle, and midbrain areas such as the deep mesencephalic nucleus, the red nucleus and the substantia nigra reticulata. The GU uptake in the lateral habenula was, however, decreased by many dopaminergic agonists.

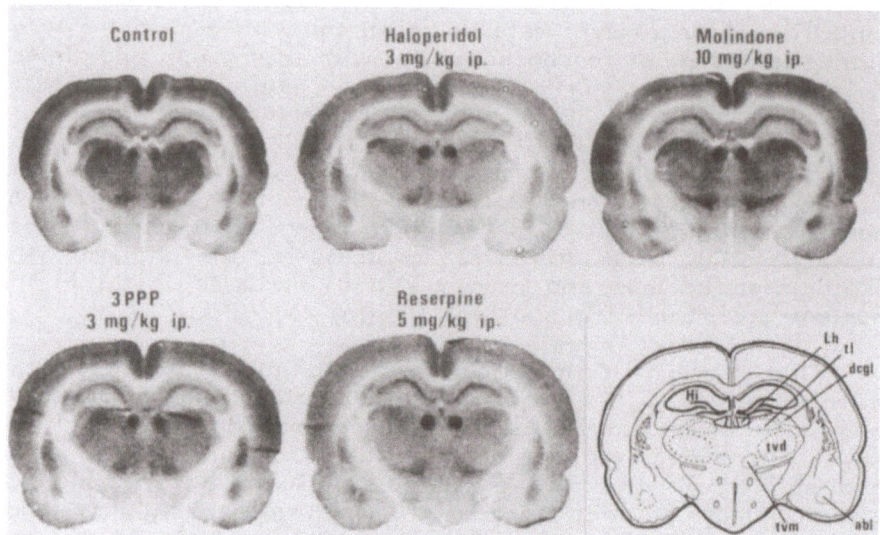

Figure 2 The "neuroleptic 2–deoxyglucose–fingerprint". The pictures are photomicrographs from 2-^{14}C–DG autoradiograms of coronal sections at a midthalamic level, from brains of control (vehicle injected) and rats treated with haloperidol (a "typical" neuroleptic) molindone (an "atypical" neuroleptic) (±) 3PPP (a "presynaptic" agonist) and reserpine (a "depletor" of endogenous catecholamines). The schematic drawing shows the main anatomical structures seen in the autoradiograms. Note the elevated optical density in the nucleus of the lateral habenula of the rats treated with the different drugs. Abbreviations are: (lh) lateral habenular nucleus, (dcgl) dorsal lateral geniculate nucleus, (tl) lateral thalamic nucleus, (tvd) ventroposterior thalamic nucleus, (tvm) ventromedial thalamic nucleus, (abl) basolateral amygdaloid nucleus, (Hi) hippocampus.

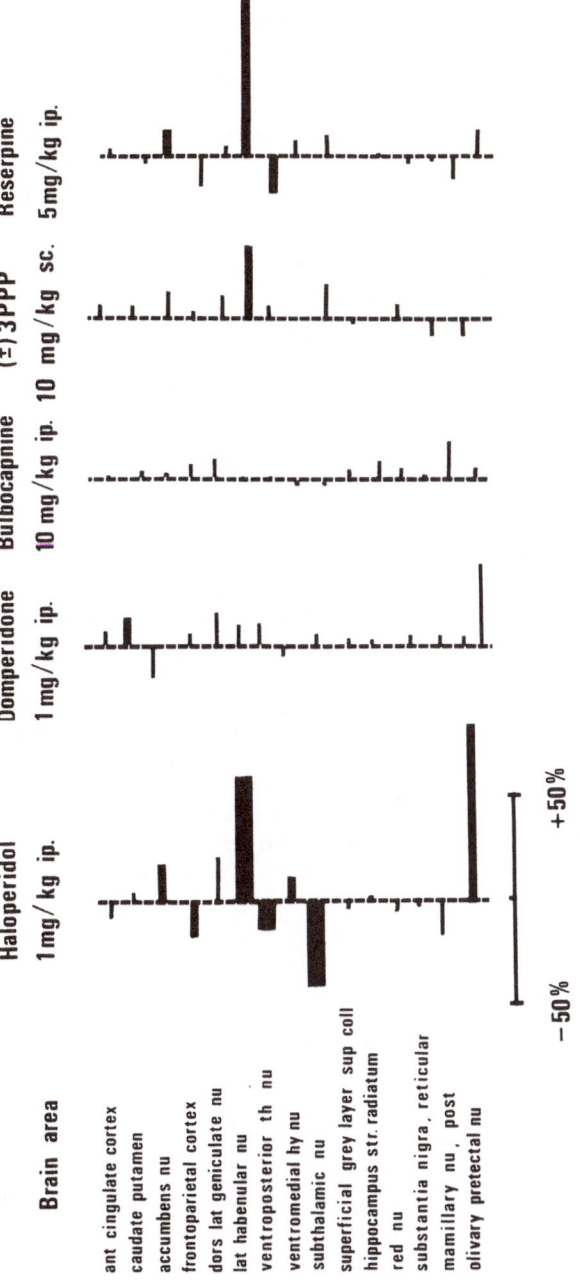

Figure 3 Semiquantitative determination of the effects of several dopaminergic blockers in 2DG uptake in several rat brain regions. Changes with respect to controls are represented by the length of the bar which is proportional to the % modification. The direction of the bar indicates increase or decrease with respect to control. The degree of statistical signification is indicated by the width of the bar. Note the lack of effects of domperidone, a peripheral DA-D_2-blocker and bulbocapnine, a relatively selective DA-D_1-antagonist.

As seen previously with the neuroleptics, the "DA–agonist 2DG fingerprint" presented the characteristics of a receptor mediated effect. Thus, the GU modifications were dose–dependent, they showed a maximal effect and the order of activity of the different agents tested correlate with that seen for these agents in other tests reflecting DA–receptor activation. The lack of effects of the DA–D_1 agonist SKF 38–393 (Fig.4 and 5) further supports the concept that glucose metabolism in brain is modified via DA–D_2 receptors but not DA–D_1, as also seen with the neuroleptics.

Treatment with indirectly acting agonists such as the DA uptake blocker nomifensin, the DA releaser amphetamine, or the precursor L–DOPA resulted in a "DA–agonist 2DG fingerprint" (Fig.4 and 5). Thus, drugs acting through different mechanisms that result in the same final effect, generate a similar "2DG fingerprint".

While other pharmacological systems have not been yet examined as extensively as the dopaminergic system preliminary results from our and other laboratories (McCulloch, 1982) indicate that drugs acting through other receptors generate different "2DG–fingerprints". These results support the utility of the 2DG technique in the study and characterization of centrally acting drugs.

In conclusion, quantitative autoradiographic techniques such as receptor autoradiography or the 2DG technique appear to be useful tools in the search for new drugs. The techniques are helpful in the investigation of the site(s) of action and the brain pathways activated as a consequence of their action in the brain. They provide valuable information about areas or physiological effects where a more specific effect of a drug is likely to be seen.

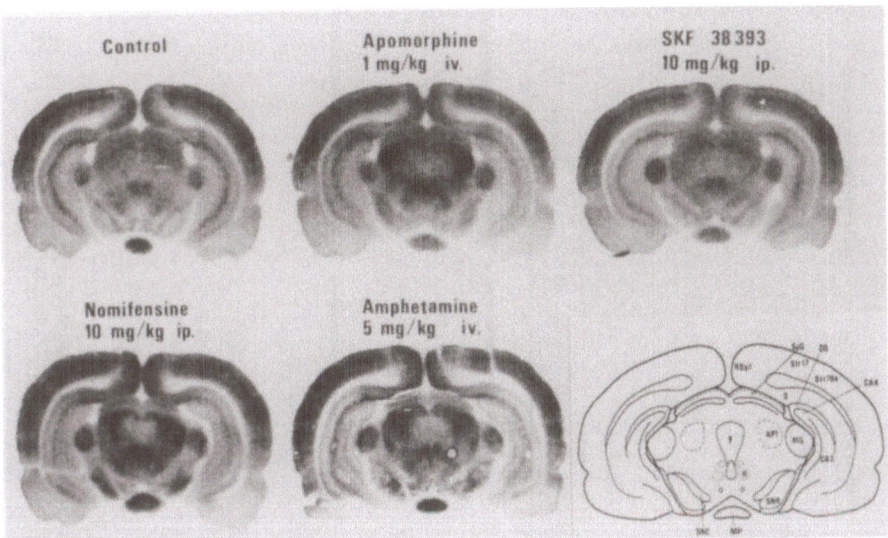

Figure 4 Effects of several dopaminergic agents on the 2DG uptake the "DA-agonist 2DG fingerprint". The pictures are photographs from autoradiograms from a midbrain coronal section from rats treated with vehicle (control) the DA-agonist apomorphine, the DA-D$_1$-agonist SKF 38-393; the DA-uptake inhibitor nomifensine and the DA-releasing drug amphetamine. Note the increased 2DG uptake in the anterior pretectal area (APT) and the substantia nigra pars reticulata (SNR) in the animals treated with apomorphine, nomifensine, and amphetamine but not SKF 38-393. Abbreviations in the schematic drawing are: (SuG) superficial grey layer sup. colliculus, (DG) dentate gyrus, (RSpl) retrosplenial cortex, (Str17) striate cortex, area 17, (Str18a) striate cortex, area 18a, (S) subiculum, (CA4) field CA4 of Ammon's horn, (ApT) anterior pretectal area, (MG) medial geniculate nucleus, (CA3) field CA3 of Ammon's horn, (R) red nucleus, (SNC) substantia nigra, compact, (MP) mammilary nucleus posterior.

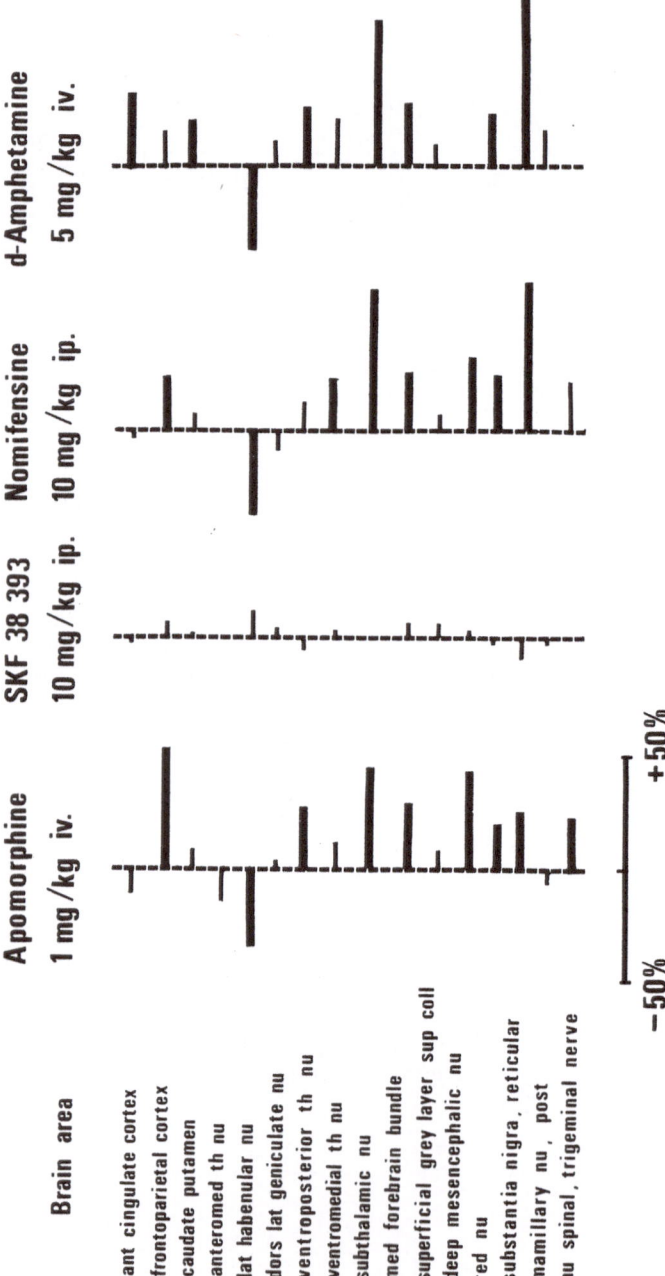

Figure 5 Schematic representation of the effects of the DA ergic drugs shown in Fig.4 in the GU in several regions. As in Fig.4, changes with respect to control are represented by the bars. Data were measured and calculated as described in "Methodology".

REFERENCES

Anderson, R.E. and Hollyfield, J.G. (1981) Light stimulates the incorporation of inositol into phosphatidylinositol in the retina. Biochem.Biophys. Acta, 665, 619–622.

Cortés, R., Palacios, J.M. and Pazos A., Visualization of multiple serotonin receptors in the rat brain by autoradiography. Brit.J.Pharmacol., in press.

Cuello, A.C., Milstein, C. and Palacios, J.M. (1984) ^{3}H–film radiohistochemistry of substance P using internally labeled monoclonal antibodies. (in preparation)

Goodman, R.R., Snyder, S.H., Kuhar, M.J. and Young, W.S. (1980) Differentiation of delta and mu opiate receptor localizations by light microscopic autoradiography. Proc.Natl.Acad. Sci. (USA) 77, 6239–6243.

Hammer, R., Berrie, C.P., Birdsall, N.J.M., Burgen, A.S.V. and Hulme, E.C. (1980) Pirenzepine distinguishes between different subclasses of muscarinic receptors. Nature 283, 90–92.

Hunt, S.P. and Mantyh, P.W. (1984) Radioimmunocytochemistry with [3H]Biotin. Brain Research, 291 203–217.

Kimes, A.S., Sweeney, D., London, E.D. and Rapoport, S.I. (1983) Palmitate incorporation into different brain regions in the awake rat. Brain Research, 274, 291–301.

Kuhar, M.J. (1982) Localization of drug and neurotransmitter receptors in brain by light microscopic autoradiography, in: Handbook of Psychopharmacology, Vol.15, edited by L.L. Iversen, S.D. Iversen and S.H. Snyder. Plenum Press, New York, pp. 299–320.

Laduron, P. (1983) More binding, more fancy. Trends in Pharmacol.Sci. 4, 333–335.

McCulloch, J. (1982) Mapping functional alterations in the CNS with (^{14}C)Deoxyglucose, in: Handbook of Psychopharmacology, Vol.15, (edited by L.L. Inversen, S.D. Iversen and S.H. Snyder) Plenum Press, New York, pp. 321–410.

Meibach, R.C., Glick, S.D., Ross, D.A., Cox, R. and Maayani, S. (1980) Intraperitoneal administration and other modifications of the 2–deoxy–d–glucose technique, Brain Res., 195, 167–176.

Murphy, K.M.M., Gould, R.J. and Snyder, S.H. (1982) Autoradiographic visualization of [^{3}H]nitrendipine binding sites in rat brain: localization to synaptic zone, Europ.J. Pharmacol, 81, 517-519.

Palacios, J.M. (1983) Quantitative receptor autoradiography: application to the study of multiple serotonin receptors in rat cortex, in: CNS Receptors - From Molecular Pharmacology to Behavior (edited by P. Mandel and F.V. DeFeudis) Raven Press, New York, pp. 455-463.

Palacios, J.M. and Wiederhold, K.H. (1984) Acute administration of 1-N-methyl-4-phenyl-1,2,3,6-tetrahydropyridine(MPTP), a compound producing parkinsonism in humans, stimulates [2-^{14}C]deoxyglucose uptake in the regions of the catecholaminergic cell bodies in the rat and guinea pig brains. Brain Res., in press.

Rainbow, Th.C., Biegon A. and McEwen, B.S. (1982) Autoradiographic localization of imipramine binding in rat brain. Eur.J.Pharmacol. 77, 363-364.

Seeman, P. (1981) Brain Dopamine Receptors, Pharmacol.Rev., 32, 229-313.

Smith, C.B., Davidson, L., Deibler, G., Patlak, C., Pettigrew, C. and Sokoloff, L. (1980) A method for the determination of local rates of protein synthesis, Trans.Amer. Soc.Neurochem. 11, 94.

Snyder, S.H. and Goodman, R.R. (1980) Multiple neurotransmitter receptors. J.Neurochem. 35, 5-15.

Sokoloff, L., Reivich, M., Kennedy, C., Des Rosiers, M.H., Patlak, C.S., Pettigrew, K.D., Sakurada, C. and Shinohara, M. (1977) The (^{14}C) deoxyglucose method for the measurement of local cerebral glucose utilization: theory, procedure and normal values in the conscious and anesthetized albino rat, J.Neurochem. 28, 13-36.

Sokoloff, L. (1980) The relationship between function and energy metabolism: its use in the localization of functional activity in the nervous system, Neurosci.Res.Prog.Bull. 19, 159-210.

Stumpf, W.E. (1971) Autoradiography techniques and the localization of estrogen, androgen and glucocorticoid in the pituitary and brain, Amer.Zool. 11, 725-739.

Tobler, H.J. and Engel, G. (1983) Affinity spectra: a novel way for the evaluation of equilibrium binding experiments. Naunyn-Schmiedeberg's Arch.Pharmacol. 322, 183-192.

Unnerstall, J.R., Niehoff, D.L., Kuhar, M.J. and Palacios, J.M. (1982) Quantitative receptor autoradiography using [^{3}H]Ultrofilm: application to multiple benzodiazepine receptors, J.Neurosci. Methods 6, 59-73.

Wamsley, J.K. and Palacios, J.M. (1983) Apposition techniques of autoradiography for microscopic receptor localization, in: Current methods in cellular neurobiology, Vol.I, edited by J.L. Barker and J.F. McKelvy. John Wiley & Sons, New York, pp. 241-268.

Wamsley, J.K. Zarbin, M.A., Birdall, N.J.M. and Kuhar, M.J. (1980) Muscarinic cholinergic receptors: autoradiographic localization of high and low affinity agonist binding sites. Brain Res. 200, 1-12.

Yamamura, H.I., Wamsley, J.K., Deshmukh, P. and Roeske, W.R. (1983) Differential light microscopic autoradiographic localization of muscarinic cholinergic receptors in the brainstem and spinal cord of the rat using ^{3}H-pirenzepine, Eur.J.Pharmac. 91, 147-149.

Young, W.S. and Kuhar, M.J. (1979) Autoradiographic localization of benzodiazepine receptors in the brain of humans and animals. Nature 280, 393-395.

QUANTITATIVE AUTORADIOGRAPHY AS A TOOL TO STUDY RECEPTORS IN NEURAL TISSUE. STUDIES ON 3H–OUABAIN BINDING SITES AND CORRELATION WITH SYNAPTIC PROTEIN PHOSPHORYLATION IN DIFFERENT BRAIN AREAS

FABIO BENFENATI[1], LUIGI F. AGNATI[1], KJELL FUXE[2],
MARLO CIMINO[1], NINO BATTISTINI[1], EMILIO MERLO PICH[1],
CONSTANZA FARABEGOLI[1] and ISABELLA ZINI[1]

[1]Department of Human Physiology, University of Modena, Modena, Italy
[2]Department of Histology, Karolinska Institutet, Stockholm, Sweden

INTRODUCTION

During recent years quantitative receptor autoradiography has been extensively used in studies on the distribution of transmitter receptors in the central nervous system (see Unnerstall et al. 1981, 1982; Palacios et al. 1981; Penney et al. 1981; Rainbow et al. 1982; Fuxe et al. 1983). Quantitative studies have been performed both by counting silver grains on nuclear emulsion and by computerized densitometry with tritium sensitive film (Unnerstall et al. 1981, Palacios et al. 1981). Saturation analysis can also be performed by quantitative receptor autoradiography as first shown by Penney et al. (1981), who determined the Kd and B_{max} values for $3H$-muscimol binding sites in brain sections using $14C$-plastic standards, which had been calibrated to tritium. The problem in quantitative receptor autoradiography has been to obtain suitable standards in view of the quenching, taking place in brain tissue and which is different in white matter and grey matter due to differences in self absorption (Alexander et al. 1981; Geary and Wooten, 1983). Previously, both tritium containing tissue paste standards and tritium containing plastic standards have been used. The $3H$-plastic standards have been calibrated by coexposing plastic squares, containing a range of tritium concentrations, and hemisections from tritium containing brains. The tissue radioactivity in the contralateral half was then determined in various regions by scintillation spectroscopy.

In the present paper we have further improved these procedures by introducing a new form of biological standard and a new logistic functional model to establish the standard curve. In addition, the distribution of $3H$-Ouabain binding sites has for the first time been mapped out using quantitative receptor autoradiography. Ouabain binds to the $Na^+ K^+$ sensitive ATPas in the nerve cell membranes, and therefore $3H$-Ouabain marks areas of the nerve cell membranes, in which action potentials are generated via changes in the permeability for sodium and potassium ions. The regional distribution of the $3H$-Ouabain binding sites has been correlated

with the regional distribution of cAMP and Ca^{2+} induced protein
phosphorylation, which takes place at the synaptic level, in order
to map out which neuronal networks are dominated by action potential induced release of transmitters and comodulators, and which
are dominated by electrotonically induced release of such compounds
(local circuit interactions).

METHODOLOGICAL ASPECTS

Quantitation of ^{3}H-radioligand binding sites in brain sections

     In order to quantify receptor autoradiograms performed with
LKB 3H-Ultrofilm, the first and most important step is the definition of the film response curve to increasing amounts of radioactivity. This response, which is not linear, links the entity of
blackening in the film to the real amount of radioligand bound to
the section. However, since tissue exerts a significant quenching
of tritium derived beta particles, it is important to determine the
emulsion response, using standards quenched by neural tissue in a

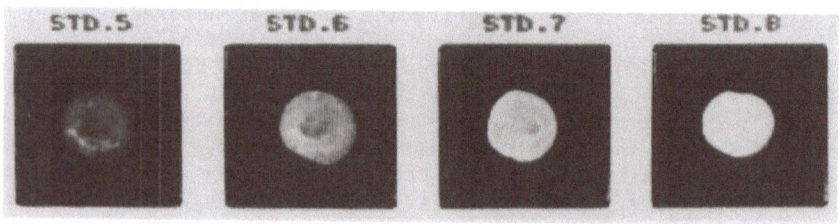

Fig. 1. A. Example of quenched standard emulsion spots, obtained
after an exposure of 15 days (3H-Ultrofilm) to different amounts
(85, 130, 420 and 730 dpm/sq.mm, respectively) of tritiated
Ouabain.
B. Example of an image analyzer printout concerning the density
measurement in a sampled area outlined by means of a light pen
device. On the left the tone histogram is shown and on the right
the mean grey value is given together with other statistical
parameters.

way which as closely as possible mimics those of the real autoradiogram. To this aim, 2 ul drops of different solutions containing a nonvolatile tritiated compound (3H-Spiperone; a D2 DA receptor blocking agent or 3H-Ouabain) in water (activity range: 10-
12000 dpm/sq.mm) were put onto gelatinized glass slides, dried and
covered by 14 µm frozen brain sections, allowing a complete diffusion of the isotope within the melting tissue. Standards were
exposed to 3H-Ultrofilm for different time periods according to
those needed for receptor autoradiography. After development of the
film (Gevaert G150 developer), the optical density of standard
emulsion spots as well as of any other autoradiographic areas was
measured using the digital image analyzer Tesak VDC 501 (Agnati et
al. 1984). The Tesak system codifies density values on a digital
scale ranging from 0 (full absorption) to 255 (full transmittance).
For a given area measurement, it gives the mean grey value together
with the relative tone distribution histogram (fig. 1). After the
measurement of the sampled area (A) and of the relative background

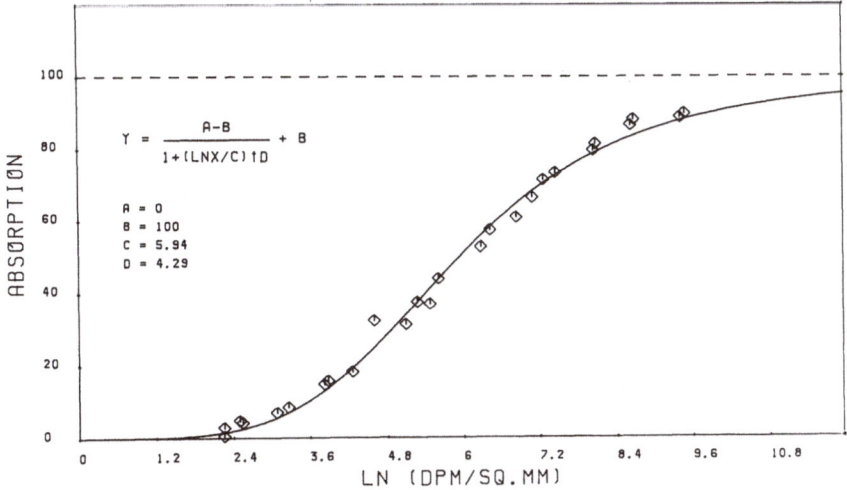

Fig. 2. 3H-Ultrofilm response curve to varying 3H-Ouabain standard
concentrations (range 10-12000 dpm/sq.mm) after 15 days of exposure. On the X axis the natural logarithm of the radioactivity
(dpm/square mm) is given. On the Y axis the respective absorption
values are reported, calculated following digital image analysis as
described in text. On the left the general form of the logistic
function used to fit the experimental points is shown together with
the final best fit parameters.

(B), the final absolute grey value was expressed as follows:
Absorption (%) = 100 - (A/B * 100)
i.e. 100 minus the ratio between sampled area and background
transmittances. This value was demonstrated to be the most reliable
density parameter in our system. In the analysis of the standard
curve, absorption values were plotted on the Y axis versus the
natural logarithm of the respective standard radioactivity (expressed in dpm/sq.mm). The calibration curve was fitted by means of an
iterative least-square non-linear regression analysis according to
the following logistic functional model:

$$Y = \frac{A - B}{1 + (\ln X/C)\ D} + B$$

where Y is the % Absorption, X the source activity and the parameters A,B,C,D are the lower plateau value, the upper plateau
value, the ED50 value and the slope factor value, respectively
(fig. 2). Once this function has been defined, the absorption

FILM RESPØNSE CURVE

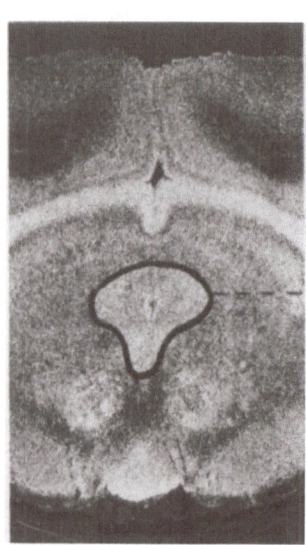
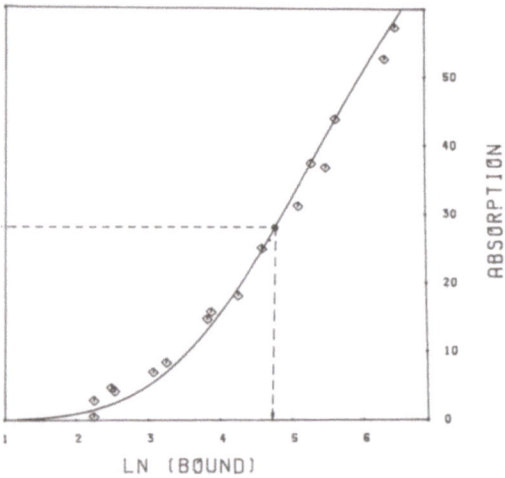

Fig. 3. Example of the interpolation of an absorption value,
obtained from an area sampled out from a receptor autoradiogram
(panel on the left: 3H-Ouabain binding at mesencephalic level;
sampled area: periaqueductal grey) in the quenched standard curve
(given on the right; for further details, see legend to fig. 4).
From one absorption value, the corresponding bound value (expressed
in dpm/sq.mm) is obtained. Notice the high labelling of the outer
layers of the superior collicle and of the interpeduncular nucleus.

values (calculated from receptor-positive and background areas in the autoradiograms by means of computerized densitometry) can then be interpolated in the quenched standard curve, obtained under the same experimental conditions, achieving the corresponding bound values expressed in fmol/sq.mm (fig. 3).

Obviously, the interpolation with non-linear functions intro-duces a potential error in the transformation of normally distri-buted observations, which can be particularly marked in the non-rectilinear parts of the logistic function. In order to assess this phenomenon, we performed a Montecarlo simulation by introducing, for various absorption levels, random errors in grey value measure-ments and analyzing variance in the transformed data. From these studies, it was apparent that variance and percent variance (i.e. variance divided by the respective mean value of the replications) are very low in an absorption range 10-40% (working range), whereas the calculated bound levels are less reliable when absorption levels are considered outside this range (fig. 4).

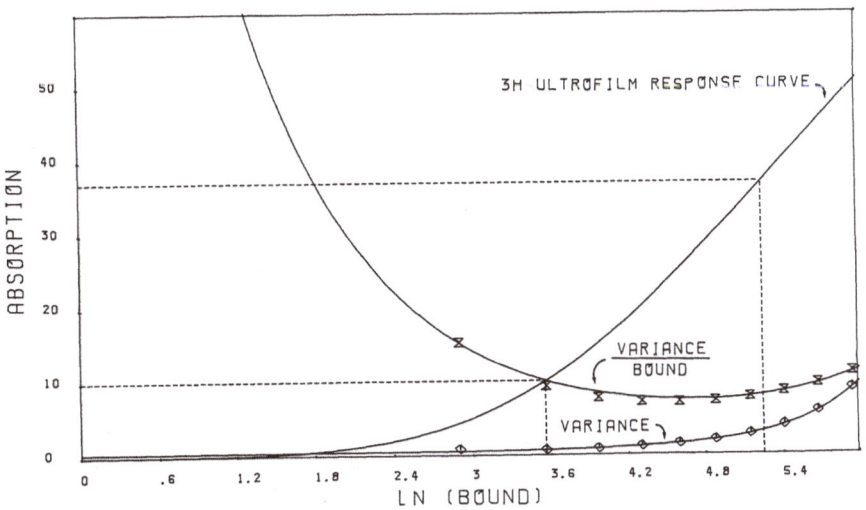

Fig. 4. Changes of variance (square symbols) and percent variance (variance/bound, double triangles) for different replications simulated with random errors and performed at various absorption levels. The 3H-Ultrofilm response curve is also shown: The part included between the two broken lines represents the working range (for more details, see text). On the X axis, the natural logarithm of bound values (in dpm/sq.mm) is given. On the Y axis absorption values are reported. Variance and percent variance are expressed in absolute values (scales not shown).

When a saturation study is performed, involving a separate interpolation of total and nonspecific binding measurements for the different ligand concentrations used, it is possible to calculate the binding parameters Kd (in nM) and B_{max} (in dpm/sq.mm). The latter value can be further changed to fmol/sq.mm or even to fmol/mg of protein, since protein contents of the sections in the standard curve can easily be determined. The values obtained, e.g., for D2 dopamine receptors, labeled by ^{3}H-spiperone in rat striatum, are in full agreement with those obtained in membrane preparations from the striatum using standard radioligand binding techniques (see Creese et al. 1977) (fig. 5).

DISTRIBUTION OF ^{3}H-OUABAIN BINDING SITES IN THE CNS AND ITS RELATIONSHIP TO SYNAPTIC PROTEIN PHOSPHORYLATION

Material and methods

Normal male Spraque-Dawley rats, kept under standard lighting

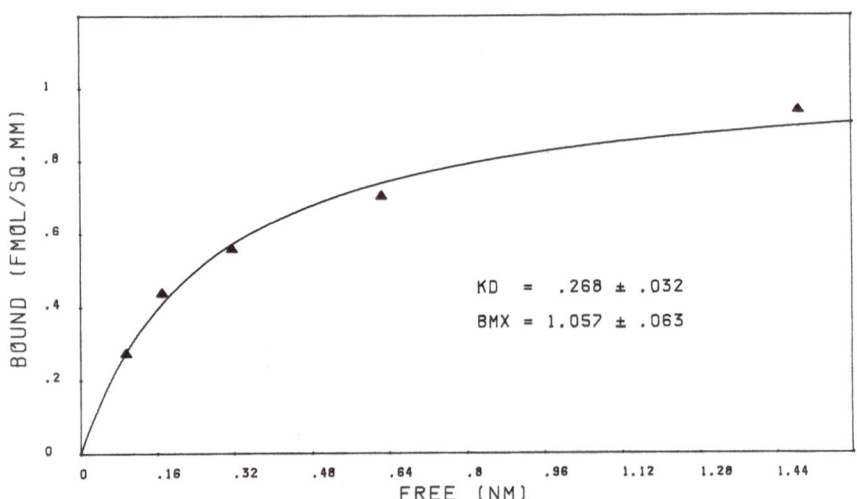

Fig. 5. Non linear fitting analysis of specific binding isotherms of 3H-Spiperone to striatal dopamine receptors. Free ligand concentrations (in nM) used in the assay and specific bound values (in fmoles (sq.mm) are reported on the X axis and on the Y axis, respectively. Final binding parameters (means ± s.e.m.) are shown.

conditions (lights on at 6 a.m. and off at 8 p.m.) were used throughout the study.

^{3}H-Ouabain was used as a tool to visualize the distribution of Na$^+$, K$^+$-ATPase in the CNS. The analysis was carried out at different CNS levels from the frontal pole to the spinal cord on 14 μm cryostat sections of rats perfused with 0.1% paraformaldehyde in PBS. ^{3}H-Ouabain binding was performed following the procedure of Silbergeld et al. (1982) with slight modifications. Briefly, sections were incubated at room temperature for 2 hours with 50 nM ^{3}H-Ouabain (NEN, Spec. Activity, 18 Ci/mmol) in Tris-HCl buffer (50mM, pH 7.5) containing 100 mM NaCl, 2 mM MgCl2 and 2 mM ATP. Nonspecific binding was defined as the binding occurring in presence of 100 μM cold Ouabain. After incubation, the sections were washed twice in ice-cold Tris buffer for 4 minutes, rinsed in distilled water and dried with a stream of cold and dry air. The sections were then transferred to X-ray film cassettes together with standard curve samples and exposed to 3H-Ultrofilm at -80°C for 15 days. Silica gel containing bags were present to ensure a dry atmosphere in the cassettes. In the evaluation of the ^{3}H-Ouabain distribution in the sections, neuroanatomical structures were identified and characterized by means of Nissl staining performed on the same section after exposure or on adjacent 14 μm thick sections. Grey value measurements were then carried out as mentioned above.

For the protein phosphorylation assay, the rats were killed by decapitation, and brain areas rapidly dissected out and homogenized. The P2 fraction, obtained following the procedure of De Robertis et al. (1967), was lysed in 50 mM HEPES buffer (pH 7.4 containing EGTA 0.1 mM, MgCl$_2$ 5 mM, DTT mM). The final protein concentration was 1 mg/ml. When cyclic nucleotides were included in the incubation mixture a phosphodiesterase inhibitor (IBMX 0.5 mM) was also added. The reaction mixture (final volume 120 μl) contained a 100 μg protein suspension plus cyclic AMP (cAMP, 0.1-1000 μM) or CaCl$_2$ (1.1 mM). The samples were preincubated at 30°C for 2 min, and the phosphorylation reaction was started by the addition of 32P-γ-ATP (4-5 μCi/sample, 3 μM final concentration) following the procedure of Greengard and colleagues Ueda et al. 1973; Schulman and Greengard 1978). After incubation at 30°C for 10 sec, the reaction was stopped by the addition of an equal volume of a stop solution (containing 4% SDS, 10% 2-mercaptoethanol, 20% glycerol, 0.002% bromophenol blue in a 0.12 mM Tris-HCl pH 6.8) and the samples were boiled for 2 min. The proteins were separated by one-dimensional SDS/polyacrylamide gel electrophoresis according to the method of Laemmli (1970). The gels were stained, destained, dried and subjected to autoradiography.

The intensity of the darkness of the bands in the film was measured by a Joyce-Loebl scanning microdensitometer.

F. Benfenati, et al.

Fig. 6. Upper part: 3H-Ouabain receptor autoradiography performed in rat brain at level A8380 (according to the Koenig and Klippel Atlas 1974). Lower part: Autoradiography of phosphorylated proteins from striatal synaptosomal fractions after separation by SDS-polyacrylamide gel electrophoresis. The phosphorylation is shown under basal conditions and after stimulation with cyclic AMP (5 μM) and CaCl2 (1.1 mM). The molecular weights of the phosphorylated proteins are shown in thousands of daltons. The proteins phosphorylation pattern in the dorsal cerebral cortex was closely similar to that found in striatum.

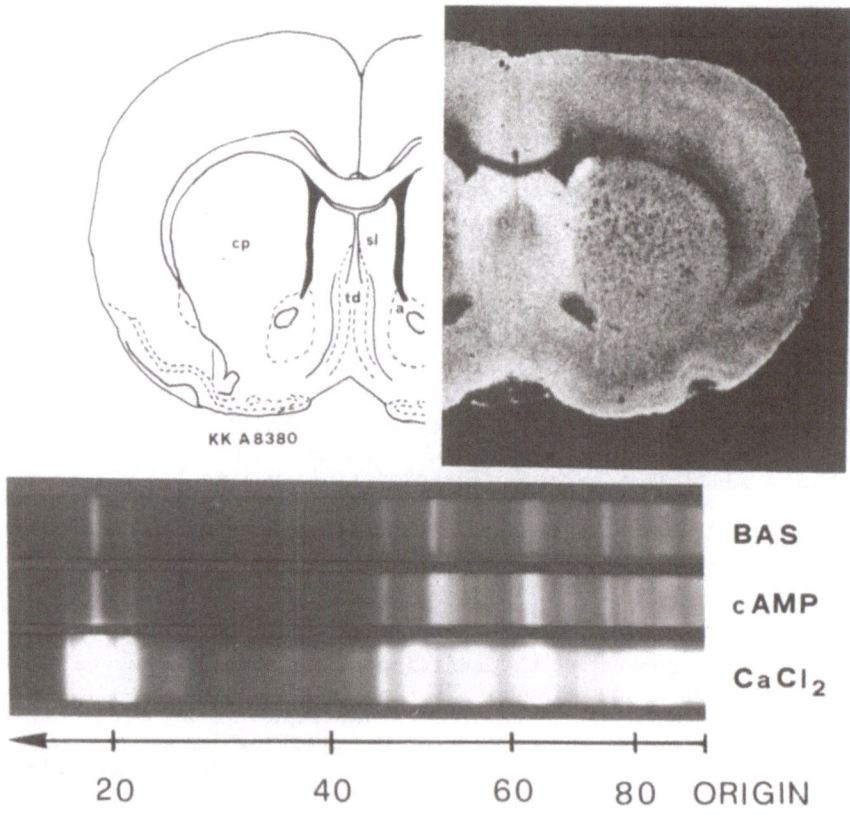

Fig. 7. Upper part: 3H-Ouabain receptor autoradiography performed at level A 2580 (Koenig and Klippel 1974). Autoradiography of phosphorylated proteins from synaptosomal fractions of the rat hippocampus. For further details, see legend to fig. 6.

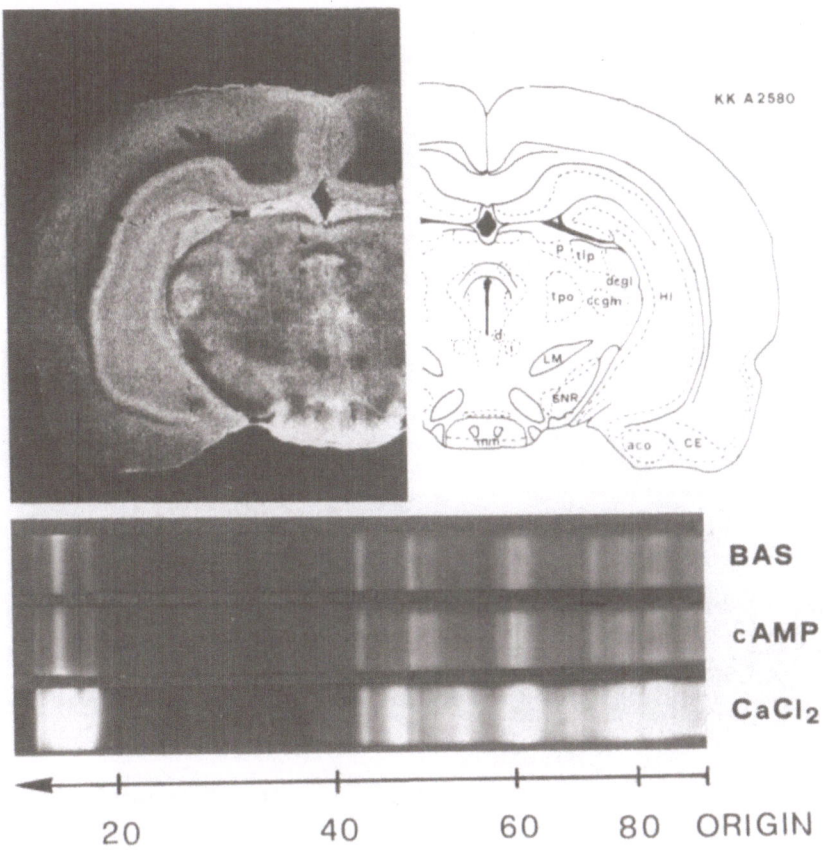

F. Benfenati, et al.

Fig. 8. Upper part: 3H-Ouabain receptor autoradiography performed at level B -10.3 (according to Pellegrino et al. (1979). Lower part: Autoradiography of phosphorylated proteins carried out in synaptosomal fractions prepared from rat cerebellum. For further details, see legend to fig. 6.

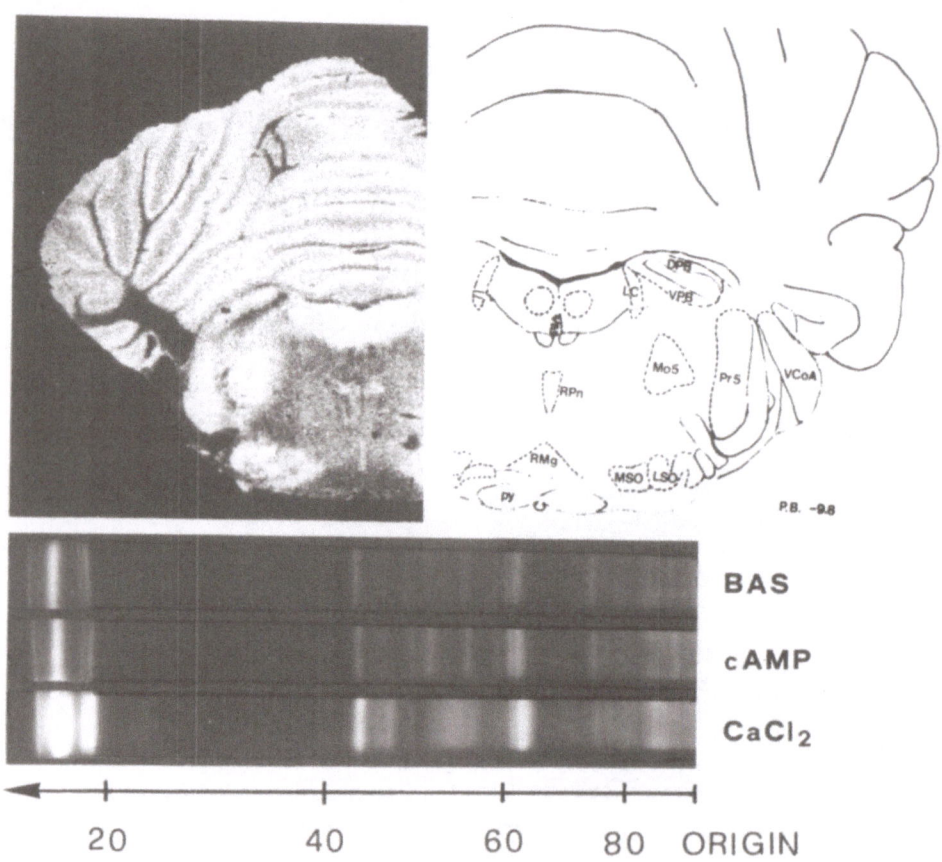

Fig. 9. Upper part: 3H-Ouabain receptor autoradiography in rat
thoracic (left) and cervical (right) spinal cord level in the
normal rat. Lower part: Autoradiography of phosphorylated proteins
from spinal cord synaptosomal fractions. For further details, see
legend to fig. 6.

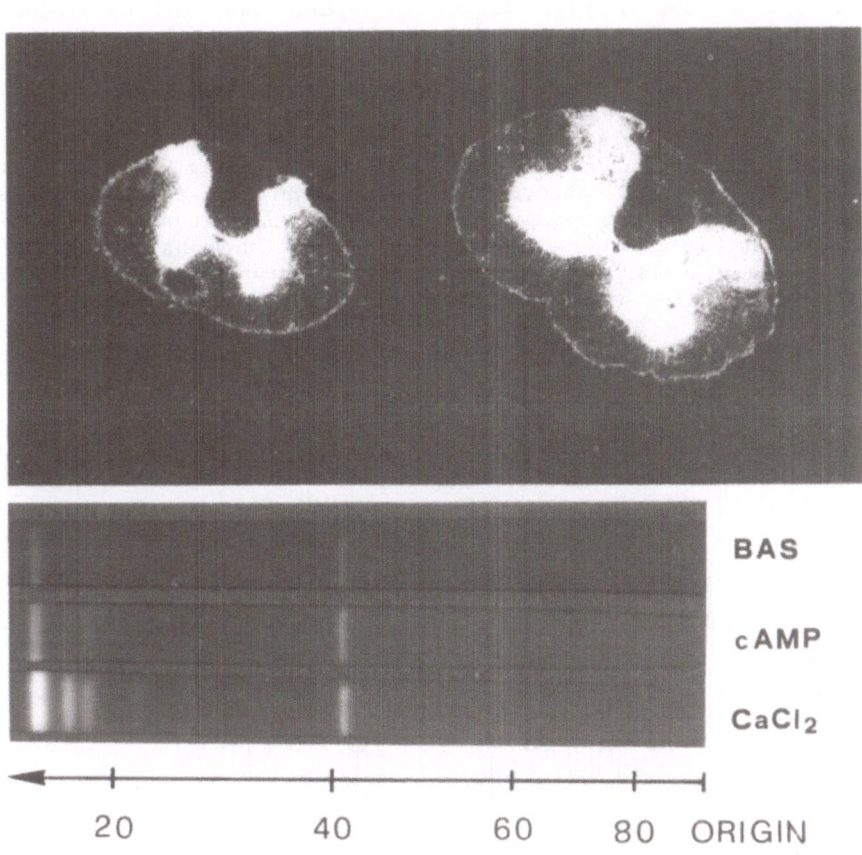

Results and discussion

The results obtained with ^{3}H-Ouabain receptor autoradiography and on the cyclic AMP and Ca^{++}-induced protein phosphorylation obtained in the telencephalon, in the midbrain, in the cerebellum, and in the spinal cord are illustrated in figs. 6-9. The highest degree of cyclic AMP and Ca$^+$-induced protein phosphorylation in the areas analyzed was found in striatal synaptosomal preparations (see fig. 6). A similar high degree of cyclic AMP and Ca$^+$-induced protein phosphorylation was observed in synaptosomal preparations from the dorsal cerebral cortex (mainly frontoparietal cortex). As seen in fig. 6, both these brain areas showed only a weak labelling by ^{3}H-Ouabain. Instead, a strong labelling was observed in the lateral septal nucleus (sl) and the anterior cingulate cortex following incubation with ^{3}H-Ouabain (50 nM). In fig. 7 it is shown that Ca$^+$-induced but not cyclic AMP-induced protein phosphorylation is also fairly strong within synaptosomal preparations from the rat hippocampus. It is also shown in fig. 7 that ^{3}H-Ouabain moderately labels the granular layer of the dentate gyrus and the pyramidal layer of the gyrus hippocampi together with adjacent parts of the

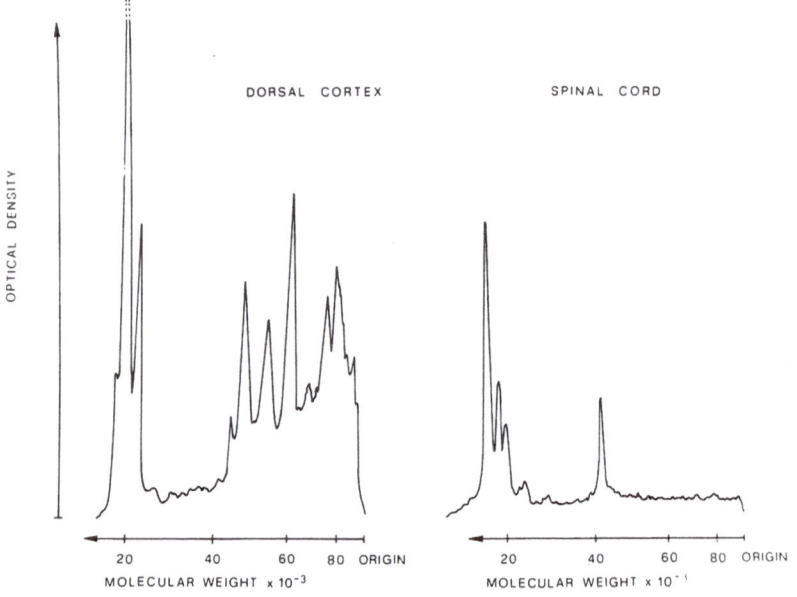

Fig. 10. Calcium-stimulated protein phosphorylation in synaptosomal fractions of dorsal cerebral cortex (mainly frontoparietal cortex) and spinal cord. The results of the densitometric analysis of phosphoprotein bands are shown. On the X-axis the protein molecular weight in thousands of daltons is given. On the Y-axis the optical density of the autoradiographic band is reported in arbitrary units.

overlying molecular and oriens layer, respectively. ^{3}H-Ouabain strongly labels the most superficial layer of the superior collicle (most anterior part) (fig. 7). In addition, ^{3}H-Ouabain moderately labels the periventricular grey, parts of the lateral geniculate body and the subiculum region, while the remaining part of the cerebral hemisphere with the entorhinal cortex and the amygdaloid cortex is very weakly labeled. In synaptosomal preparations of the cerebellum only a moderate degree of Ca^+-induced protein phosphorylation is observed (fig. 8). Cyclic AMP-induced protein phosphorylation cannot be very clearly observed, except in a few bands corresponding to a molecular weight of 60.000. In contrast, ^{3}H-Ouabain strongly labels the entire molecular layer and the adjacent Purkinje cell layer of the cerebellar cortex, and the granular layer is very weakly labeled, if at all (fig. 8). ^{3}H-Ouabain also strongly labels the entire periventricular grey of the pons, the motor nucleus of the trigeminal nerve (nucleus motorius nervi trigemini), the motor nucleus of the facial nerve, the parabrachial nucleus (lateral and medial part) and the ventral cohlear nucleus. As shown in fig. 9, ^{3}H-Ouabain very strongly labels the entire grey matter of the spinal cord. Instead, as also seen in fig. 9, Ca^+-induced protein phosphorylation is very weak and cyclic AMP-induced protein phosphorylation cannot safely be demonstrated in these experiments. The profound difference in Ca^+-induced protein phosphorylation in synaptosomal fractions of the cerebral cortex and the spinal cord is illustrated in fig. 10. Thus, within the dorsal cerebral cortex a large number of phosphoproteins are observed, and they are also more strongly labelled than those found in synaptosomal preparations of the spinal cord. In fig. 11 an inverse correlation is demonstrated between ^{3}H-Ouabain binding and Ca^+-induced protein phosphorylation based on the analysis of five different regions of the central nervous systems.

The present studies on the distribution of the ^{3}H-Ouabain binding sites in the central nervous system in relation to the cyclic AMP and Ca^+-induced protein phosphorylation in various brain areas are of substantial interest, since they indicate that in the phylogenetically old parts of the brain, most brain regions have large numbers of ^{3}H-Ouabain binding sites and a relatively low degree of cyclic AMP and Ca^+-induced protein phosphorylation. Instead the opposite is true for the phylogenetically recent parts of the brain (striatum and especially cerebral cortex). Thus, it seems possible that the areas of the spinal cord and the brain stem mainly operate via action potential induced release of transmitters and comodulators, since the presence of ^{3}H-Ouabain binding sites should label those domains of the nerve cell membranes, where sodium and potassium channels exist. It is wellknown that the opening and closure of such channels underly the propagation of the action potential. Instead, within the striatum and within the cortical areas, especially the neocortex, calcium ions and cyclic AMP induce a powerful phosphorylation of large numbers of proteins, while ^{3}H-Ouabain only weakly labels these areas. Thus, it seems reasonable that the higher, phylogenetically recent centers of the

F. Benfenati, et al.

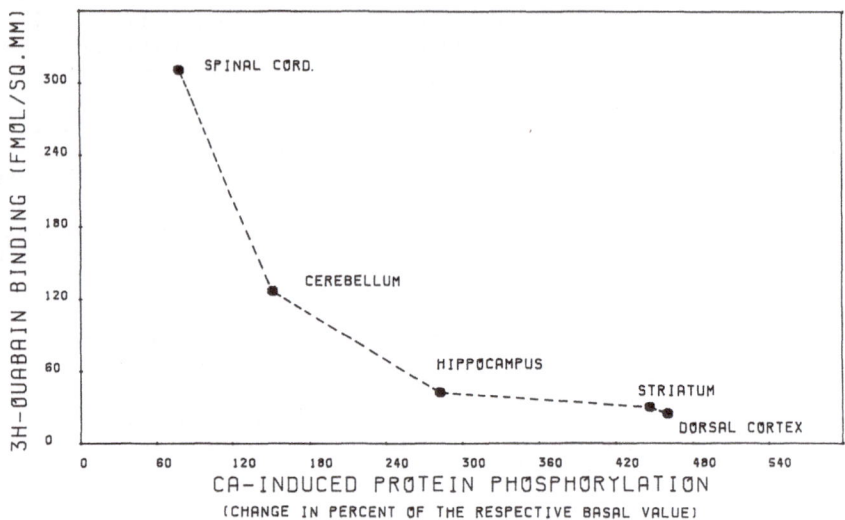

Fig. 11. Correlation between calcium-induced protein phosphorylation (shown on the X-axis as percent changes with respect to the
basal value) and 3H-Ouabain binding (expressed in fmoles/sq.mm on
the Y-axis) for the different CNS areas investigated.

brain mainly perform their integrative activity via electrotonically induced release of transmitters and comodulators in the
local circuits of these centers. Thus, in order to perform more
sophisticated functions, the brain has developed a number of
proteins which can be phosphorylated on synaptic activation, These
phosphoproteins can control both ion permeability and metabolic
functions. They may in this way markedly enhance information
storage and processing (see Nestler and Greengard, 1984).

ACKNOWLEDGEMENTS

    This work has been supported by a grant (MH25504-10) from the
National Institute of Health and by a CNR International grant. For
excellent technical assistance we are grateful to Mrs. B. Tinner
and for excellent secreterial assistance to Mrs. A. Edgren, Mrs.
Ulla-Britt Wedin and Mrs. G. Castellano.

REFERENCES

Agnati, L.F., Fuxe, K., Benfenati, F., Zini, I., Zoli, M., Fabbri, L. and Härfstrand, A. (1984). Computer assisted morphometry and microdensitometry of transmitter identified neurons with special reference to the mesostriatal dopamine pathway. I. Methodological aspects. Acta Physiol. Scand., Suppl. 532, 5-36.

Alexander, G.M., Schwartzman, R.J., Bell, R.D., Yu, J. and Renthal, A. (1981). Quantitative measurement of local cerebral metabolic rate for glucose using tritiated 2-deoxyglucose. Brain Res., 223, 59-67.

Creese, I., Schneider, R., Snyder, S.H. (1977). ^{3}H-spiroperidol labels dopamine receptors in pituitary and brain. Eur. J. Pharmacol., 46, 377-382.

De Robertis E., De Lores Arnaiz, G.R., Alberici, M., Butcher, R.W. and Sutherland, E.W. (1967). Subcellular distribution of adenyl cyclase and cyclic phosphodiesterase in rat brain cortex. J. Biol. Chem., 242, 3487-3493.

Fuxe, K., Calza, L., Benfanati, F., Zini, I., Agnati, L.F. (1983). Quantitative autoradiographic localization of ^{3}H imipramine binding sites in the brain of the rat: Relationship to ascending 5-hydroxytryptamine neuron systems. Proc. Natl. Acad. Sci., 80, 3836-3840.

Geary, W.A. and Wooten, G.F. (1983). Quantitative film autoradiography of opiate agonist and antagonist binding in rat brain. J. Pharmacol. Exp. Ther., 225, 234-240.

König, J.F.R. and Klippel, R.A. (1974). The rat brain. Robert E. Krieger Publishing Co. Inc., New York.

Laemmli, U.K. (1970). Cleavage of structural proteins during the assembly of the head of bacteriophage T4. Nature 227, 680-685.

Nestler, E.J. and Greengard, P. (1984). Protein Phosphorylation in the Nervous System, John Wiley & Sons, Inc., Toronto.

Palacios, J.M., Niehoff, D.H. and Kuhar, M.J. (1981). Receptor autoradiography with tritium-sensitive film: Potential for computerized densitometry. Neurosci. Lett., 25, 101-105.

Pellegrino, L.J., Pellegrino, A.S. and Cushman, A.J. (1979). A Stereotaxic Atlas of the Rat Brain, 2 nd edition, Plenum Press, New York.

Penney, J.B., Pan, H.S., Young, A.B., Frey, K.A. and Dauth, G.W. (1981). Quantitative autoradiography of (^{3}H) muscimol binding in rat brain. Science (Wash. DC) 214, 1036-1038.

Rainbow, T.C., Biegon, A. and Bleisch, W.V. (1982). Quantitative autoradiography of neurochemicals. Life Sci., 30, 1769-1774.

Rainbow, T.C., Bleisch, W.V., Biegon, A. and McEwen, B.S. (1982). Quantitative densitometry of neurotransmitter receptors. J. Neurosci. Methods, 5, 127-138.

Silbergeld, E.K., Anderson, S.M. and Morris, S.J. (1982). Interactions of erithrosin B with rat cortical membranes. Life Sci., 31, 957-969.

Schulman, H. and Greengard, P. (1978). Ca-dependent protein phosphorylation system in membranes from various tissues, and its activation by calcium dependent regulator. Proc. Natl. Acad. Sci., USA, 75, 5432-5436.

Ueda, T., Maeno, H. and Greengard, P. (1973). Regulation of endogenous phosphorylation of specific proteins in synaptic membrane fractions from rat brain by adenosine 3:5-monophosphate. J. Biol. Chem. 248, 8295-8305.

Unnerstall, J.F., Kuhar, M.J., Niehoff, D.J. and Palacios, J.M. (1981). Benzodiazepine receptors are coupled to a subpopulation of γ-aminobutyric acid (GABA) receptors: Evidence from a quantitative autoradiographic study. J. Pharmacol. Exp. Ther., 218, 797-804.

Unnerstall, J.R., Niehoff, D.L., Kuhar, M.J. and Palacios, J.M. (1982). Quantitative receptor autoradiography using (^{3}H) Ultrofilm: Application to multiple benzodiazepine receptors. J. Neurosci. Methods, 6, 59-73.

Young, W.S., III and Kuhar, M.J. (1979). A new method for receptor autoradiography: (^{3}H) opioid receptors in rat brain. Brain Res., 179, 255-270.

QUANTITATIVE AUTORADIOGRAPHY OF DRUG BINDING SITES: METHODOLOGICAL CONSIDERATIONS

G.F. WOOTEN and W.A. GEARY, II

Department of Neurology, Box 394, University of Virginia Medical Center, Charlottesville, Virginia 22908, USA

INTRODUCTION

The goal of this current line of work was to develop quantitative autoradiographic methods to regionally quantify the binding properties (i.e. Kd and Bmax) of opiate ligands in morphine dependent and naive rat brain. Such regional studies were not possible utilizing binding techniques to brain membrane preparations because of the limited availability of tissue from dissection of small brain regions. Even with regional dissection techniques, the fine anatomical resolution available with autoradiography was sacrificed. With quantitative autoradiographic methods, one could ask whether the development of opiate tolerance/dependence occurred as a result of regulation of the affinity or concentration of opiate binding sites in any of one or several regions rich in opiate binding sites.

Opiate ligand binding in brain crude membrane preparations has been extensively studied (for review - Simon and Hiller, 1978; Snyder and Childers, 1979). Qualitative autoradiographic procedures employing fresh frozen sections and both in vivo and in vitro labeling techniques have been used to describe the regional distribution of opiate binding sites (Atweh and Kuhar, 1977a, b; Young and Kuhar, 1979). In addition, opiate ligand binding to tissue sections was shown to be saturable, stereospecific, and high affinity using in vitro labeling techniques (Young and Kuhar, 1979).

Quantitative studies of silver grains on nuclear emulsion for single concentrations of radioligands have been described by Unnerstall, et al. (1981). Quantitative film autoradiography of the kinetic features of radioligand ([^{3}H] muscimol) using ^{14}C-plastic standards was first described by Penney et al. (1981). Palacios et al. (1981) suggested the potential for computerized densitometric analysis of film autoradiographs of radioligands. By using a different but analogous method, Unnerstall et al. (1982) have characterized features of benzodiazepine binding in rat brain with tritium-containing tissue paste standards. We

397

reported the first quantitative data concerning Kd and Bmax for
opiate ligands in specific nuclei of rat brain using tritium-
impregnated plastic standards and densitometric assay (Geary and
Wooten, 1983).

In this paper, we wish to review several methodological
aspects of quantitative receptor ligand autoradiography and
present data on the effects of the development of opiate
dependence and withdrawal on regional opiate ligand binding
properties.

METHODS

Radioligand Binding to Brain Sections

Rat brains were rapidly removed and immersed in liquid Freon
for 3-4 minutes. Serial 20 μ sections were cut at -22°C, thaw-
mounted on gel-coated coverslips, and stored at -35°C. For
experimentation, sections were preincubated for 5 minutes at 4°C
in 500 ml of phosphate buffered saline. For saturation, studies
alternating adjacent sections were incubated for 30 minutes with
either increasing concentrations of ^{3}H-naloxone (0.5-6.6 nM) or
radioligand plus excess unlabeled ligand (200 nM naloxone-HCl or
150 nM etorphine-HCl). Sections were then rinsed twice (3-5
seconds) in PBS and once in deionized water (all at 4°C), dried
at room temperature, and exposed to LKB Ultrofilm along with ^{3}H-
plastic standards. The developed films were then analyzed
densitometrically by region and densitometric measurements minus
non-specific binding were converted to actual ligand concentration using a standard curve. The resulting data were analyzed by
Eadie Hoffstee and Hill Plots. Quantitative estimates of Kd and
Bmax were calculated according to Zivin and Waud (1982).

[^{3}H] Plastic Standards

Tritium containing plastic standards were prepared according
to Alexander et al. (1981) with minor modifications. Varying
amounts of ^{3}H-imipramine in absolute ethanol were added to 1 ml
of liquid methacrylate monomer. After thorough mixing, the
monomer was added to 2 g of powdered methacrylate polymer in
Petri dishes (5 cm in diameter), mixed, and allowed to harden.
Resultant discs were approximately 1 mm thick. Discs were cut
into 3 mm squares, exposed to LKB Ultrofilm, and analyzed for
homogeneity by densitometry. Within group O.D.'s exhibited a
range of 5% or less. The calibration of standards was
accomplished by co-exposing a series of plastic standards
containing a range of tritium concentrations and 20 μ coronal
hemisections from tritium-containing brains (labeled in vivo with
varying concentrations of ^{3}H-2-deoxyglucose) where contralateral
tissue radioactivity content had been determined in regional

brain homogenates by scintillation spectroscopy (Reivich et al., 1969). Calibration curves (tissue O.D. vs. tissue µCi/mg) were generated by linear regression of log O.D. vs. log tissue µCi/mg by method of least squares. Plastic tritium standards were calibrated by interpolation (see Figs. 1 & 2).

To determine if commercially available carbon-14 standards could be used instead of tritium standards for quantitative tritium autoradiography, we compared O.D.'s of carbon-14 and tritium plastic standards for different times of exposure to LKB Ultrofilm from 1-4 weeks. There was a marked lack of covariance between carbon-14 and tritium standards exposed for different

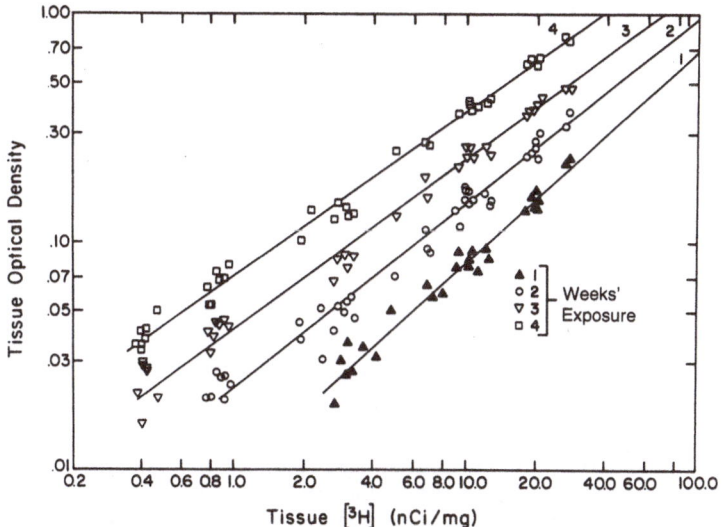

Fig. 1. Calibration curve for tritium tissue content. Rats were injected with 100-1000 µCi of ^{3}H-2-deoxyglucose (N=6 concentrations), killed 50 minutes later and their brains hemisected. One hemisphere was immediately frozen and prepared for autoradiography. Quadruplicate 20 µ sections from 10 selected brain regions were exposed serially to LKB Ultrofilm for 1-4 weeks. Corresponding regions from the contralateral hemisphere were dissected and their tritium content determined by liquid scintillation spectroscopy. Tissue sections were co-exposed with tritium-impregnated plastic standards. Tissue equivalent (TE) values for standards were interpolated from these curves. There was a positive correlation (P<0.01) between tissue optical density and tissue tritium content for each exposure time.

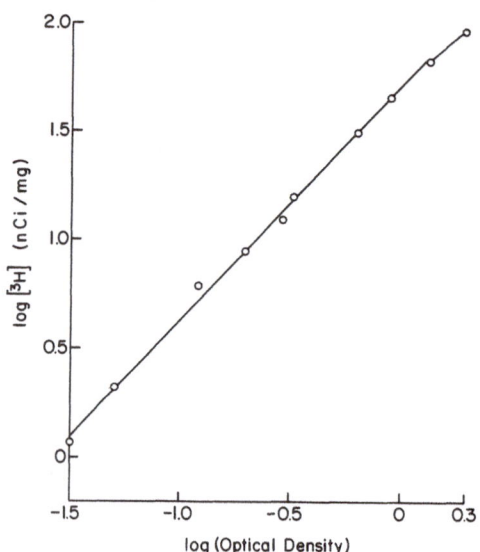

Fig. 2. Typical standard curve for two week exposure time. Ordinate represents log tissue equivalent tritium concentration and abscissa represents corresponding log optical density for each of ten plastic standards.

periods of time with carbon-14 standards producing a more rapid increase in O.D. with longer exposures than tritium (Fig 3). Only when carbon-14 standards are calibrated for the same length of exposure time as is utilized in the experiments using tritium images can carbon-14 standards be accurately employed. In contrast, tritium-containing plastic standards calibrated for any given exposure time will produce invariable equivalent tissue concentrations for any other exposure time within the linear O.D. response range of LKB Ultrofilm (approximately 0.08 to 0.8 O.D. units).

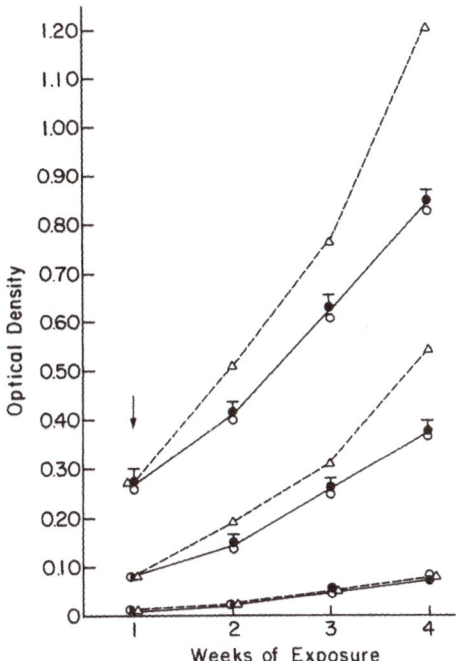

Fig. 3. Non-covariation of tritium vs. carbon-14 for different
exposure times. Tritium in tissue sections (filled circles),
tritium in plastic standards (open circles), and carbon-14 in
plastic standards (open triangle) of equal optical densities at
one week were co-exposed for 2, 3, and 4 weeks to LKB Ultrofilm.
Tritium in tissue and in plastic co-varied over the 4 week
exposure time. While carbon-14 demonstrated a faster rate of
film saturation than either tritium source.

Correction for Tissue Quenching of Tritium Image
_____

        Because of the greater physical density of white matter
compared to gray matter, the same concentration of tritium in a
white matter structure might be expected to produce a weaker
image than from a gray matter structure. Paired sections of rat
brains labeled in vivo with ^{3}H-2-deoxyglucose were exposed to LKB
Ultrofilm after one of each pair of sections had been exposed to
chloroform for 2, 5, or 10 minutes at room temperature to extract

the tissue lipid. This extraction resulted in a 30% reduction in
tissue dry weight but did not alter the tritium content per
section. In eight gray structures choroform produced a 28.7±5.6%
increase in mean O.D. compared to nonextracted controls (range
22-38%); while in three white matter structures chloroform
extraction produced and mean increase in O.D. of 115.9±29.3%
above control (i.e. more than doubled O.D.) (Fig. 4). Using this
approach correction factors for regional tissue tritium
autoabsorption can be estimated and applied to regional binding
data to maximize its accuracy.

Production of Opiate Dependence and Withdrawal

     Opiate dependence was produced by twice daily SC injections
of morphine sulfate for 11 days beginning with 10 mg/kg and
ending with 100 mg/kg. Dependent rats were killed for binding
studies at 90 minutes after their final morphine injection.
Studies of the morphine withdrawal state used rats that were

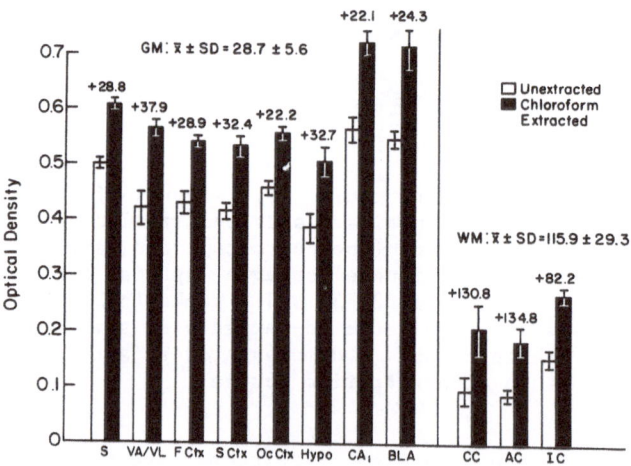

Fig. 4. Effects of chloroform extraction on regional image
quenching. Scintillation spectroscopy of extracted sections did
not differ from unextracted controls. No redistribution of
tritium was noted on the images from extracted tissue. Effects
of chloroform extraction were the same for 2, 5, and 10 minutes
of incubation in chloroform.

killed 10-12 minutes after intravenous administration of
naloxone-HCl 0.5 mg/kg; this naloxone treatment was given 10-12
hours after the final morphine dose.

RESULTS

The Kd and Bmax values for ^{3}H-naloxone binding in 13
selected brain regions of naive rats are given in Table 1.
The range of Kd values was 0.70 nM (hippocampus CA1) to 3.2
nM (whole striatum). The range of Bmax values was 30 fmol/mg
(non-cluster striatum) to 152 fmol/mg (habenula). The mean±S.D.
values for Kd and Bmax for all structures pooled were 1.9±0.9 nM
and 75±45 fmol/mg, respectively. Greater than 95% of the Kd
values of the distribution fall within two standard deviations of
the mean suggesting a continuous, single population phenomenon.
ANOVA tests for heterogeneity among structures confirmed
differences between Kd values for structures at either extreme of
the distribution.
The effects of chronic morphine treatment (dependent state)
and chronic morphine followed by naloxone treatment (withdrawal
state) on ^{3}H-naloxone binding are given in Tables 2 and 3.
No differences were found in either the affinity or capacity
of ^{3}H-naloxone binding among any of the 3 treatment groups for
any of the eight structures.

Table 1. ^{3}H-Naloxone Binding in Naive Rats:
           Regional Saturation Studies

Structure	Kd	Bmax
N. accumbens (rostral)	1.9±0.5(nM)	42±7 (fmol/mg)
N. accumbens (caudal)	2.1±0.5	47±6
Central amygdala	1.6±0.5	106±13
Habenula	2.5±0.5	152±25
Striatum (non-cluster)	1.8±0.2	30±5
Striatum (whole)	3.2±0.3	57±7
Interpeduncular N.	0.7±0.1	143±18
Superior colliculus (SG)	1.4±0.3	35±5
Median raphe	2.7±0.6	80±11
Cort. N. amygdala	2.6±0.7	103±24
Hippocampus CA1	0.7±0.1	35±8
Cingulate Cortex	1.4±0.1	38±1
Med. Frontal Cortex	1.4±0.4	36±6

N>4 rats for each structure
Data are mean±S.D.

Table 2.   Effects of Opiate Treatment on Regional
Naloxone Binding Constants
(Comparion of K_d's; units = nM)

Structures	Control	Dependent	Withdrawal
Cent. Amyg.	1.6±0.5	1.1±0.3	1.2±0.7
Habenula	2.5±0.5	2.4±0.3	2.7±0.5
Striatum (NC)	1.8±0.2	1.6±0.2	1.5±0.2
Striatum (W)	3.2±0.3	3.0±0.3	3.0±0.2
IPN	0.7±0.1	0.8±0.2	0.8±0.1
S.C.	1.4±0.3	1.4±0.2	1.3±0.3
Cort. N. Amyg.	2.6±0.7	2.7±0.3	2.4±0.4
CA1	0.7±0.1	0.6±0.1	0.7±0.1

$N \geq 4$ rats for each structure
Data are mean±S.D.

Table 3.   Effects of Opiate Treatment on Regional
Naloxone Binding Constants
(Comparison of B_{max}'s; units=fmol/mg tissue)

Structure	Control	Dependent	Withdrawal
Cent. Amyg.	106±13	95±19	97±20
Habenula	152±25	164±15	167±46
Striatum (NC)	30±5	36±6	33±3
Striatum (W)	57±7	65±5	53±11
IPN	143±18	166±23	175±20
S.C.	35±5	35±5	37±5
Cort. N. Amyg.	103±24	99±14	91±12
CA1	35±7	38±7	30±5

$N \geq 4$ rats for each structure
Data are mean±S.D.

DISCUSSION

   Quantitative film autoradiography is a powerful method that combines fine anatomical resolution with the capacity for quantitative estimates of regionally selective molecular properties of tissue. The use of calibrated tritium standards, whether they be composed of tissue paste or plastic, allows for the accurate estimation of ligand concentration simultaneously in multiple brain regions by densitometric measurements. Calibrated standards of different isotopes may be used (e.g. carbon-14) but only when calibrated for the precise exposure time to be used in the experiment because carbon-14 and tritium sources do not co-vary over different exposure times. Recently, the Amersham Corporation (Arlington Heights, Illinois, USA) has produced commercially available tritium plastic standards that we have calibrated, thereby facilitating the general use of these quantitative methods.

   Regional differences in brain lipid content may result in regionally selective differences in autoabsorption of low energy tritium emissions. Regional correction factors obtained by chloroform extraction of [^{3}H]-2-deoxyglucose impregnated tissue should allow for substantial resolution of this potential source of error in tritium quantitative autoradiography.

   Our results with quantitative ^{3}H-naloxone autoradiography in naive, opiate dependent and opiate withdrawal state rats indicate that neither regulation of naloxone binding site affinity nor concentration in any of several brain regions is the molecular event underlying the development of opiate tolerance/dependence. Therefore, this regulation must occur either at a post-binding transduction site or in other neurotransmitter neuronal systems.

   The quantitative autoradiographic methods using radiolabeled ligands can also be applied to the measurement of regional enzyme activities with radiolabeled, irreversibly bound suicide substrates, for mapping messages for specific proteins with radiolabeled cDNA probes, for radioimmunohistochemistry using tritium labeled primary or secondary antibodies, and for high resolution 2-deoxy [^{3}H] glucose autoradiography.

REFERENCES

Alexander, G.M., Schwartzman, R.J., Bell, R.D., Yu, J. and Renthal, A. (1981). Quantitative Measurement of Local Cerebral Metabolic Rate for Glucose Using Tritiated 2-Deoxyglucose. Brain Res., 223, 59-67.

Atweh, S.F. and Kuhar, M.J. (1977a). Autoradiographic Localization of Opiate Receptors in Rat Brain (II). Brainstem. Brain Res., 129, 1-12.

Atweh, S.F. and Kuhar, M.J. (1977b). Autoradiographic Localization of Opiate Receptors in Rat Brain (III). The Telencephalon Brain Res., 134, 393-405.

Geary, W.A. and Wooten, G.F. (1983). Quantitative Film
Autoradiography of Opiate Agonist and Antagonist Binding in Rat
Brain. J. Pharmacol. Exp. Ther., 225, 234-240.

Palacios, J.M., Niehoff, D.H., and Kuhar, M.J. (1981). Receptor
Autoradiography with Tritium-Sensitive Film: Potential for
Computerized Densitometry. Neurosci. Lett., 25, 101-105.

Penney, J.B., Pan, H.S., Young, A.B., Frey, K.A., and Dauth, G.W.
(1981). Quantitative Autoradiography of [^{3}H] Muscimol Binding in
Rat Brain. Science (Washington, D.C.), 214, 1036-1038.

Reivich, M., Jehle, J.W., Sokoloff, L. and Kety, S.S. (1969)
Measurement of Regional Cerebral Blood Flow with Antipyrine-^{14}C
in Awake Cats. J. Appl. Physiol., 27, 296-300.

Simon, E.J. and Hiller, J.M. (1978). The Opiate Receptors. Ann.
Rev. Pharmacol. Toxicol., 18, 371-394.

Snyder, S.H. and Childers, S.R. (1979). Opiate Receptors and
Opioid Peptides. Ann. Rev. Neurosci., 2, 35-64.

Unnerstall, J.R., Kuhar, M.J., Niehoff, D.L. and Palacios, J.M.
(1981). Benzodiazepine Receptors are Coupled to a Subpopulation
of Gamma-Aminobutyric Acid (GABA) Receptors: Evidence from a
Quantitative Autoradiographic Study. J. Pharmacol. Exp. Ther.,
218, 797-804.

Unnerstall, J.R., Niehoff, D.L., Kuhar, M.J. and Palacios, J.M.
(1982). Quantitative Receptor Autoradiography using [^{3}H]
Ultrofilm: Applications to Multiple Benzodiazepine Receptors.
J. Neurosci. Methods, 6, 59-73.

Young, W.S. and Kuhar, M.J. (1979). A New Method for Receptor
Autoradiography: ^{3}H Opioid Receptors in Rat Brain. Brain Res.,
179, 155-170.

Zivin, J.A. and Waud, D.R. (1982). How to Analyze Binding,
Enzyme, and Uptake Data: The Simplest Case, a Single Phase.
Life Sci., 30, 1407-1422.

RECENT ADVANCES IN THE VISUALIZATION AND QUANTIFICATION OF BENZODIAZEPINE RECEPTORS

GRAYSON RICHARDS and HANNS MÖHLER

Pharmaceutical Research Department, F. Hoffmann–La Roche & Co Ltd, CH–4002 Basel, Switzerland

INTRODUCTION

The anxiolytic, hypnotic, anticonvulsant and muscle relaxant effects of the benzodiazepines are mediated by the inhibitory neuro-transmitter GABA (γ-aminobutyric acid). The benzodiazepines mod-ulate, in a facilitatory way, GABAergic transmission in a variety of principle output neurons and interneurons in the mammalian CNS (see Haefely, 1983 for a current review). These are receptor-mediated effects. Specific, high-affinity, saturable binding sites for the benzodiazepines form part of an oligomeric complex with a GABA receptor and its associated chloride channel protein (Möhler and Richards, 1983 ; Braestrup and Nielsen, 1983). Recent studies of the pharmacology and binding characteristics of the benzodia-zepine receptor ligands have revealed a whole spectrum of compounds with a high affinity for the receptor but with different functional consequences (Haefely et al., 1984). Thus, three basic types of ligands can now be distinguished : *agonists,* which produce the char-acteristic therapeutic effects of the benzodiazepines ; *inverse agonists,* which produce diametrically opposite effects to those of agonists ; and *competitive antagonists,* which have no major intrin-sic activity per se but prevent or abolish receptor-mediated pharma-cological effects of agonists and inverse agonists (Figure 1). The receptor ligands could influence, allosterically, either the affinity of GABA for its receptor or the coupling between the GABA receptor and the chloride channel. Mixed agonists/antagonists (so-called partial agonists) are novel receptor ligands whose potential anxioselectivity is now being clinically evaluated.

The adaptation of quantitative autoradiographic techniques to study the sites and mechanisms of drug action (see Kuhar, 1982 ; Wamsley and Palacios, 1983 for details) has opened up numerous options for research on the benzodiazepine receptors. In this brief review, we will highlight the recent advances in the visualization and quantification of these receptors and also discuss future pros-pects in this area of research. Since the methodology which we have

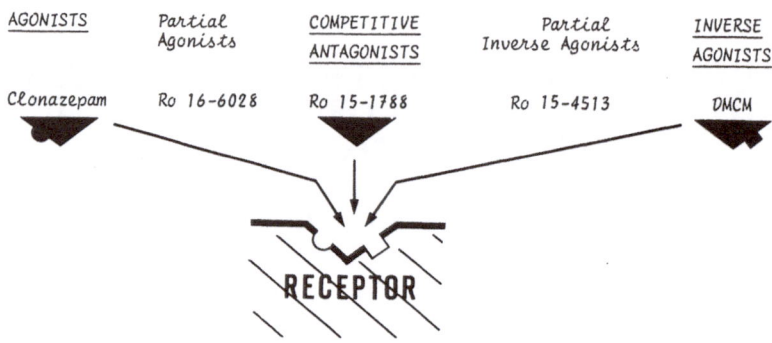

Figure 1. A simple schematic representation of three basic types of benzodiazepine receptor ligands with different modes of receptor interaction : clonazepam (a benzodiazepine), Ro 15-1788 (an imidazodiazepine) and DMCM (a β-carboline). Ro 16-6028 (an imidazobenzodiazepinone) is a partial agonist and Ro 15-4513 (an azide derivative of Ro 15-1788) a partial inverse agonist.

used is essentially similar to that described and discussed in the foregoing chapters (Palacios and Wiederhold ; Wooten), we will restrict ourselves to some useful applications of quantitative autoradiography and immunohistochemistry to the study of benzodiazepine receptors (see Richards and Möhler, 1984b for a current review).

RADIOLIGAND BINDING STUDIES IN VITRO

The presence of high-affinity, saturable and stereospecific binding sites for the benzodiazepines in a crude synaptic membrane preparation was first revealed in an in vitro radioligand binding assay using ^{3}H-diazepam (Möhler and Okada, 1977 ; Squires and Braestrup, 1977). The benzodiazepines were found to have a higher affinity for their receptors than for any other membrane binding sites particularly when using highly radioactive ligands in low concentrations which favours specific over non-specific binding.

Based on this in vitro binding assay, autoradiographic studies (Young and Kuhar, 1980 ; Richards and Möhler, 1984a for a review) have since revealed the precise localization of the benzodiazepine binding sites in the mammalian CNS with a macroscopic (Figure 2) and microscopic resolution. Unnerstall et al. (1981) have described a quantitative approach to receptor autoradiography in which the brain sections are co-exposed with brain paste standards which are subsequently used to calibrate the optical densities measured microdensitometrically and thereby to express the binding in fmol/mg tissue. We are currently setting up a computerized image analysis system (ASBA, Wild-Leitz) for such quantitative studies.

The in vitro autoradiographic approach has now been extended
to study : the ontogenetic development of these binding sites in
the rat brain (Schlumpf et al., 1983), the magnitude of the GABA-
induced increase in the affinity of benzodiazepine receptors
(Unnerstall et al., 1981), the heterogeneity of benzodiazepine
receptors (Young et al., 1981) and, more recently, human postmortem
brain tissue (Niehoff and Whitehouse, 1983 ; Palacios, personal
communication). Future application of the in vitro technique to
the study of human postmortem brain to include various pathological
states affecting benzodiazepine receptors could help to further
our knowledge of drug therapy in anxiety, epilepsy and insomnia.

RADIOLIGAND BINDING STUDIES IN VIVO

Specific binding of the benzodiazepines in the CNS can be
demonstrated not only in vitro but also in vivo (Chang and Snyder,
1978 ; Williamson et al., 1978). In fact, a single intravenous dose
of a receptor ligand, e.g. ^{3}H-Ro 15-1788 (2-5 minutes before decapitation) leads to an autoradiographic distribution very similar to
that observed in vitro (Richards, in preparation ; Figure 3). In
this way, it might be possible to study the degree of receptor occu-
pancy in various regions of the CNS.

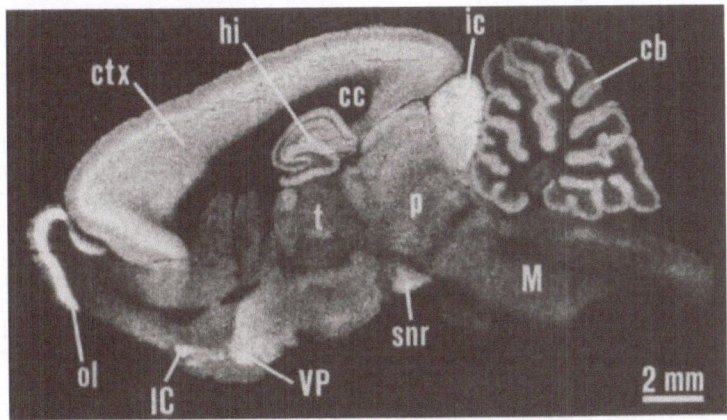

Figure 2. Autoradiogram (LKB film) of total binding of 1 nM ^{3}H-clo-
nazepam in vitro in a parasagittal rat brain section. White areas
correspond to the presence of the radioligand. Note the uneven distribution of receptors : highest densities occur in the glomerular
and external plexiform layers of the olfactory bulb (ol), in the
cerebral cortex (ctx), islets of Calleja (IC), ventral pallidum
(VP), hippocampus (hi), inferior colliculus (ic), substantia nigra-
zona reticulata (snr), and cerebellum (cb), and lowest densities
in the corpus callosum (cc), parts of the thalamus (t), pons (p)
and medulla (M).

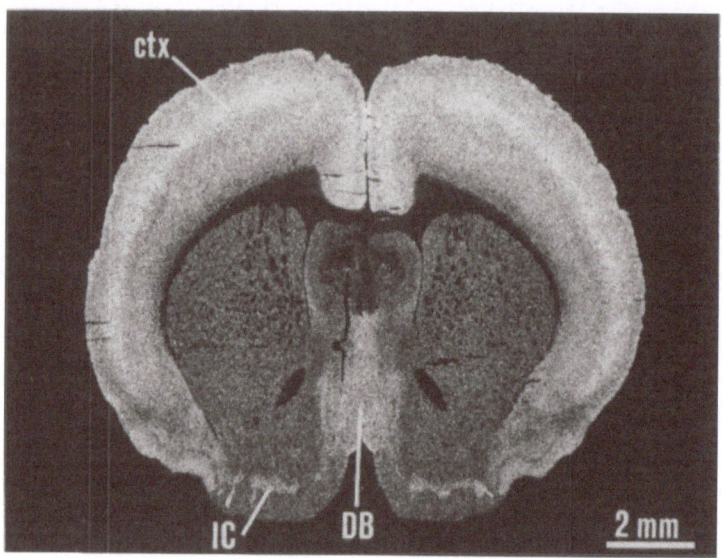

Figure 3. Autoradiogram (LKB film) of total binding of ^{3}H-Ro 15-1788 in a frontal section of rat brain in vivo. White areas correspond to the presence of the radioligand. The distribution of the radiolabel (300 µCi/kg i.v., 5 min) is very similar to that observed with in vitro binding. This suggests that most of the ligand is specifically bound to its receptor. ctx, cerebral cortex ; DB, nucleus of the diagonal band of Broca ; IC, islands of Calleja.

Radioligand binding studies in vivo have been combined with the technique of photolabelling to demonstrate the subcellular localization of the benzodiazepine receptors by electron microscopical autoradiography (Möhler et al.,1980). When ^{3}H-flunitrazepam is injected i.v. to rats and brain slices are exposed to weak UV light and then processed for autoradiography, radiolabelled receptors can be identified in areas of synaptic contacts, i.e. regions consisting of nerve terminals, postsynaptic membranes and adjacent glial structures (the autoradiographic resolution does not allow a more precise localization of the label). After immunohistochemical staining with antiserum to glutamate decarboxylase, a marker enzyme for GABAergic neurons, approximately one third of the photolabelled nerve terminals could be identified as GABAergic (Möhler et al., 1981). Thus, at least some GABAergic synapses are primary target sites for the pharmacological effects of the benzodiazepines.

Photolabelling with ^{3}H-flunitrazepam results in the irreversible (covalent) binding of the ligand to only 25 % of the total number of binding sites present in the tissue. In contrast, with ^{3}H-Ro 15-4513, an azide derivative of the benzodiazepine antagonist Ro 15-1788, all binding sites can be photolabelled (Möhler et al., in press). The regional brain distribution of binding sites for

^{3}H-Ro 15-4513 and ^{3}H-Ro 15-1788 was virtually identical (Figure 4). The proteins photolabelled by ^{3}H-Ro 15-4513 or -flunitrazepam were identical with the exception of the cerebellum where an additional protein was moderately labelled by the former ligand. Since the granular layer of the cerebellum contained (in autoradiograms) a high density of ^{3}H-Ro 15-4513 binding which could not be displaced by diazepam, the additional protein is probably the site of this non-specific binding. Ro 15-4513 is a novel tool for the visualization and quantification of benzodiazepine receptors in future studies at a subcellular level.

The imaging of psychopharmaceuticals in the human brain in vivo is now possible by positron emmission tomography (PET), e.g. using ^{11}C-flunitrazepam and ^{11}C-Ro 15-1788 (Comar et al., 1981 ; Mazière et al., 1983). Although the present resolution of this technique (8 mm) limits the interpretation of the positron images, it will certainly have immense value in future research on anxiety, epilepsy and insomnia in vivo.

IMMUNOHISTOCHEMICAL STUDIES WITH MONOCLONAL ANTIBODIES TO THE ISOLATED RECEPTOR COMPLEX

In order to obtain a fuller understanding of the molecular mechanisms of GABA receptor function and its regulation by benzodiazepine receptor ligands, the receptor complex has now been isolated and partially purified (Sigel et al., 1983 ; Schoch and Möhler, 1983). The purified receptor preparation contained not only the high-affinity neuronal binding sites for the benzodiazepines but also high and low affinity sites for GABA. Thus, all three sites are localized on the same physical entity.

Monoclonal antibodies were raised against the purified receptor complex in collaborative studies with colleagues in our immunology department (Drs. P. Häring, B. Takacs and Ch. Staehli). The hybridoma technique yielded a total of 20 monoclonal cultures secreting antibodies against the receptor complex. The immunoprecipitated complex showed all the binding properties of the isolated receptor. The distribution of immunoreactivity to the monoclonal antibodies (Schoch et al., in preparation) was demonstrated in cryostat sections of perfusion-fixed rat brain and immersion-fixed bovine brain using the indirect peroxidase-antiperoxidase (PAP) technique. Figure 5 illustrates the distribution of antigenic sites in bovine hippocampus. In the rat and bovine brain the regions of intense staining are known to contain a high density of GABAergic efferents and neuronal type benzodiazepine receptors. Similar results were obtained radioimmunohistochemically using ^{125}I-labelled monoclonal antibodies which therefore offers a further approach to quantify, densitometrically, the regional distribution of these receptors. Colloidal gold-labelled antibodies are currently being used to determine the subcellular localization of antigenic sites with high resolution.

Figure 4. Autoradiogram (LKB film) of a parasagittal rat brain section photolabelled with 1 nM ^{3}H-Ro 15-4513 in vitro. White areas
correspond to the presence of the radioligand. Note the similarity
in the distribution of binding sites to that illustrated in Fig. 2.
Photolabelling in the granular layer of the cerebellum was not displaceable by diazepam and is therefore considered to be non-
specific.

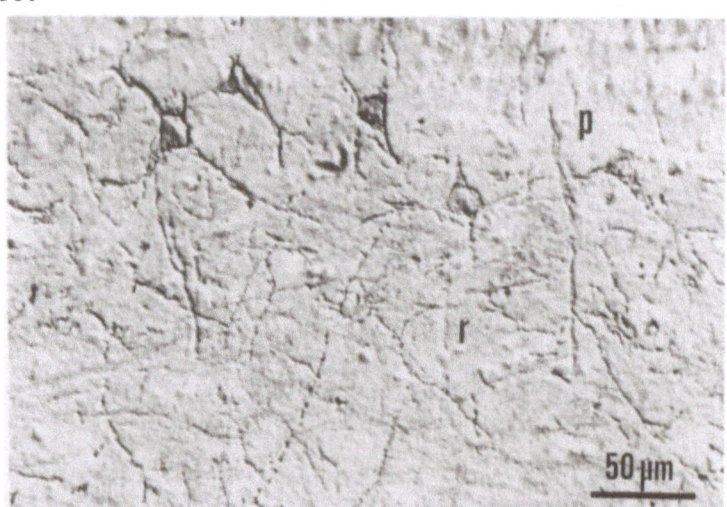

Figure 5. Immunohistochemical localization (PAP technique) of
antigenic sites in a cryostat section of bovine hippocampus
(interface between the stratum pyramidale (p) and the stratum
radiatum (r)). Note the immunoreactive cells and their processes
(in this interference-contrast image) revealing the distribution
of the receptor complex.

FUTURE RESEARCH PROSPECTS

The foregoing discussion has highlighted the recent advances in the visualization and quantification of the benzodiazepine receptors in the mammalian CNS.

Future trends and prospects in this area of research will be determined by the new techniques described. They provide the tools to answer several questions concerning the mechanisms of action of these drugs e.g.
- What is the relationship between the magnitude of effect and receptor occupation ? -in vivo binding studies.
- Are there regional alterations of benzodiazepine receptors in human brain in cases of chronic anxiety, epilepsy and insomnia ? - in vitro binding and PET studies.
- Are there regional adaptive changes of receptors in response to stress situations or to chronic drug treatment ? - in vivo binding studies.
- Are the neuronal antigenic sites (receptors), identified by monoclonal antibodies,pre- or postsynaptic ? Specifically which cell types are stained ? - immunocytochemistry with gold-labelled antibodies.

REFERENCES

Braestrup, C. and Nielsen, M. (1983) Benzodiazepine receptors. In Handbook of Psychopharmacology, Vol. 17, (eds. L. Iversen, S.D. Iversen and S.H.Snyder). Plenum Press, New York.

Chang, R.S.L and Snyder, S.H. (1978) Benzodiazepine receptors: labeling in intact animals with [^{3}H]flunitrazepam. Eur. J.Pharmac. 48, 213-218.

Comar, D., Mazière, M., Cepeda, C., Godot, J.-M., Menini, C. and Naquet, R. (1981) The kinetics and displacement of [^{11}C]flunitrazepam in the brain of the living baboon. Eur. J. Pharmacol. 75, 21-26.

Haefely, W. (1983) The biological basis of benzodiazepine actions. The benzodiazepines today - two decades of research and clinical experience. J. Psychoact. Drugs 15, 19-39.

Haefely, W., Kyburz, E., Gerecke, M. and Möhler, H. (1984) Recent advances in the molecular pharmacology of benzodiazepine receptors and in the structure-activity relationship of their agonists and antagonists. Advances in Drug Res. (in press).

Kuhar, M.J. (1982) Localization of drug and neurotransmitter receptors in brain by light microscopic autoradiography. In Handbook of Psychopharmacology, Vol. 15, (eds. L. Iversen, S.D. Iversen and S.H. Snyder). Plenum Press, New York.

Mazière, M., Prenant, Ch., Sastre, J., Crouzet, M., Comar, D., Hantraye, P., Kaisima, M., Guibert, B. and Naquet, R. (1983) ^{11}C-Ro 15-1788 and ^{11}C-flunitrazepam, deux coordinats pour l'étude par tomographie par positrons des sites de liaison des benzodiazépines. C.R. Acad. Sc. Paris. T. 296, Série 111, 871-876.

Möhler, H. and Okada, T. (1977) Benzodiazepine receptors : demonstration in the central nervous system. Science 198, 849-851.

Möhler, H. and Richards, J.G. (1983) Receptors for anxiolytic drugs. In Anxiolytics : Neurochemical, Behavioural and Clinical Perspectives. (eds. J.B. Malick, S.J. Enna and H.I. Yamamura). Raven Press, New York.

Möhler, H., Battersby, M.K. and Richards, J.G. (1980) Benzodiazepine receptor protein identified and visualized in brain tissue by a photoaffinity label. Proc. Natl. Acad. Sci. USA 77, 1666-1670.

Möhler, H., Richards, J.G. and Wu, J.-Y. (1981) Autoradiographic localization of benzodiazepine receptors in immunocytochemically identified γ-aminobutyrergic synapses. Proc. Natl. Acad. Sci. USA 78, 1935-1938.

Möhler, H., Siegart, W., Richards, J.G. and Hunkeler, W. (1984) Photoaffinity labeling of benzodiazepine receptors with a partial inverse agonist. Eur. J. Pharmacol. (in press).

Niehoff, D.L. and Whitehouse, P.J. (1983) Multiple benzodiazepine receptors : autoradiographic localization in normal human amygdala. Brain Res. 276, 237-245.

Richards, J.G. and Möhler, H. (1984a) Benzodiazepine receptors. Neuropharmacology 23, 233-242.

Richards, J.G. and Möhler, H. (1984b) Benzodiazepine receptors and their ligands. In Mechanisms of Drug Action. (ed. G.N. Woodruff). Macmillan Press (in press).

Schlumpf, M., Richards, J.G., Lichtensteiger, W. and Möhler, H. (1983) An autoradiographic study of the prenatal development of benzodiazepine binding sites in rat brain. J. Neurosci. 3, 1478-1487.

Schoch, P. and Möhler, H. (1983) Purified benzodiazepine receptor retains modulation by GABA. Eur. J. Pharmacol. 95, 323-324.

Schoch, P., Richards, J.G., Häring, P., Takacs, B., Stähli, C., Staehelin, T., Haefely, W. and Möhler, H. (1984) A purified GABA/benzodiazepine receptor : immunohistochemistry and structural analysis with monoclonal antibodies. (in preparation).

Sigel, E., Stephenson, F.A., Mamalki, C. and Barnard, E.A. (1983)
A γ-aminobutyric acid/benzodiazepine receptor complex of bovine
cerebral cortex : purification and partial characterization. J.
Biol. Chem. 258, 6965-6969.

Squires, R.F. and Braestrup, C. (1977) Benzodiazepine receptors in
rat brain. Nature 266, 732-734.

Unnerstall, J.R., Kuhar, M.J., Niehoff, D.L. and Palacios, J.J.
(1981) Benzodiazepine receptors are coupled to a subpopulation of
γ-aminobutyric acid (GABA) receptors : evidence from a quantitative
autoradiographic study. J. Pharm. Exp. Ther. 218, 797-804.

Wamsley, J.K. and Palacios, J.M. (1983) Apposition techniques of
autoradiography for microscopic receptor localization. In Current
methods in cellular neurobiology, Vol. 1 (eds. J.L. Barker and
J.F. McKelvy). John Wiley, New York.

Williamson, M.J., Paul, S.M. and Skolnick, P. (1978) Demonstration
of [3H]diazepam binding to benzodiazepine receptors in vivo. Life
Sci. 23, 1935-1940.

Young, W.S. and Kuhar, M.J. (1980) Radiohistochemical localization
of benzodiazepine receptors in rat brain. J. Pharm. Exp. Ther. 212,
337-346.

Young, W.S., Niehoff, D., Kuhar, M.J., Beer, B. and Lippa, A.S.
(1981) Multiple benzodiazepine receptor localization by light
microscopic radiohistochemistry. J. Pharm. Exp. Ther. 216, 425-430.

INDEX

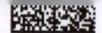